RREPS13 AND Meghri13

International symposium on Radiation from Relativistic Electrons in Periodic Structures (RREPS13) and the 3rd International Conference on Electron, Positron, Neutron and X-ray Scattering under External Influences (Meghri13)

Journal of Physics: Conference Series Volume 517

Lake Sevan, Armenia
23 – 28 September 2013

ISBN: 978-1-63439-298-3
ISSN: 1742-6588

Printed from e-media with permission by:

Curran Associates, Inc.
57 Morehouse Lane
Red Hook, NY 12571

Some format issues inherent in the e-media version may also appear in this print version.

Printed by Curran Associates, Inc. (2014)

For permission requests, please contact the Institute of Physics
at the address below.

Institute of Physics
Dirac House, Temple Back
Bristol BS1 6BE UK

Phone: 44 1 17 929 7481
Fax: 44 1 17 920 0979

techtracking@iop.org

Additional copies of this publication are available from:

Curran Associates, Inc.
57 Morehouse Lane
Red Hook, NY 12571 USA
Phone: 845-758-0400
Fax: 845-758-2634
Email: curran@proceedings.com
Web: www.proceedings.com

RREPS13 AND Meghri13

International symposium on Radiation from Relativistic Electrons in Periodic Structures (RREPS13) and the 3rd International Conference on Electron, Positron, Neutron and X-ray Scattering under External Influences (Meghri13)

Journal of Physics: Conference Series Volume 517

Lake Sevan, Armenia
23 – 28 September 2013

ISBN: 978-1-63439-298-3
ISSN: 1742-6588

IOP Institute of Physics

TABLE OF CONTENTS

PARAMETRIC X-RADIATION

DIFFRACTION RADIATION AND SMITH-PURCELL EFFECT

COHERENT BREMSSTRAHLUNG AND CHANNELING RADIATION

X-RAY SCATTERING WITHOUT AND BY ACOUSTIC SUPERLATTICES

INTERACTION OF PARTICLE BEAMS WITH ARTIFICIAL STRUCTURES

APPLICATION OF RADIATION BEAMS

Preface

These Proceedings are published as a recollection of contributions presented at the X International Symposium on "Radiation from Relativistic Electrons in Periodic Structures" (RREPS-13) merged with III International Conference "Electron, Positron, Neutron and X-ray Scattering under External Influences" (Meghri-13), which was held at Lake Sevan, 23–28 September, 2013, Armenia. RREPS-13 and Meghri-13 were co-organized by Tomsk Polytechnic University (Russia) and Institute of Applied Problems of Physics (Armenia). The main goal of the symposium was to bring together the scientists from around the world who work on designs of new radiation sources and their applications.

There were 89 participants from 12 countries. The website of the symposium is available at http://rreps.tpu.ru/

The scientific program of the symposium consisted of 8 sections and a satellite Workshop on Terahertz Radiation generation. All papers in these Proceedings refer to one from the following topics:

Section 1 General Properties of Radiation from Relativistic Particles

Section 2 Transition Radiation

Section 3 Parametric X- Radiation

Section 4 Diffraction Radiation and Smith–Purcell Effect

Section 5 Coherent Bremsstrahlung and Channeling Radiation

Section 6 X-Ray Scattering without and by Acoustic Superlattices

Section 7 Interaction of Particles Beams with Artificial Structures (Acoustic Superlattices, Metamaterials, etc.)

Section 8 Application of Radiation Beams

The published papers cover nearly all "hot" topics of current interest on investigations of monochromatic and broadband radiation sources based on accelerators and X-ray tubes. Different mechanisms of radiation emission such as Compton backscattering, Cherenkov radiation, transition radiation, diffraction radiation, Smith–Purcell effect, parametric X-ray were considered in Sections 1, 2, 3, 4 and 5. The problem of control of radiation parameters by external acoustic fields is discussed in Section 6. Several applications of electron, proton, gamma and X-ray beams are proposed in Sections 7 and 8.

We are extremely thankful to all authors for providing their valuable contributions for these Proceedings as well as the reviewers for their constructive recommendations and criticism aiding to improve the presented articles.

We are looking forward to welcoming all colleagues at the next Symposium of the biennial series RREPS-15, which will be hosted by Saint Petersburg State University in 2015. We invite all researchers interested in the field including the authors of these Proceedings.

Professor Alexander Potylitsyn
Tomsk Polytechnic University, Tomsk, Russia
Dr Pavel Karataev
Royal Holloway, University of London, Egham, United Kingdom
Professor Alpik Mkrtchyan
Institute of Applied Problems of Physics, Yerevan, Armenia

Journal of Physics: Conference Series **517** (2014) 011002 doi:10.1088/1742-6596/517/1/011002

Peer review statement

All papers published in this volume of *Journal of Physics: Conference Series* have been peer reviewed through processes administered by the proceedings Editors. Reviews were conducted by expert referees to the professional and scientific standards expected of a proceedings journal published by IOP Publishing.

Spectral properties of Compton inverse radiation: Application of Compton beams

Eugene Bulyak[1] **and Junji Urakawa**[2]
[1] NSC KIPT, Kharkov, Ukraine
[2] KEK, Tsukuba, Ibaraki, Japan
E-mail: bulyak@kipt.kharkov.ua

Abstract. Compton inverse radiation emitted due to backscattering of laser pulses off the relativistic electrons possesses high spectral density and high energy of photons – in hard x–ray up to gamma–ray energies – because of short wavelength of laser radiation as compared with the classical electromagnetic devices such as undulators.

In this report, the possibility of such radiation to monochromatization by means of collimation is studied. Two approaches have been considered for the description of the spectral–angular density of Compton radiation based on the classical field theory and on the quantum electrodynamics. As is shown, both descriptions produce similar total spectra. On the contrary, angular distribution of the radiation is different: the classical approach predicted a more narrow radiation cone. Also proposed and estimated is a method of the 'electronic' monochromatization based on the electronic subtraction of the two images produced by the electron beams with slightly different energies. A 'proof–of–principle' experiment of this method is proposed for the LUXC facility of KEK (Japan).

1. Introduction

Sources of Compton radiation, in which photons of intense laser pulse scattered off from relativistic electrons, are able to produce bright x–ray beams with narrow bandwidth. The process of Compton scattering can be treated as two–particle elastic scattering: one of the particle is represented by the electron, another one by the photon. Since the laser photon has negligibly small energy as compared with electron's, the recoil of electron is negligibly small – the photon is scattered off within a narrow cone along electron's trajectory.

Such sources have a substantial potential for applications in different areas of medicine, biology, physics, etc. This potential is emerged not only from its brightness, but tunability and 'quasimonochromaticity' of the spectrum.

The report highlights spectral properties of Compton sources and presents a potential application of these sources for x–ray angiography, which possesses a substantial advantage over the conventional methods.

2. Spectrum of collimated Compton inverse radiation

The spectrum of inverse Compton radiation is determined by two relations, (i) kinematic dependence of the energy of a scattered off quantum, E_{x} upon the crossing angle ϕ between the laser photon and the electron, energy of both the electron and the photon, and the scattering angle ψ, and (ii) the differential cross section – dependence of probability of scattering the photon at angle ψ. Within the small–angle approximation, $\psi \ll 1$ (the crossing angle $\phi = 0$ corresponds to the head–on collision), neglected the electron recoil, these relations read ([1]):

$$E_\mathrm{x} = \frac{2\gamma^2(1+\cos\phi)E_\mathrm{las}}{1+\gamma^2\psi^2}\ ; \qquad \mathrm{d}\sigma = 8\pi r_0^2 \frac{\psi\gamma^2\left(1+\gamma^4\psi^4\right)}{\left(1+\gamma^2\psi^2\right)^4}\mathrm{d}\psi\ ,$$

where $\gamma = E_\mathrm{e}/m_\mathrm{e}c^2$ is the Lorentz factor of the electron, r_0 is the classical electron radius.

Convolution of the cross section with distribution functions of both the electron bunch and the photon pulse taking into account the kinematic relation will produce a real collimated spectrum of the radiation. As it follows from the kinematics, the width of spectrum is determined by the collimation angles $\psi_i \leq \psi \leq \psi_f$, as is depicted in Fig.1. The spectral–angular density of the collimated Compton radiation for an ideal case (monoenergetic electrons and laser photons with parallel trajectories) is presented in Fig.2.

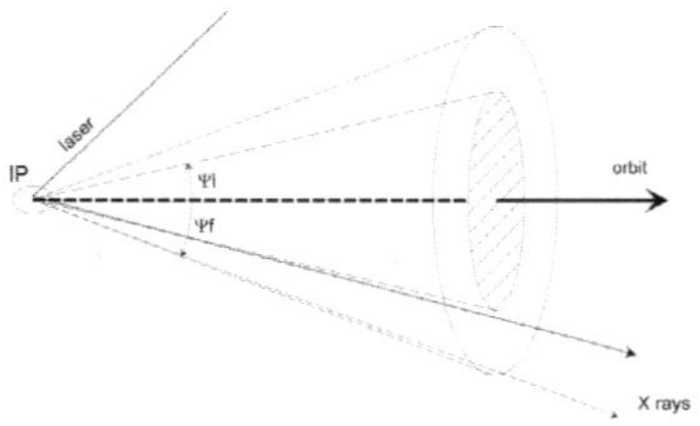

Figure 1. Scheme of the Compton radiation collimation.

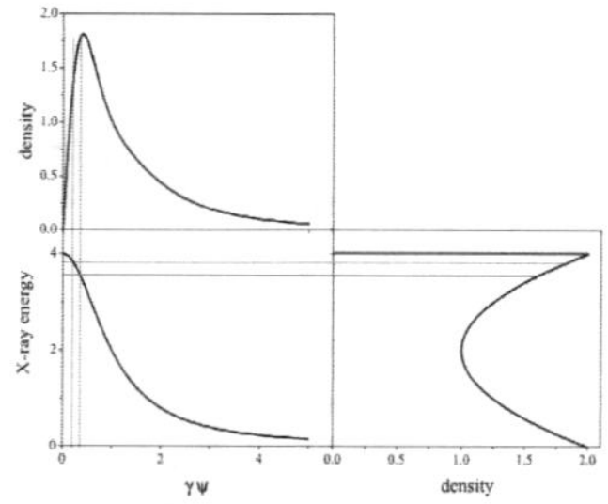

Figure 2. Spectral-angular density of the Compton radiation.

For the vast majority of practical sources the inner collimation angle is zero, $\psi_i = 0$. We will consider below this particular case of an iris collimator.

In general, the spectrum is dependent on the following factors: the collimator opening angle, the energy spread of electrons and photons and the angular spread of their trajectories at the interaction point (IP). In practical cases, the angular spread of photon 'trajectories' produces least impact on the spectrum, as it can be seen from the kinematics: Its effect is proportional to the spread squared because IP is usually set up at the laser waist, while the electron spread impact is γ^2 times larger. Also the spread of electron energy within the bunch usually much higher than that of photons. Therefore below we will consider effects of the electron bunch phase volume – spreads of energy and trajectories – upon the collimated spectrum of scattered off laser quanta.

2.1. Spread of electrons energy

The energy spread dilutes the spectral–angular density curve in Fig.2 along the energy axis. Since the energy spread in bunches of Compton sources is usually small – from a few percents down to fraction of a percent, – the corresponding partial energy spread of x–rays is equal to doubled the electrons': For the normal (Gaussian) distribution of electrons' energy in the bunch, the 'pin-hole' collimated x–ray beam has the doubled reduced dispersion: $\sigma_x/E_x \approx 2\sigma_e/E_e$. (Detailed study on the spectrum for head-on collision is presented in [2].) Analytical collimated spectra for finite energy spread of electrons with parallel trajectories are presented in Fig.3.

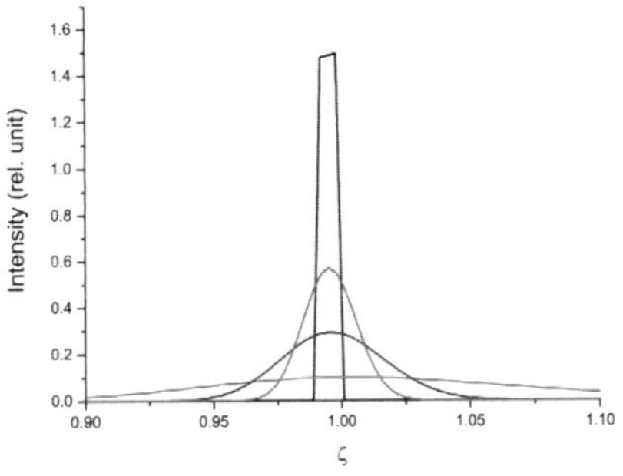

Figure 3. X-ray spectra for the electron energy spreads $\sigma_e\gamma = (0, 0.01, 0.02, 0.03)$ into collimating range $\psi\gamma = (0, 0.1)$.

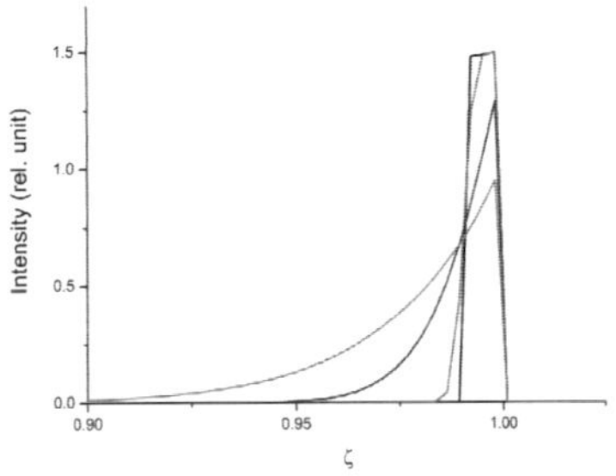

Figure 4. X-ray energy spectra for the electron trajectories spreads $\sigma_e\gamma = (0.01, 0.1, 0.5, 1.0)$ into collimating range $\psi\gamma = (0, 0.1)$.

2.2. Angular spread of electrons trajectories

The angular spread of electrons' trajectories induces much more widening into the spectrum since the radiation is emitted within narrow cone along the trajectory. (It should be noted that a quantum emitted at the angle $1/\gamma$ to the electron trajectory has a half of the energy of that emitted along the trajectory.) Also tight focusing of the bunches at the interaction point aimed to gain the yield of x rays, increases the angular spread.

A collimated spectrum with account for the angular spread has asymmetric and more complicated shape than that due to the energy spread. It is specified with steep high–energy cutoff (stemmed from the energy conservation law) and long low–energy tail, see Fig.4.

Examples of simulated spectra for different collimation opening angles is presented in Fig.5 (the parameters for simulation were taken similar to those of LUCX facility, see [3]).

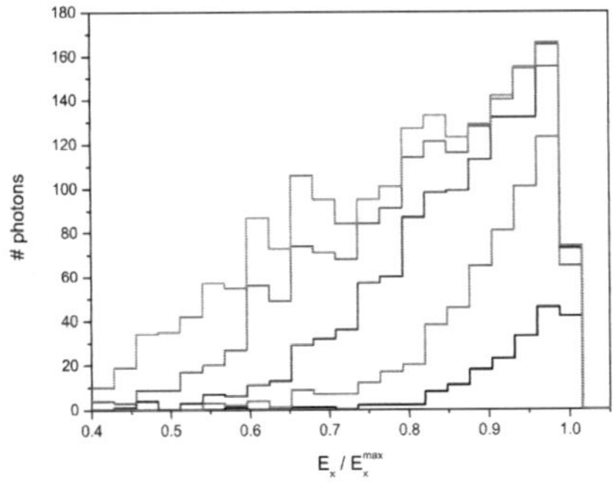

Figure 5. Spectra for collimation $\psi\gamma = 0.2, 0.4, 0.6, 0.8, 1.0$ (from the narrowest to widest profiles). Angular horizontal and vertical spreads in the bunch are $\gamma\sigma_{x,y} = 0.08, 0.32$.

From the study on the collimated x-ray spectra generated by Compton sources the points are followed:

- The spectrum width is limited by the high energy cutoff, no photons with higher energy.
- The spectrum has the maximum close to the high-energy cutoff.
- High-energy cutoff of the spectrum is steep.
- The collimator opening angle should not be more narrow than the (approximately doubled) angular spread of electron trajectories at IP. More narrow collimation will reduce the spectral maximum but not the width.

3. Application of Compton sources for angiography

Angiography, a medical x–ray based imaging technique to visualize the inside of blood vessels and organs of the body, uses a radio-opaque contrast agent injected into the blood vessel. The difference between x–ray images without and with the contrast agent in the blood produces a picture of the blood flow.

Iodine–based radiocontrast agents are most commonly used in angiography with maximum contrast energy of x–rays just above the K–edge of iodine, 33.17 keV. Hence, the x–rays within a band of ∼30–35 keV most effectively produce the images. Other energies are redundant and even detrimental since they cause additional radiation load.

The Compton x–ray sources with the tunable energy peak possess certain advantages over the conventional bremsstrahlung sources. In Fig.6, there an ideal Compton spectrum and a bremsstrahlung one are presented that produce equal fluxes within 30–35 keV energy range.

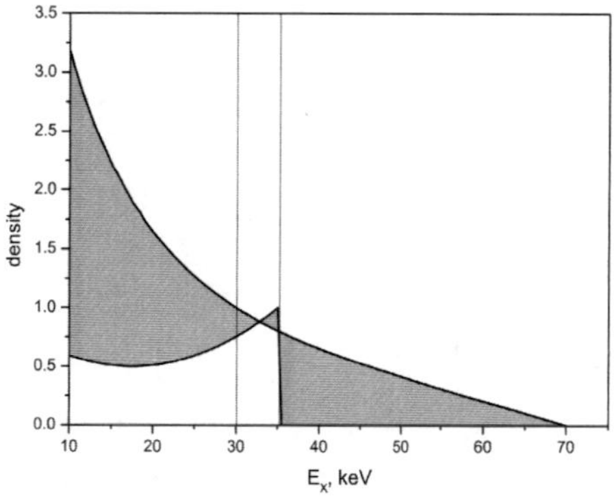

Figure 6. Bremstrahlung (70 keV electrons and tungsten radiator) and Compton spectra. the vertical lines limit a useful x-ray energy band, the gray areas beyond show redundant x–ray radiation load from the bremsstrahlung spectrum.

As it can be seen from the picture, the conventional source produces much more background x–ray quanta (the gray areas beyond 30–35 keV range).

3.1. Simulation of Compton angiography

For verification of Compton sources applicability for angiography, a digital model has been created. The models simulates the image of the tissue composed from the skeletal muscle, the bone cortical, blood and iodine components (x–ray attenuation coefficients were taken from NIST [5]). Distribution of the energy of x–ray quanta is simulated with Monte Carlo method according to the collimate ideal Compton spectrum. Distribution of the impinging quanta over the tissue face is uniform, random.

A run of simulation on the model reveals applicability of Compton x–ray source for angiography as is presented in Fig.7 (left). The model input data was: x–ray energy ranges 23.3–35 keV; number of quanta in the range 2×10^7 (total number over the spectrum 2×10^7); the sensor mesh 100 × 100 pixels (1 mm^2 pixels on 10 × 10 cm^2 tissue); the surface densities of tissue's substances are as follows (in g/cm^2): muscle 5, bone 0.5, blood 0.5, iodine 0.0125. The muscle component is uniform density over the tissue, the bone–like substance is added at the bottom ($2 < y < 4$ cm), the blood one at the top of the tissue ($7 < y < 8$ cm), iodine is placed in the left half of the blood tape ($x < 5$ cm).

To compare the left half of '35 keV' image in Fig.7 (blood with the contrast agent) with the right half (no contrast agent), one can detect location of the iodine. Total flux density for the simulated case is 4×10^5 quanta per cm^2.

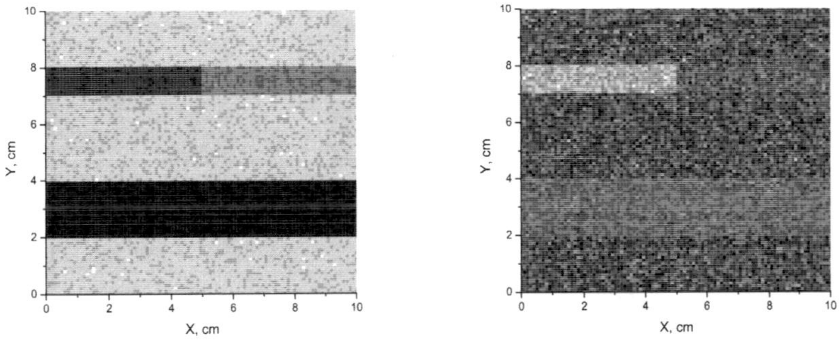

Figure 7. X–ray image with $E_x^{max} = 35\,\mathrm{keV}$ (left) and a subtracted '30 keV' minus '35 keV' image (right).

3.2. Advanced method of Compton angiography

Specific properties of the Compton x–ray radiation spectrum – steep high-energy cutoff with the maximum near it, and strong dependence of the maximal energy upon the energy of electrons, – enable us to propose an advanced method of Compton angiography.

The method consists in following. Dissimilar to the conventional angiography where location of the contrast agent is revealed as a result of subtracting the image without the agent from that with it, the both images are taken with presence of the contrast agent but one of them at the maximum energy of the spectrum tuned just below the K–edge of the agent, another – above it. Difference between these images will display the contrast agent location.

Simulations of the method have validated it. In Fig.7 (right panel) a subtracted image is presented, the image of the described above tissue at the maximum energy of Compton x–rays of 35 keV (left panel) was subtracted from the image with the maximum energy 30 keV. Since opaqueness of the agent increases with increasing the energy above K–edge while the other components become more transparent, the location of agent in the difference image is lighter than the others (see Fig.7, right panel).

Thus, the advanced method is capable to reveal the location of the contrast agent much faster than the conventional Compton angiography method with about the same flux of x–ray quanta. Speed up of the imaging is due to the fact that injection of the contrast agent into the blood vessel will take much more time than switching of the accelerator energy (we believe it will take a fraction of second).

3.3. 'Proof–of–principle' experiment proposal for LUCX

The 'proof–of–principle' experiment must verify the basic suggestion for Compton x–ray source: opaqueness of a contrast agent increases with increasing the maximal energy of spectra above the K-edge of contrast agent. Also ability of Compton sources to gather sufficient statistics revealing the location of the contrast agent should be proved.

Such an experiment is proposed to conduct at LUCX facility [3, 4]. Since the energy of Compton x–ray photons in this accelerator with YAG laser (1 eV) can not overlap the K–line of iodine, we propose to use bromine as a contrast agent with K-edge at 13.4737 keV, see [5].

We have simulated a 'proof–of–principle' experiment for LUCX. Results of the simulation is depicted in Fig.8. The sample of 20×20 cells was irradiated by 2×10^6 quanta of each energy, $E_x^{(max)} = 13, 16\,\mathrm{keV}$ (the images 'A' and 'B', respectively). The Compton spectrum was collimated to $1/\gamma$ opening angle. The sample thickness (muscle skeletal) was $0.5\,\mathrm{g/cm^2}$ with additional $0.15\,\mathrm{g/cm^2}$ in top left corner, bromine thickness was $10\,\mathrm{mg/cm^2}$ (bottom left corner).

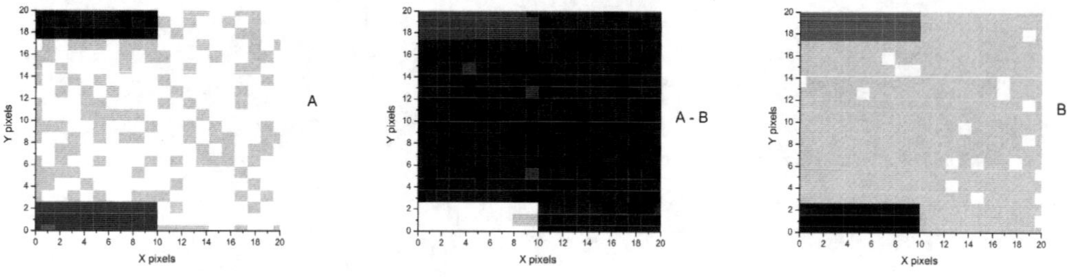

Figure 8. '13 keV' image (A, left), '16 keV' image (B, right) and the subtracted density (A-B, centre).

As is follows from the images, one cannot tell from single run (A and B images) which of the shadows induced by bromine or by an additional thickness. The bromine shadow is thinner when the spectrum does not overlap the K–peak, 'A' image. The subtracted image, 'A-B' clearly indicates which of shadows belong to the additional thickness or to the bromine – the contrast agent is more transparent for lower energy inverse to the regular behavior.

4. Summary and conclusion

X–ray radiation generated by the Compton sources has the spectrum with the steep high–energy cutoff, which is independent on the collimation opening angle. The specific shape of the spectrum with tunable maximum is able to substantially reduce radiation load for the angiography procedure.

We propose a novel angiography procedure consisting in deriving a subtracted image from the two images having been made with different maximum energy of the Compton spectrum, one with the energy below the K–edge of a contrast agent, another – with the maximum energy overlapped it.

The carried out simulations proved the suggestion of advantage the Compton subtracted scheme for angiography. Also we proposed a 'proof–of–principle' experiment that can be conducted at LUCX facility of KEK employing bromine as a radiocontrast agent.

Acknowledgments

Authors wishing to acknowledge Prof. A. Dovbnya, Drs. P. Gladkikh and V. Skomorokhov for their assistance and helpful discussions. Work is supported by the Photon and Quantum Basic Research Coordinated Development Program by the Ministry of Education, Culture, Sports, Science and Technology 'Fundamental Technology Development for High Brightness X-ray Source and the Imaging by Compact Accelerator.'

References

[1] Bulyak E and Skomorokhov V 2004 *Proc. Eur. Part. Accel. Conf. 2004* report THPKF063
[2] Sun C, Li J, Rusev G, Tonchev A P and Wu Y K 2009 *Phys. Rev. ST-AB* **12** 062801
[3] Fukuda M, Araki S, Aryshev A, Honda Y, Terunuma N, Urakawa J, Sakaue K and Washio M 2013 *Proc. NAPAC'13* report TUPMA01
[4] Sakaue K, Washio M, Fukuda M, Honda Y, Terunuma N, Urakawa J 2013 *Proc. IPAC 2013* report WEPWA017
[5] `http://www.nist.gov/pml/data/xraycoef/`

Journal of Physics: Conference Series **517** (2014) 012002 doi:10.1088/1742-6596/517/1/012002

Radiation of a laser-channeled electron

E N Frolov[1,2], A V Dik[1] and S B Dabagov[1,3,4]

[1] LPI RAS, Moscow, Russia
[2] NR TPU, Tomsk, Russia
[3] NRNU MEPhI, Moscow, Russia
[4] LNF INFN, Frascati (RM), Italy

E-mail: frol.onn@gmail.com

Abstract. The possibility of electrons trapping by the channels formed by electromagnetic waves may open new techniques for beam reflection and steering. This work describes relativistic electron dynamics in the field formed by crossed laser fields together with the emitted radiation spectrum at such propagation. The plane wave approximation for the laser field is used, and the results of both analytical and numerical analysis are provided.

1. Introduction

Nowadays channeling of both charged and neutral particles has been widely applied for beam shaping, deflection and transportation. And, moreover, the very term "channeling" that originates from specific propagation of charged particles in aligned crystals (crystal channeling) [1, 2] is getting wider today absorbing new phenomena that could be successfully described within channeling phenomenology, from X-ray quanta channeling in nanostructures [3] and propagation of electrons in plasma media [4, 5, 6] or nanotubes [7, 8, 9] to controlled particles movement in the fields of different nature [10, 11, 12, 13] that form, let say, "potential channels". Hence, suppose we have created a system to be characterized by a set of well oriented channels, in other words, by periodic potential wells with respect to the charged projectiles. Propagation of particles in such systems is to be considered as channeling, since they perform oscillations in periodic potential channels and since such oscillations are orientation-dependent with defined critical angle. Together with such similarities, the phenomena are united not only by the methods used for their description but also by the application fields. Nevertheless, some particular points about these channeling processes differ essentially.

In this work the motion of relativistic electron trapped by the potential well formed by standing electromagnetic waves is considered. The potential well (channel) could be generated either by crossing laser beams or by electromagnetic waves propagating in a waveguide. On the contrary to electron channeling in crystals, in this specific case no medium is present inside the channel, hence, no inelastic processes would take place at propagation in such a field. Besides, channeling phenomenon in crystals is known mostly for fast particles (to prevent particle diffraction in crystals), since, their large longitudinal momenta allow averaging the atomic raw/plane potential along the direction of the projectiles that form channels inside the crystal (the basics of channeling). Typically, for channeling experiments the beam alignment has to be done along the main crystallographic directions, passing nearby of which the beams of relativistic charged particles might be channeled. But in the case of, let us call it so, "laser-channeling" an averaged potential could be calculated for nonrelativistic projectiles too that was previously shown [11, 14]. The dynamics of relativistic beams, namely, relativistic laser-channeled electrons is analyzed in this work. Apart of additional interest applying the proposed technique for hadron beams shaping, which is under investigation by our group and a subject of separate publication, we have first presented below spectral characteristics of relativistic laser-channeled electrons comparing this type of radiation with channeling radiation in crystals.

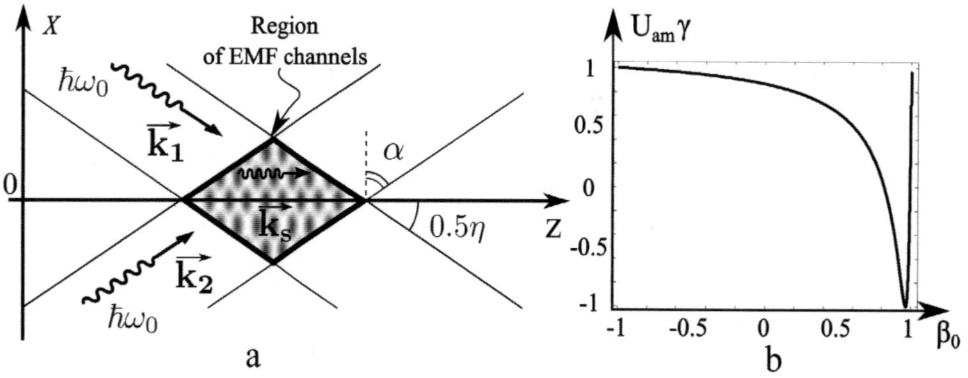

Figure 1. a: General system geometry. The region, where standing waves form the channels, is separated in the center. The peaks of transverse electrical component of the field fast propagating in the $\mathbf{k}_s = \mathbf{k}_1 + \mathbf{k}_2$ direction shapes the potential barrier. b: Dependence of the barrier height of a channel potential (normalized by the value $[e^2u_0^2k^2]/[8m\omega_0^2]$) from the particle velocity normalized by the speed of light.

2. Basics of channeling in a laser field

Channeling phenomenon takes place during the propagation of relativistic particles through a crystalline target under small angles to the main crystallographic planes (for planar channeling) or axes (for axial channeling). Due to its velocity, the particle moving along either the plane or the axis barely feels each separate atom rather than moves in an averaged field of the whole atomic plane or string. Such a concept of continuum potential approximation for the description of the inner structure potential for a monocrystalline target was introduced by Lindhard [2]. The potential well forms longitudinally equipotential channel preferable for the particle propagation. When the transverse energy of a particle is less than potential barrier height it becomes channeled and performs transverse oscillations inside the potential well together with relativistic motion along the channel. On the one hand, such a classical treatment is common for channeling of charged particles in crystals, on the other hand, similar approach could be applied to a grasp of other cases, i.e. to a particle propagating either in the field of crossed laser beams, or in a waveguiding structure together with electromagnetic waves. Being mathematically similar, these two latter situations will be described below, giving an example of electron channeled in electromagnetic field.

For simplicity let us consider laser beams wave surface intensity distribution to be uniform and the problem to be two-dimensional (shown in figure 1,a), thereby two plane electromagnetic waves of the same frequency ω_0 and amplitude u_0 are aligned with some nonzero angle to Oz-axis so that the sum of their wave vectors $\mathbf{k}_s = \mathbf{k}_1 + \mathbf{k}_2$ is co-directional with Oz (longitudinal). The magnetic components of both electromagnetic waves are aligned normal to xOz, so all the forces will be situated in the xOz-plane. Having fixed z-coordinate and measured either

x or z components of electrical field, one can reveal standing electromagnetic waves that have nodes and antinodes at some particular points. On the contrary, fixing x-coordinate at any point between nodes and measuring a component of electrical field along the line parallel to Oz, electromagnetic wave propagating along longitudinal direction could be resolved.

A charged particle in such a field could moving along a complex trajectory defined by a sum of small rapid oscillations and slow averaged movement. The former is a result of the particle interaction with both peaks and wells of electromagnetic field passing by, the latter is a result of particle motion in averaged potential of described system. As shown in [14] for nonrelativistic electrons, potential channels parallel to $\mathbf{k}_s$ appear in such systems, that results into transverse averaged oscillations near the potential channel bottom. So, there is a preferable direction for particle motion, and if its transverse energy is less than potential barrier height, the particle can be trapped and becomes channeled in a laser field.

3. Field potential and particle dynamics

Let us consider relativistic electron in a field characterized by the following potential

$$\varphi = 0; \quad \mathbf{A} = -u_0 \frac{k_z}{k_x} \cos(k_x x) \cos(\omega_0 t - k_z z)\mathbf{e}_x + u_0 \sin(k_x x) \sin(\omega_0 t - k_z z)\mathbf{e}_z \ , \tag{1}$$

where $k_{x,z}$ are the absolute values of transverse and longitudinal components of $\mathbf{k}_{1,2}$, u_0 is the vector potential amplitude, ω_0 is the characteristic field frequency. Since the terms of the right side of the expression for a vector-potential can posses comparable values for relativistic particle, we have to analyze this expression keeping all terms without any approximation at the beginning. Such a field configuration could be the result of either crossing two laser beams or propagation of electromagnetic wave inside an infinite planar waveguide parallel to Oz-axis. As above mentioned, the particle trajectory can be presented as a sum of averaged smooth component $\bar{x}_i$ and rapid oscillations χ_i, $x_i = \bar{x}_i + \chi_i$. The latter, on the contrary to the former term $\bar{x}_i$, is characterized by both small amplitude and high frequency. Solving the equation of motion for χ_i, we can reduce high oscillation part of the trajectory

$$[\chi_x(t),\, \chi_z(t)] = \frac{eu_0 k}{m\omega_0^2(1-\beta_0 \sin\alpha)^2} \left[\frac{\sin\alpha - \beta_0}{\bar{\gamma}},\, \frac{\cos\alpha}{\bar{\gamma}^3}\right] \cos\Phi_1 \sin\Phi_2 \ , \tag{2}$$

where $\Phi_1 = k\bar{x}\cos\alpha$, $\Phi_2 = \omega_0(1-\beta_0\sin\alpha)t - k_z\bar{z}_0$, e and m are the electron charge and rest mass, $\chi_{x,z}$ are the small additions to the particle averaged trajectory, $\bar{\gamma} = \left\{1 - \left[(d\bar{x}/dt)^2 + (d\bar{z}/dt)^2\right]/c^2\right\}^{-1/2}$ is the Lorentz factor, β_0 ($\approx \beta_z$) is the initial relativistic electron velocity and $k = \omega_0/c$. The averaged trajectory $\bar{x}_i$ is defined by the motion equations

$$m\frac{d}{dt}\left(\bar{\gamma}\frac{d\bar{x}}{dt}\right) + A(\omega_0, \alpha, \beta_0)\sin(2k\bar{x}\cos\alpha) = 0 \quad \text{and} \quad m\frac{d}{dt}\left(\bar{\gamma}\frac{d\bar{z}}{dt}\right) = 0, \tag{3}$$

where α is the angle of the wave vector $\mathbf{k}$ with respect to the Ox-axis. Neglecting small rapid oscillations and, in the first approximation, assuming γ to be defined only by z-velocity, $\gamma \approx 1/\left(1-\beta_0^2\right)^{-1/2}$, we can write down the averaged potential

$$U_{eff} = U_0 - \frac{e^2u_0^2k^2(-\cos(2\alpha) - 2\beta_0\sin\alpha + (1+\cos^2\alpha)\beta_0^2)}{8\gamma m\omega_0^2(1-\beta_0\sin\alpha)^2\cos^2\alpha}\cos(2kx\cos\alpha) \tag{4}$$

The channel potential depth (or height) is a function of the system geometry and particle characteristics: $U_{am} = U_{am}(\alpha, u_0, \omega_0, \beta_0, m)$. And it defines the effective potential $U_{eff} =$

$U_0 - U_{am}\cos(2kx\cos\alpha)$. Having fixed all described parameters except β_0, we can get the dependence of a channel potential depth on the particle velocity (shown in figure 1,b).

In dependence with the particle velocity β_0, from the analysis we can reveal the change in a channel spatial position. Herein it is important to remind that the electron z-velocity might be either co-directed or opposite directed with respect to high-frequency sum laser field propagation; that is the reason of a β_0 sign change. As seen from the dependence in figure 1,b, the potential depth curve suffers a qualitative change, from slowly decreased part at positive values to sharply felt down to the minimum at negative values. This feature corresponds to the potential change from attracting one to scattering one. For attracting potential the β_0-parameter is in the region of positive U_{am}, and the potential channels' axes lie at $x_n = (0.5+n)\pi/k_z$ while the peaks of the field are at $x_n = n\pi/k_z$. On the contrary, for scattering potential (for β_0 near the minimum of potential amplitude U_{am}) the potential channels' axes lie at $x_n = \pi n/k_z$ and the field peaks - at $x_n = (0.5+n)\pi/k_z$.

In the applicability range of analytical solutions for the system shown above the computer simulations are in rather good agreement with analytically derived results, allowing us to go beyond the range of small transverse energies. Supposing the particle oscillations in the field of this potential near the channel bottom, the averaged component of its trajectory is expressed by harmonic function of a fixed frequency ω_1. The total energy of channeled electron in such a case is much less than the height of the channel potential barrier. But higher total transverse energy of a particle results in less linear process of these oscillations and the trajectory becomes such as shown in figure 3,a with larger averaged oscillation period and amplitude.

Summing up, a charged particle moving in the described configuration of electromagnetic field could be trapped and successfully channeled. The dynamics of such a particle could be treated with the described methodology, approved by the numerical experiments conducted. Of course, the radiation of such a channeled electron is of great interest either for diagnostic purposes, or for future applications. First results of its analysis is presented below.

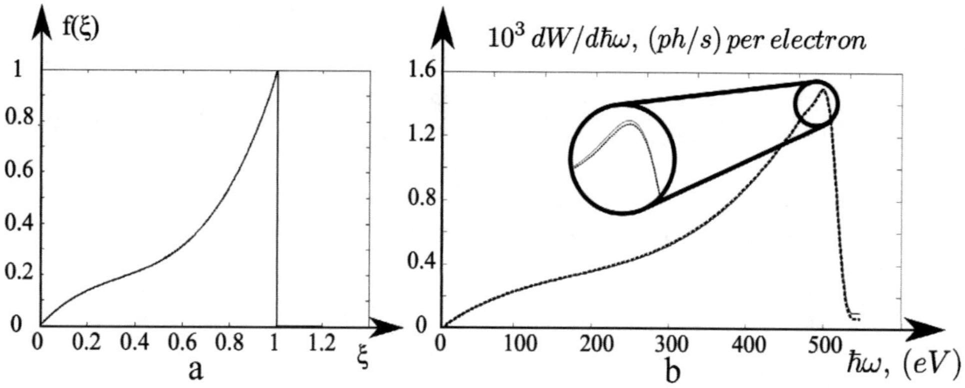

Figure 2. a: Dimensionless spectral power distribution $f(\omega) = (c^3/e^2(\omega_2)^3 a_x^2\gamma^2)dW/d\omega$ versus dimensionless frequency $\xi(\omega) = \omega/2\gamma^2\omega_2$ for a large number of small oscillations. b: Matching analytical and numerical calculations for the photon yield from 20 rapid oscillations of one relativistic electron initially positioned near channel bottom with $\beta_0 = 0.9997$, zero initial transverse velocity channeled in a field of laser beams of frequency $w_0 = 7.5 \cdot 10^{15}\ Hz$ and intensity 300 MV/m crossed at the angle of $\eta = 30°$.

4. Radiation of a relativistic laser-channeled particle

The analytical spectral intensity distribution was found for electron moving close to the bottom of the potential channel. For the large smooth oscillations the spectral intensity distribution was calculated numerically. Let consider the case of small-amplitude averaged oscillations and particle velocity, close to the speed of light, directed along the channel. Then we can rewrite the trajectory equations in the assumption of absence of longitudinal particle oscillations, since they are γ^2-times less than transverse ones for the ultra-relativistic case

$$z(t) \approx v_0 t \quad \text{and} \quad x(t) \approx a_1 \sin(\omega_1 t) + a_2 \sin(\omega_2 t)\ , \tag{5}$$

where $\omega_1^2 = \left(2eu_0 k^2 \cos\alpha\right)^2 \left(\gamma m \omega_2\right)^{-2} \left|-\cos(2\alpha) - 2\beta_0 \sin\alpha + (1+\cos^2\alpha)\beta_0^2\right|$ and $\omega_2 = \omega_0(1 - \beta_0 \sin\alpha)$. Thus, the radiation by the electron at large distances is defined by the expressions

$$[A_x, A_z, \varphi] = -\left[\frac{(1-\beta_0\cos\theta)}{\sin\theta\cos\phi}, \beta_0, 1\right] \frac{e}{r(1-\beta_0\cos\theta)} \sum_{n,m} e^{-i\omega_{n,m}(t-r/c)} J_n(\nu a_1)\, J_m(\nu a_2) \tag{6}$$

where $\omega_{n,m} = (n\omega_1 + m\omega_2)/(1-\beta_0\cos\theta)$ and $\nu = (\omega(n,m)\sin\theta\cos\phi)/c$. Assuming the frequency of averaged oscillations is a multiple of the frequency of rapid oscillations $\omega_2 = N\omega_1$ (which is surely possible for some strict system and particle parameters) and using expressions (6) it becomes possible to express the expansion of the radiation field potentials in spherical harmonics, and, hence, to write down the spectral intensity distribution[1] [15, 16]

$$\frac{d^2W}{do d\omega} = \frac{\omega_1}{2\pi c N_0} \sum_{n=1} n^2 \frac{\sin^2\left(\pi N_0 \left(\frac{\omega(1-\beta_0\cos\theta)}{\omega_1} - n\right)\right)}{\pi^2 \left(\frac{\omega(1-\beta_0\cos\theta)}{\omega_1} - n\right)^2} \left(|A_n|^2 - |\varphi_n|^2\right) \tag{7}$$

Here N_0 is the number of full averaged oscillations. For small values θ we deal with small-amplitude averaged oscillations with respect to the width of potential well and the field, where $\omega_2/\omega_1 = N \gg 1$. Hence, the radiation spectrum peaks are located at frequencies $\Omega_1 {=} \omega_1/(1-\beta_0\cos\theta)$, $\Omega_2 {=} \omega_2/(1-\beta_0\cos\theta)$. And taking into account that the radiation by ultra relativistic particle fans out in a narrow forward direction cone and performing integration by the angle, we can write down the expression of radiation spectral power distribution for relativistic laser-channeled electron

$$\frac{dW}{d\omega} = \sum_{i=1}^{2} \frac{e^2\omega_i^3 a_i^2 \gamma^2}{c^3} \zeta_i \left(1 - 2\zeta_i + 2\zeta_i^2\right) \Theta\left(\pi N_i (1-\zeta_i)\right) \tag{8}$$

where N_i is the numbers of ω_i-frequency oscillations, $\zeta_i = \omega/(2\gamma^2\omega_i)$, $\Theta(x) = 0.5 + \mathrm{si}(2x)/\pi - \sin^2(x)/\pi x$. The expression (8) shows that for small-amplitude averaged oscillations the full radiation spectrum includes two curves of a similar form (see figure 2,a) and different absolute amplitudes as well as maximum frequencies.

The numerical calculation of spectral radiation distribution for relativistic particle can be performed by integration of well-known expressions [16]. The simulation results match analytically calculated curves for the case of small averaged oscillations radiation (see figure 2,b). Taking into account rather good correspondence of analytical and numerical results, shown in figure 2,b, the code has been used to calculate the radiation by relativistic electron in the case of large-amplitude averaged oscillations shown in figure 3,a, which goes beyond analytical treatment. The radiation spectrum for this trajectory is shown in figure 3,a. Obviously, several peaks in figure 3,b characterize spectral structure of nonlinear $\bar{x}(t)$ function harmonic expansion. The less is averaged oscillations amplitude, the more sinusoidal the electron trajectory $\bar{x}(t)$ becomes and the closer the results of figure 3,b gets to the ones of figure 2,a.

[1] This condition is not necessary for $N \gg 1$ case, while for other values of N it is important

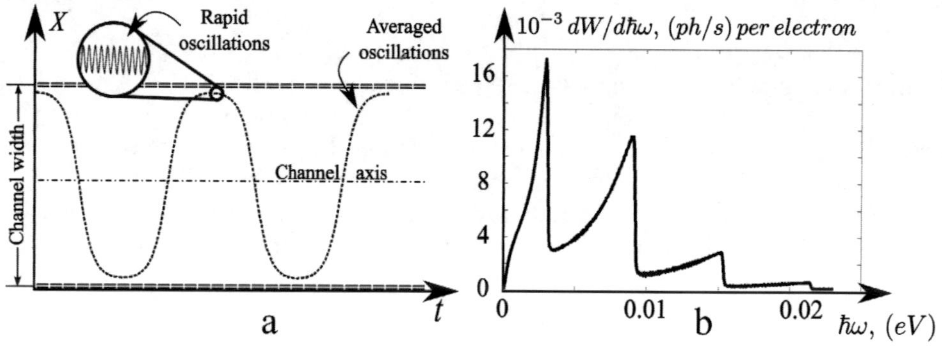

Figure 3. a: Numerical simulation for the large amplitude averaged oscillations of relativistic channeled electron with total transverse energy close to the height of the potential barrier. b: Numerical results for the photon yield from 20 large-amplitude averaged oscillations of the same electron initially positioned near channel border in the same fields system.

Conclusions

The dynamics of relativistic electron in electromagnetic field inherent for the system of two crossed laser beams or for the infinite biplanar waveguiding structure carrying the electromagnetic wave was described. The possibility and characteristics of relativistic electron channeling regime in such a configuration was shown. Analytical expressions describing the electron dynamics in the system were supported by fully coincident numerical experiment results. The averaged potential function (4) presents the way for manipulating channel width and height by varying the parameters of laser beams or electromagnetic wave traveling the waveguide. This channeling regime of a charged particle in electromagnetic field could be interesting for future applications on accelerators for beams steering and shaping; especially, with curved axial waveguiding structures implementation. The second interesting point described is relativistic electron radiation in the considered system, which looks similar to channeling radiation: it is shifted to the high energy region due to the Doppler effect and has a tunable spectral peak frequency depending on both system field and particle parameters. Deeper studies are needed to answer the question on a new radiation source based on electron channeling in a crossed-laser field. Comparatively small photon yield obtained could be, for instance, compensated by the number of electrons, since the channel width for an IR laser is $\sim 10^{-5}$ cm.

This work was supported by the Ministry of Education and Science of the Russian Federation in the frames of Competitiveness Growth Program of National Research Nuclear University MEPhI, Agreement 02.A03.21.0005.

References

[1] Gemmel D S 1974 *Rev Mod Phys* **46 (1)** 129
[2] Lindhard J 1965 *K. Dan. Vidensk. Selsk. Mat.-Fys. Medd.* **34 (14)**
[3] Dabagov S B 2003 *Phys. Usp.* **46** 1053
[4] Mangles S P D, Walton B R, Tzoufras M *et al* 2005 *Phys. Rev. Lett.* **94 (24)** 24500
[5] Pukhov A, and Ter Vehn J M 2002 *Appl. Phys.* B **74** 4
[6] Dik A V, Ligidov A Z, and Dabagov S B 2013 *Nucl. Instr. Meth.* B **309C** 210
[7] Artru X, Fomin S P, Shul'ga N F, Ispirian K A, and Zhevago N K 2005 *Phys. Rep.* **412** 89
[8] Dabagov S B, and Zhevago N K 2008 *Riv. Nuovo Cimento* **31** 491
[9] Karabarbounis A, Sarros S, and Trikalinos Ch 2013 *Nucl. Instr. Meth. Phys. Res.* B **316** 160
[10] Kapitza P L, and Dirak P A M 1933 *Math. Proc. Cambridge* **29 (02)** 297
[11] Andreev A V, and Akhmanov S A 1990 *Zh. Eksp. Teor. Fiz.* **99** 1668
[12] Balykin V I, Subbotin M V, and Letokhov V S 1996 *Opt. Commun.* **129** 177
[13] Artru X, Ispirian K A, and Ispiryan M K 2007 *ArXiv e-prints* 0707.0148
[14] Frolov E N, Dik A V, and Dabagov S B 2013 *Nucl. Instr. Meth.* B **309C** 157
[15] Landau L D, and Lifshitz E M 1988 *Field Theory* Moscow, Nauka
[16] Bagrov V G, Bisnovatyi-Kogan G S, Bordovitsyn V A *et al* 2002 *Radiation theory of relativistic particles* Moscow, Fizmatlit

RREPS13 and Meghri13 IOP Publishing
Journal of Physics: Conference Series **517** (2014) 012003 doi:10.1088/1742-6596/517/1/012003

X-Ray Cherenkov Radiation as a Source for Transverse Size Diagnostics of Ultra-relativistic Electron Beams

A S Konkov[1], P V Karataev[2], A P Potylitsyn[1] and A S Gogolev[1]

[1] Tomsk Polytechnic University, Lenin Ave. 30, Tomsk, 634050, Russian Federation

[2] John Adams Institute at Royal Holloway, University of London, Egham, Surrey, TW20 0EX, UK

E-mail: anatoliy.konkov@gmail.com

Abstract. Development of advanced non-invasive methods for charged particle beam diagnostics is crucial in modern and future accelerators. Therefore, it is essential to develop non-invasive methods to generate EM radiation which properties depend on the beam parameters. In this paper we propose to use X-ray Cherenkov radiation (XCR) which can be generated by ultrarelativistic charged particles in the vicinity of the absorption edges of several materials. Due to a resonant nature of the refractive index near the absorption edges the generated radiation is quasi-monochromatic. On the other hand its properties depend on the parameters of the incident particle beam. In this paper we represent an approach for calculating the XCR generated along with transition radiation and diffraction radiation. The effect of the transverse electron beam size on the XCR characteristics is discussed.

1. Introduction

Polarization radiation (PR) generated by relativistic charged particles has been intensively studied as a source for various applications. It appears as a result of dynamic polarization of a medium [1]. Easily detectable photon yield resulted in development of precise methods for transverse beam size diagnostics using optical transition radiation (OTR) [2]. Due to a very low energy loss of the particles such PR types as diffraction radiation (DR) [3] and Smith-Purcell radiation [4] give opportunities to develop non-invasive methods for charged particle beam diagnostics. If the radiation wavelength is longer than the bunch length the radiation is emitted coherently, i.e. the photon yield is proportional to the squared number of particles in the bunch [1], enabling scientists to use the PR mechanism as an intense source of THz radiation with known polarization and spectral-spatial characteristics [5].

Transverse beam size and emittance diagnostics are one of the most challenging tasks in modern and future accelerators such as electron-positron linear colliders [6] or X-ray free electron lasers [7]. OTR and optical DR (ODR) based methods are the most promising ones. However, they benefit from the fact that the radiation is incoherent, i.e. each electron emits radiation independently. However in the state-of-the-art X-ray FELs the bunch length is very short and the longitudinal structure is such that some small fraction of the beam might generate radiation coherently in optical wavelength range [8] excluding OTR/ODR phenomena from the list of potential candidates for the beam size monitoring. Moreover, the diagnostics requirements are so strict that the resolution of the OTR/ODR techniques is limited by diffraction.

Keeping in mind that the transition radiation (TR)/DR spectra are very broad then one of the solutions is to use shorter wavelength of the generated radiation to reduce both the diffraction limit and the coherency effects. For example, the EUV TR monitor is being developed in Mainz (Germany) [9]. A similar approach was proposed for UV/X-DR monitor in Cornell electron synchrotron Test Accelerator (CesrTA, Cornell) [10]. However, the UV or even X-ray frequency is comparable to or shorter than the plasmon frequency of metals. It means that the photon yield is either significantly supressed or requires a very special generation and observation geometry. In [11] authors have proposed to use the parametric X-ray radiation for transverse beams size measurements. In this paper we propose to use X-ray Cherenkov radiation (XCR) as a promising source for extremely low emittance charged particle beam diagnostics.

Cherenkov radiation (CR) is generated when a charged particle moves in a medium with velocity larger than the phase velocity of light in that medium, i.e. $v > c/n$, where n is the real part of the refractive index. Since the refractive index is a function of wavelength and is reduced towards shorter wavelength, the CR in almost entire X-ray region is either significantly suppressed or absent because the real part of the refractive index is smaller than unity. However, in [12] the authors have realized that in the vicinity of the atomic absorption edges the refractive index at high frequencies experiences a resonant increase. In this case the CR can be generated in X-ray region. Moreover, due to the fact that the refractive index has a sharp peak in the dispersion curve, CR spectrum is quasi-monochromatic with the bandwidth defined by the dispersion resonance.

In [13] the XCR at the K-edge of carbon generated by 1 GeV electrons was observed experimentally. Later the XCR generated by moderately relativistic electrons in titanium and vanadium foils at 5 MeV [14] and in silicon foil at 10 MeV [15] and at 75 MeV [16] was measured and investigated.

In this paper we represent a theoretical approach for calculating the characteristics of the XCR along with transition/diffraction radiation generated from the target surface. We performed a detailed analysis of the spectral-spatial properties and their distortion due to the beam size effect.

2. Theoretical formalism

The theoretical approach is based on the method of *polarization currents* [17–20]. The field of the PR emitted by medium atoms excited (polarized) by the external field $\mathbf{E}^0$ of the passing particle with energy $\gamma = E/(mc^2) = 1/\sqrt{1-\beta^2}$ moving rectilinearly and with constant velocity $\mathbf{v} = \beta c$ in a substance (or in its vicinity) can be represented as a solution of *vacuum* macroscopic Maxwell equations. Here γ is the charged particle Lorentz-factor, c is the speed of light, and m is the mass of the particle. For a non-magnetic medium the density of the polarization currents in a right hand side of the equations can be written as

$$\mathbf{j}_{pol} = \sigma(\omega)\Big(\mathbf{E}^0 + \mathbf{E}^{pol}(\mathbf{j}_{pol})\Big), \tag{1}$$

where $\mathbf{E}^0 \equiv \mathbf{E}^0(\mathbf{r},\omega)$ and $\mathbf{E}^{pol} \equiv \mathbf{E}^{pol}(\mathbf{r},\omega)$ are the Fourier image of the particle filed in vacuum and the field generated by the currents induced in the medium. The medium conductivity $\sigma(\omega)$ is related to the dielectric permittivity $\varepsilon(\omega)$ as

$$\sigma(\omega) = \frac{i\omega}{4\pi}(1-\varepsilon(\omega)), \tag{2}$$

here $\omega = 2\pi c/\lambda$, with λ being the radiation wavelength. Solving the Maxwell equations in a wave-zone for a target of finite volume V_T one can derive an expression for magnetic field $\mathbf{H}^{pol}$

of the PR emitted by the medium atoms excited by the passing particle filed in the form

$$\mathbf{H}^{pol}(\mathbf{r},\omega)=\nabla\times\frac{1}{c}\int_{V_T}\sigma(\omega)\mathbf{E}^0(\mathbf{r}',\omega)\frac{\exp\Big(i\sqrt{\varepsilon(\omega)}|\mathbf{r}'-\mathbf{r}|\omega/c\Big)}{|\mathbf{r}'-\mathbf{r}|}d^3r'. \tag{3}$$

In this model we assume that the energy loss by the particle is negligibly small in comparison to its total energy. One should note that (3) is an exact solution of Maxwell equations which allows us to avoid solving the differential equation (1). When we took into account the second term in (1), the wave number ω/c in vacuum was simply replaced by $\sqrt{\varepsilon(\omega)}\omega/c$. Such replacement describes the *renormalization* of the particle field inside the medium due to the field of the polarization currents [18,21]. Despite of a simple form equation (3) describes all types of PR generated in a medium of an arbitrary conductivity and an arbitrary inhomogeneity (i.e. in a target of an arbitrary shape), which is the main advantage of the method.

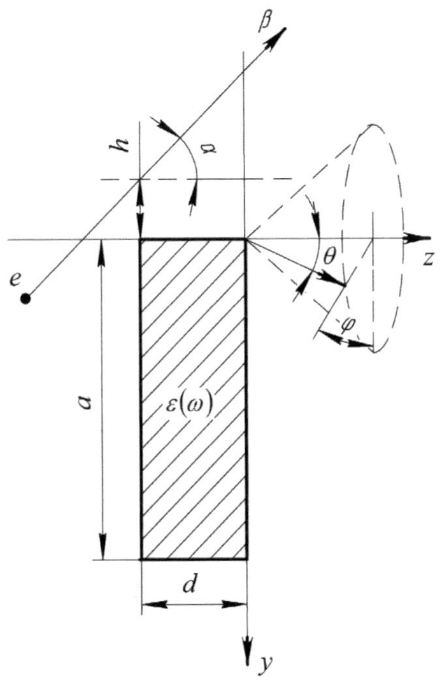

Figure 1. The calculation geometry.

Due to the fact that the polarization currents are induced in a finite volume determined by the target dimensions (see, for instance, Fig. 1), the solution of equation (3) can be found in the form:

$$\mathbf{H}^{pol}(\mathbf{r},\omega)=\frac{2\pi i}{c}\frac{\exp(ikr)}{r}\mathbf{k}\times\int_0^a dy'\int_{-d}^0\frac{\sigma(\omega)\mathbf{E}^0(k_x,y',z',\omega)}{\exp\Big(i(k_y y'+k_z z')\Big)}dz', \tag{4}$$

where $\mathbf{k}=\mathbf{e}\sqrt{\varepsilon}\omega/c$ is the wave vector, $\mathbf{e}=\mathbf{r}/\mathrm{r}$ is the unit vector along the photon emission. Taking into account the Snell's law of refraction the unit vector is

$$\mathbf{e}=\Big\{\sin\theta_m\sin\varphi;\sin\theta_m\cos\varphi;\cos\theta_m\Big\}=\frac{1}{\sqrt{\varepsilon}}\Big\{\sin\theta\sin\varphi;\sin\theta\cos\varphi;\sqrt{\varepsilon-\sin^2\theta}\Big\}. \tag{5}$$

Here θ_m and θ are the polar angles in the medium and in vacuum respectively, and φ is the azimuthal angle.

In (4) the integration is performed over the area confining the currents. In x direction, which is perpendicular to the plane of Fig. 1, we assume the screen is infinite. To solve the equation (4) we have to know the Fourier component of the charged particle field. Knowing a complete Fourier image of the field in the form [17–19]

$$\mathbf{E}^0(\mathbf{k},\omega) = \frac{4\pi i}{\omega}\frac{\mathbf{j}^0(\mathbf{k},\omega)(\omega/c)^2 - \mathbf{k}\Big(\mathbf{k}\mathbf{j}^0(\mathbf{k},\omega)\Big)}{\mathbf{k}^2-(\omega/c)^2} \tag{6}$$

we can derive an expression for the Fourier component as

$$\mathbf{E}^0(k_x, y', z', \omega) = \int_{-\infty}^{+\infty} dk_y \int_{-\infty}^{+\infty} \mathbf{E}^0(\mathbf{k},\omega)\exp\Big(i(k_y y' + k_z z')\Big)dk_z. \tag{7}$$

In (6) the Fourier transform of the particle currents density is

$$\mathbf{j}^0(\mathbf{k},\omega) = \frac{e\mathbf{v}}{(2\pi)^3}\delta(\mathbf{k}\mathbf{v}-\omega)\exp(ik_y h), \tag{8}$$

where e is the particle charge, $\mathbf{v} = v\{0; -\sin\alpha; \cos\alpha\}$ is the particle velocity vector, $\delta(\mathbf{k}\mathbf{v}-\omega)$ is the Dirac delta-function, h is distance from the particles trajectory to the edge of the target.

Substituting equations (2) and (6)–(8) in (4) and integrating over the target volume we can derive an expression for the field strength of PR.

3. Dielectric properties of Carbon (C)

According to [22] the dielectric permittivity of a medium is defined by the following equation

$$\varepsilon(\omega) = 1 - \delta(\omega) + i\eta(\omega). \tag{9}$$

This equation is a function of radiation frequency. For metals and many other opaque materials in X-ray wavelength range the real part is smaller than unity. In this case generation of CR is impossible because the Cherenkov condition can not be fulfilled. However, in the vicinity of absorption edges the real part of (9) experiences a resonant rise reaching values above unity, therefore the Cherenkov condition might be fulfilled for relativistic particles. An example of

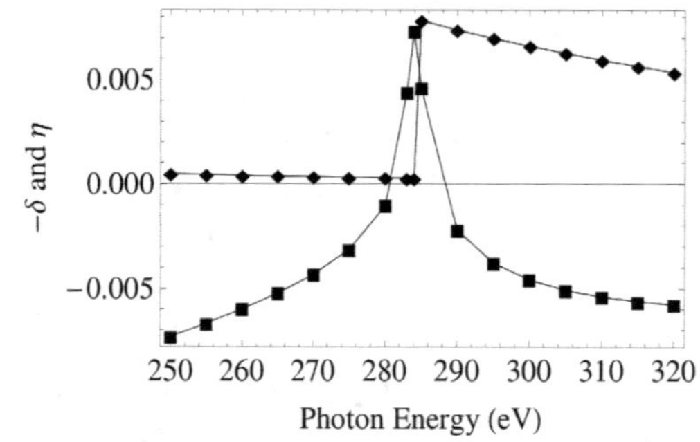

Figure 2. Dependence of the real $(-\delta)$ – ■ and imaginary (η) – ♦ parts of the dielectric constant of C in the vicinity of K-edge of absorption.

the such a rise in the vicinity of the K-edge of absorption of C is demonstrated in Fig. 2. The peak is rather sharp. One would expect the CR to be monochromatic. This phenomenon makes the CR attractive both for generating intense monochromatic X-ray radiation beams and its application for charged particle beam diagnostics.

4. Analysis of the radiation properties

In book [1] the authors demonstrated that in X-ray wavelength range when the particle moves in the vicinity of the target (diffraction Cherenkov radiation (CDR) geometry shown in Fig. 1) the CR has two maxima. Figure 3 illustrates the dependence of the PR intensity versus polar angle calculated for different incidence angles. The asymmetry appearing due to a small difference in the angle of incidence gives an opportunity to determine the angular jitter of the particle beam. In this geometry the calculation model is limited by the critical angle α_{cr}, i.e. the particle must not directly interact with the target. Its values will be determined by the following equation [19]:

$$\alpha_{cr} = \tan^{-1}\left(\frac{h}{d}\right). \tag{10}$$

If the particle actually crosses the screen (the geometry of transition Cherenkov radiation (CTR)) the model described by us in [20] can be used. In this case the asymmetry is smaller, but the CR intensity is higher than the TR intensity as it is shown in Fig. 3(*b*)). Moreover averaging over the beam angular divergence one may use the smoothing of the CTR peak as an indicator of the degree of the angular divergence.

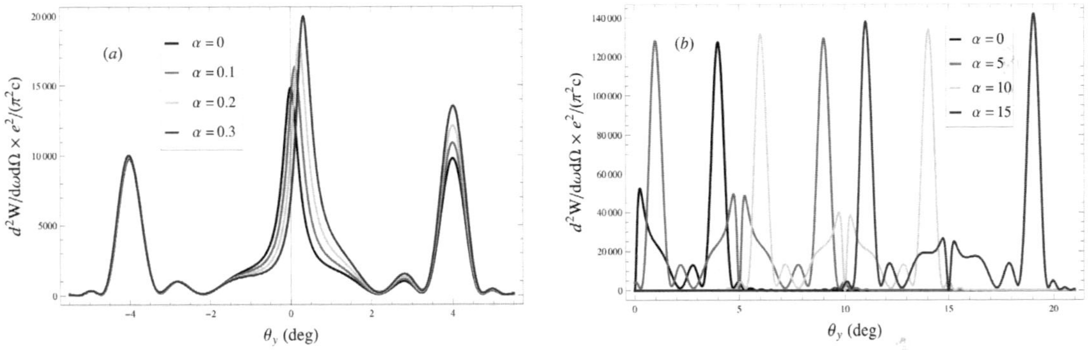

Figure 3. Dependence of the PR intensity on the projection observation angle $\theta_y = \theta\cos\varphi$ for different incidence angles α (degree): (a) for DR geometry; (b) for TR geometry. Parameters: $\gamma = 240$, $a = 70$ mm, $d = 5$ μm, $h = 0.1$ μm, $\hbar\omega = 284$ eV (K-edge of Carbon), $\varphi = 0, \pi$.

The photon yield can be optimized if we use the targets made of different materials. For example, Figure 4 presents dependences of the photon yield angular distributions calculated for Titanium ($\hbar\omega = 453.8$ eV), Silicon ($\hbar\omega = 99.8$ eV), and Aluminum ($\hbar\omega = 72.6$ eV) in the vicinity of L-edge of absorption. To show all these curves on a single plot the intensities were scaled accordingly. The scaling factors are shown on the graph in brackets. The Si targets (see Fig. 4(a)) are promising because even electrons with moderate relativistic energies are able to generate an easily detectable CR photon yield ($\sim 6\cdot 10^3$ ph/(el eV sr) in the distribution maximum). The use of CTR will lead to a significant increase of the intensity by 2–3 orders of magnitude (see Fig. 4(b)). The use of targets with large Z allows to significantly increase the CTR yield (compare, for example, the solid lines in Figs. 4(a) and 4(b)).

The X-ray CDR is very sensitive to the photon and electron energy. The latter property is the clearest in the electron energy range from keV to a few tens of MeV. The dependence on the photon energy is connected with the resonant behaviour of the dielectric permittivity in the vicinity of the absorption edges where the CR condition is fulfilled. Therefore the CDR spectrum represents a monochromatic line [14] with FWHM (Full Width at Half Maximum) of order of 1 – 1.5 eV. Figure 5(a) illustrates how strong the dependence on the photon energy is. Changing the photon energy by just a few eV the CR intensity goes from maximum to nothing

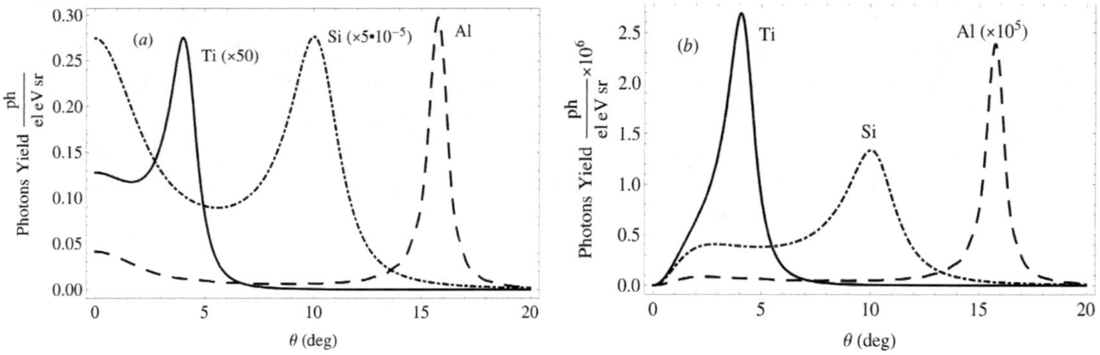

Figure 4. Dependence of the PR photons yield on the polar observation angle θ for different target materials: (a) for DR geometry; (b) for TR geometry. Parameters: $\gamma = 11$, $a = 70$ mm, $d = 5$ μm, $h = 0.1$ μm, $\varphi = \pi$.

whereas the DR peak remains. One may see a strong dependence of the position of the CDR peak in the angular distribution on the photon energy. For example, for given electron energy the change of the CDR photon energy on 1 eV leads to the peak shift on 1 degree.

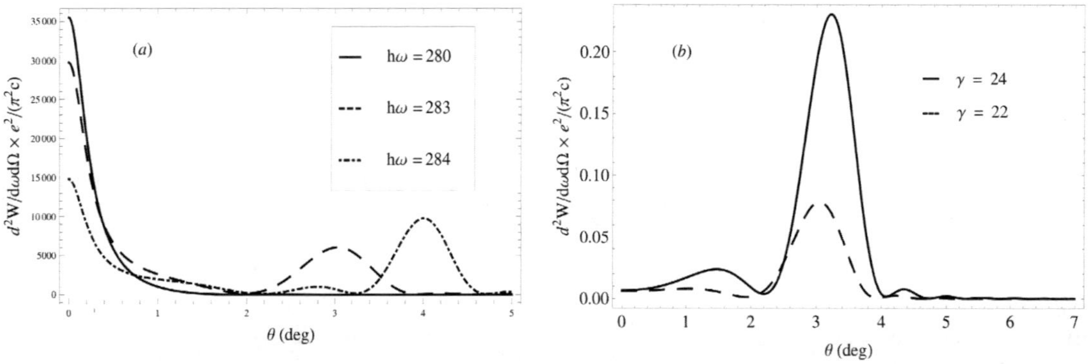

Figure 5. Dependence of the CDR intensity on the polar observation angle θ: (a) for different photon energies ($\gamma = 240$); (b) for different electron energies ($\hbar\omega = 284$ eV (K-edge of Carbon)). Parameters: $\alpha = 0$, $a = 70$ mm, $d = 5$ μm, $h = 0.1$ μm, $\varphi = \pi$.

In the range of moderately relativistic energies the CDR spectral-angular distribution allows to monitor the $< 10\%$ change in the electron energy (see, for instance, Fig. 5(b). Even a small change in the electron energy leads to a shift of the CDR peak position and intensity. With a larger change in the electron energy the shape of the distribution changes (compare, for example, dash-dotted line in Fig. 5(a) and solid line in Fig. 5(b)). Nevertheless, for a energy resolution of 1% or less an experimental verification is necessary.

5. Conclusion

In this paper we have described the model for calculating PR characteristics in the X-ray range from a target of an arbitrary shape and arbitrary dielectric permittivity. For calculations we used a recently developed method of polarization currents. In the limiting cases of normal incidence the expressions obtained for the radiation intensity coincide with the results of papers [12–15] for TR geometry and book [1] for DR geometry.

We have demonstrated that CDR characteristics in X-ray region significantly depend on the energy of the emitted photons, because the CDR is only generated in the frequency region in the vicinity of the atomic absorption edges. This peculiarity can be explained by resonance behavior of the permittivity in the given frequency range. Due to strong non-linear dependence of the photon yield on the photon energy, the CR in the vicinity of the absorption edge is highly monochromatic. The CR photon yield is comparable, and for many cases significantly higher, then the TR and DR photon yields.

For the moderate relativistic energies $\gamma < 100$ the variation of the energy significantly influences the CDR characteristics. Such peculiarity can be used to measure the beam energy via the analysis of the spectral-angular distribution shape and intensity. In the vicinity of moderate relativistic energies one can increase the CDR intensity by increasing the angle of incidence. However, the spectral angular CDR density becomes very sensitive to the angle of the particle incidence with increasing the charged particle beam energy. Such peculiarity can be used to measure the particle beam angular jitter essentially for accelerator operation.

Summarizing the analysis and the peculiarities of the CDR in the vicinity of the absorption edges open a wide range of possibilities for charged particle beam diagnostics. Nevertheless, the model is still under analysis. A more detailed proposal devoted to the diagnostics will be presented later.

Acknowledgments

Authors are grateful to M V Shevelev for stimulating criticism and useful discussions. The work was partially supported by Russian Ministry of Science and Education within the grant No. 14.B37.21.0912 and the RFBR grants No. 14-02-31642-mol_a, No. 14-02-01032-A.

References

[1] Potylitsyn A P, Ryazanov M I, Strikhanov M N and Tishchenko A A 2010 *Diffraction Radiation from Relativistic Particles* (*Springer Tracts in Modern Physics* vol 239) (Berlin: Springer) p 278
[2] Karataev P, Aryshev A, Boogert S, Howell D, Terunuma N and Urakawa J 2011 *Phys. Rev. Lett.* **107** 174801
[3] Chiadroni E, Castellano M, Cianchi A, Honkavaara K, Kube G, Merlo V and Stella F 2011 *Phys. Rev.* ST–AB **14** 102803
[4] Blackmore V, Doucas G, Perry C, Ottewell B, Kimmitt M F, Woods M, Molloy S and Arnold R 2009 *Phys. Rev.* ST–AB **12** 032803
[5] Jin Z *et al* 2011 *Phys. Rev. Lett.* **107** 265003
[6] ILC Technical Design Report 2013 http://www.linearcollider.org/ILC
[7] XFEL Technical Design Report 2006 (*DESY*) 097 http://xfel.desy.de
[8] Loos H *et al* 2008 *Proc. FEL 08 (Gyeongju, South Korea)* p 485
[9] Sukhikh L G, Bajt S, Kube G, Popov Yu A, Potylitsyn A P and Lauth W 2012 *Proc. IPAC 12 (New Orleans, Louisiana, USA)* **MOPPR019** p 819
[10] Bobb L, Billing M, Chritin N, Karataev P and Lefevre T 2012 *Proc. IBIC 12 (Tsukuba, Japan)* **MOCC01** p 24
[11] Gogolev A, Potylitsyn A and Kube G 2012 *J. Phys: Conf. Ser.* **357** 012018
[12] Bazylev A V, Glebov V I, Denisov E I, Zhevago N K and Khlebnikov A S 1976 *JETP Lett.* **24** 371
[13] Bazylev A V, Glebov V I, Denisov E I, Zhevago N K, Kumakhov M A, Khlebnikov A S and Tsinoev V G 1981 *Sov. Phys. - JETP* **54** 884
[14] Knulst W, van der Wiel M J, Luiten O J and Verhoeven J 2003 *Appl. Phys. Lett.* **83** 4050
[15] Knulst W, van der Wiel M J, Luiten O J and Verhoeven J 2001 *Appl. Phys. Lett.* **79** 2999
[16] Moran M J, Chang B, Schneider M B and Maruyama X K 1990 *Nucl. Instrum. Methods* B **48** 287
[17] Karlovets D V and Potylitsyn A P 2009 *JETP Lett.* **90** 368
[18] Karlovets D V 2011 *Sov. Phys. - JETP* **113** 27
[19] Kruchinin K O and Karlovets D V 2012 *Russ. Phys. J.* **55** 9
[20] Konkov A S, Gogolev A S, Potylitsyn A P and Karataev P V 2013 *Proc. IBIC 13 (Oxford, UK)* **WEPF36** p 1
[21] Ginzburg V L and Tsytovich V N 1990 *Transition Radiation and Transition Scattering* (Bristol: A Hilger) p 433
[22] Henke B L, Gullikson E M and Davis J C 1993 *At. Data Nucl. Tables* **54** 181

Journal of Physics: Conference Series **517** (2014) 012004 doi:10.1088/1742-6596/517/1/012004

Polarization Radiation in a Teflon Target

G Naumenko, A Potylitsyn, M Shevelev, V Bleko and V Soboleva
Tomsk Polytechnic University, Lenina str. 2, Tomsk, 634050, Russia

E-mail: naumenko@tpu.ru

Abstract. Contribution of Cherenkov radiation and Diffraction radiation when electrons are moving close to a dielectric prism for different angles of the prism face in respect to the electron beam direction was experimentally investigated. The comparison of experimental results with calculations using the known theoretical model shows the significant discrepancy in angular dependence for different orientation angles. The experimental analysis of the causes of discrepancy was performed. The experiment was performed on the relativistic electron beam with energy 6.2 MeV in Tomsk Polytechnic University.

1. Introduction

In [1] the characteristics of Cherenkov radiation (ChR) from the Teflon triangular prism surface, which is in parallel to the relativistic electron beam, and diffraction radiation (DR) from the upstream surface of prism, when electrons are moving close to the prism were experimentally investigated. In this paper we present recent investigations on the nature of these types of radiation, focusing on the radiation from the prism surface nearest to the electron beam. Namely, have studied the angular distribution of the radiation for different values of the angle ψ (see figure 1) by rotating the prism around the vertical axis. As it will be shown later, the structure of this radiation is rather complicated. To exclude the DR contribution from the upstream surface of the target (see [1]), the surface was covered by an absorber.

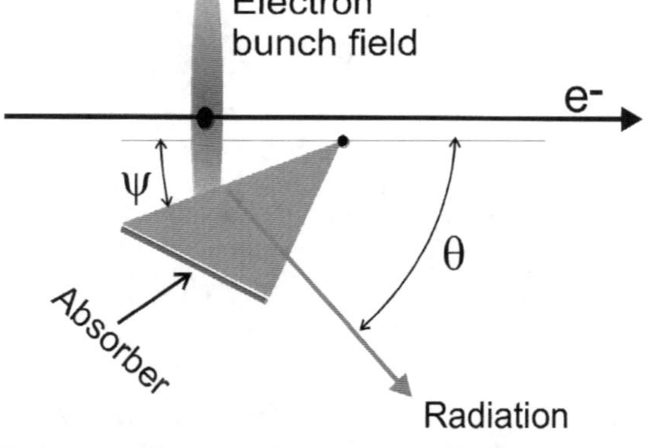

Figure 1. Principle of the measurement.

This configuration allows us to investigate the radiation emitted only from the surface nearest to the electron beam.

2. Experimental set-up and procedure

The experiment was carried out in the extracted electron beam of the Physical and Technical Institute of Tomsk Polytechnic University microtron with beam parameters presented in Table 1.
The beam is extracted from the vacuum chamber through a 50 μm thick beryllium foil.

Table 1. Electron beam parameters.

Electron energy	6.1 MeV ($\gamma = 12$)	Bunch period	380 psec
Train duration	$\tau \approx 4\,\mu$sec	Bunch population	$N_e = 6\cdot 10^8$
Bunches in a train	$n_b \approx 1.6\cdot 10^4$	Bunch length	$\sigma \approx 1.9$~2.4 mm

The radiation was measured using a room-temperature detector based on the antenna with low-threshold diode, produced by Tomsk Semi-conductive Devices Institute (Russia). The high frequency limit of the wavelength interval is limited both by the spectral efficiency of the detector and by the bunch form-factor due to the coherency of radiation ($\lambda_{\min} \approx 9\,\text{mm}$ for $\sigma \approx 2.2\,\text{mm}$). In figure 2 is shown the instrumental electron bunch pseudo-photon spectrum measured using this detector in circuit of the interferometer, which is placed directly on the electron beam.

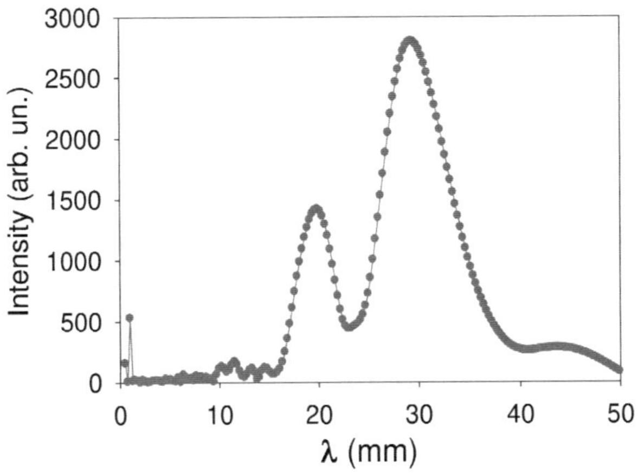

Figure 2. The instrumental electron bunch pseudo-photon spectrum.

To exclude the prewave zone effect [2] a parabolic telescope (see figure 4) was used for the angular distribution measurement. This method was suggested and tested in [3] and gives the same angular distribution as in the far field zone ($R \gg \gamma^2\lambda$). The layout of experimental set-up is shown in figure 3. The target used in the experiment is a triangular prism made out of Teflon with dimensions of 175x175x74 mm. The target was placed at a distance of 220 mm from the extraction window; the distance from the target to the electron beam (impact parameter) was 20 mm. The refractive index of Teflon target was $n = 1.41 \pm 0.1$ in the wavelength range from 10 to 30 mm. To keep the same impact-parameter the target was rotated around the point a for the positive angle ψ (see figure 3), and around the point b for negative values of ψ (see insertion in figure 3).

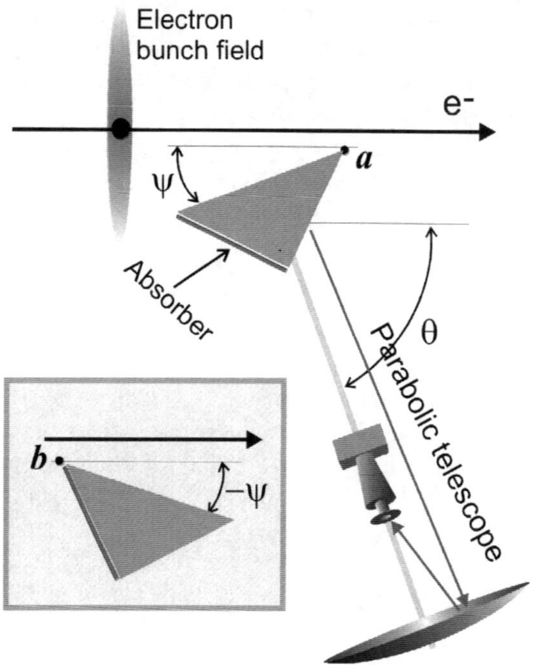

Figure 3. Experimental layout including the radiation monitor for the negative angle ψ.

3. Measurements

The intensity of the Coherent radiation was measured by displacing the detector along the angle θ for different values of angle ψ . The angular distribution obtained for $\psi = 0$ is shown in figure 4.

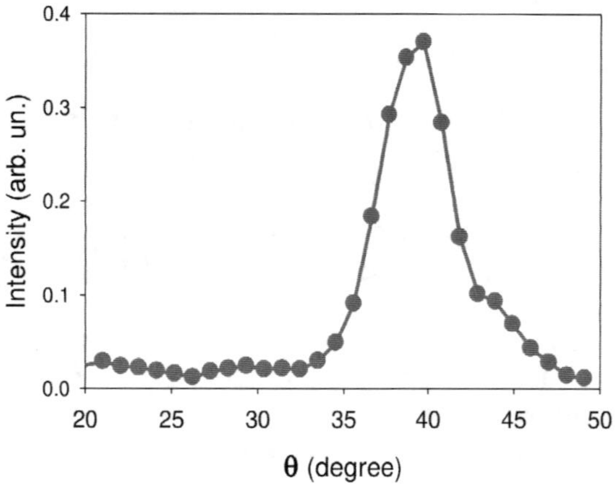

Figure 4. Angular distribution of the coherent radiation for $\psi = 0$.

In the next paragraph we will consider the position and the intensity of the peak of the radiation angular distribution for each value of angle ψ .

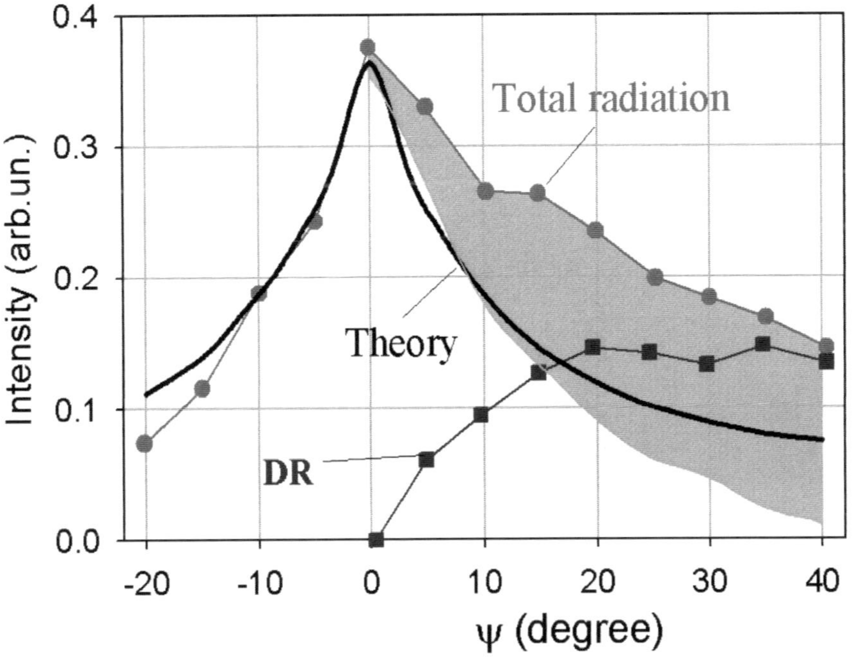

Figure 5. Peak intensity for different values of angle ψ.

In figure 5 the measured coherent radiation intensity is shown as red circles for different values of angle ψ.

ChR is defined as the result of interference of radiation, induced by the electron field in target, from different points of electron trajectory. So, if the electron trajectory is screened, like it is shown in figure 6, the contribution of the ChR in the total radiation should be excluded. The absorber on the copper screen was used to avoid multiple reflections of radiation from screen and target surface.

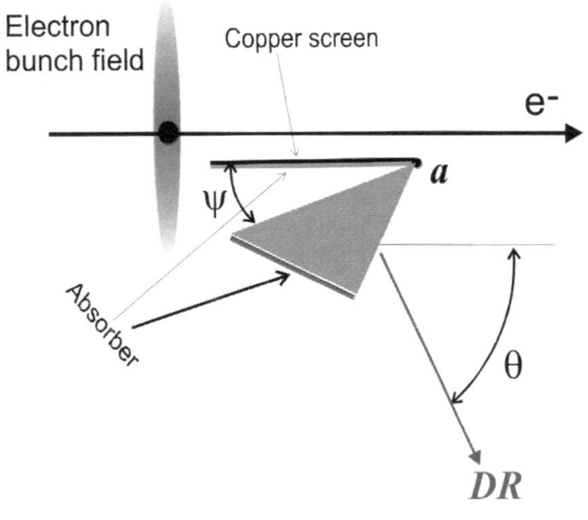

Figure 6. Experimental layout showing the screening of the Cherenkov radiation.

The result of the measurements taken in these conditions is shown in figure 5 by square points (DR). The filled area depicts the part of DR intensity in the total radiation assuming the simple approach of sum of radiation intensities.

4. Discussion

In order to compare the measurements with theory, let's use the polarization currents method, developed in [4]. The calculations for these experimental conditions, normalized to the experimental radiation intensity acquired for $\psi = 0$ are depicted in figure 5 by solid line (Theory). We see a good agreement of the theoretical calculations with experimental data for $\psi < 0$. However, in the case of $\psi > 0$ a large discrepancy is observed. A possible cause of the discrepancy can be proposed by considering in more detail the second part of experiment, where ChR is screened by copper screen. We will use the pseudo-photons viewpoint for analyses of an interaction of the relativistic electron electromagnetic field with matter. The "pseudo-photon" method proposed by Fermi [5] and developed by Williams [6] is widely used for theoretical studies of electromagnetic processes (see for instance [7] and [8]). According to this approach, the field of a charged particle may be replaced by a field of photons, which in this case are called pseudo-photons (in [9] is used the term "virtual quanta"; this term should differ from the term "virtual quanta" in quantum mechanics). This approach provides a good accuracy for the ultra-relativistic particles when the particle velocity is close to the light velocity ($v_p \approx c$) (see [7]) and when the longitudinal electric field component of the particle is negligible. In this case the particle field has the same properties as the field of real photons.

According to the pseudo-photons viewpoint, the electron field, cutted by the copper screen, will continue to propagate as real photons. It corresponds to the DR in traditional notion. Further this DR will be refracted on the target surface and propagate as a usual radiation. So, the total radiation consist from ChR and DR. Unfortunately we can not use the simple subtraction of DR from total radiation due to interference between these types of radiation, but we see that qualitatively this difference may be close to the theoretical curve.

Finally we can conclude, that in calculations using the model of polarization currents, DR is not taken into account.

Acknowledgment

This work is particularly supported by grant of Russian MES Program "Nauka" number 2456.

References

[1] Shevelev M V, Naumenko G A, Potylitsyn A P *et al J. Phys.: Conf. Ser.* **357** 012020
[2] Verzilov V A 2000 *Phys. Lett. A* **273** 135
[3] Kalinin B N, Naumenko G A, Potylitsyn A P *et al* 2006 *JETP Letters* **84** 3 110-114.
[4] Karlovets D V and Potylitsyn A P 2009 *JETP Lett.* **90** 5 326-331[5] Fermi E 1924 *Zh. Phyz* **29** 315
[6] Williams E J 1935 *Kgl. Danske Videnskab. Selskab. Mat.-fys. Medd.* **13** 4
[7] Ter-Mikaelian N L 1972 *High-Energy Electromagnetic Processes in Condensed Media* (Wiley, New York)
[8] Ter-Mikaelian N L 1969 *The Influence of the Medium on High Energy Processes at High Energies* (Publishing House of Academy of Science of Armenia, Yerevan)
[9] Jackson J D 1998 *Classical Electrodynamics, 3rd ed.* (J. Willey&Sons, New-York)

General solutions for the fields generated by a relativistic electron beam propagating in a rectangular dielectric waveguide

A M Altmark[1], **S S Baturin**[1], **I L Sheinman**[1] **and A D Kanareykin**[1,2]
[1]St. Petersburg Electrotechnical University, St. Petersburg, 5 Prof. Popov St., 197376, St. Petersburg Russia
[2]Euclid Techlabs LLC 400 Professional Dr. Ste. 220 Gaithersburg, MD 20879, USA

E-mail: aaltmark@mail.ru

Abstract. Dielectric lined waveguides are under extensive study as accelerating structures that can be excited either by Cherenkov radiation from electron beams or by external X-band or THz rf sources. Rectangular dielectric structures can be used both in proof of principle experiments for new accelerating schemes but also for studying the electronic properties of the structure loading material. General solutions for the fields generated by a relativistic electron beam propagating in a rectangular dielectric waveguide have been derived using the mode expansion method for the transverse operators of the Helmholtz equation, a new approach presented here for the first time. We derived an expression for the Green function of the combined Cherenkov and Coulomb fields obtained in terms of a superposition of *LSM* and *LSE*-modes. The resulting solution was incorporated into the previously developed BBU-3000 particle simulation code for modelling transverse beam dynamics. The parameter space essential for the beam stability defined by the Cherenkov radiation and the local Coulomb fields has also been obtained.

1. Introduction

Dielectric lined structures can be used for the generation of high power X-band or THz radiation when excited by a relativistic electron beam [1–4]. These radiation fields can also used for acceleration of a witness electron beam passing through the same structure at a specific delay with respect to the drive beam [5]. While the original work in this area focused on cylindrical structures, rectangular structures have recently come under intensive scrutiny in recent years both in testing new methods of acceleration and for studying the electronic properties of the dielectric itself [6, 7].

In this paper, we obtained a general solution for the field inside the electron bunch propagating in a rectangular dielectric waveguide (figure 1a). Analytic expressions for the fields were obtained using a modification of the transverse operator method [6]. This approach allows us to obtain a complete analytical solution for the eigenfunctions and therefore solve the problem of the generation of radiation in a rectangular dielectric loaded structure. In these studies we considered the Cerenkov field, which is sufficient for the calculation of the wakefield that is the mechanism for particle acceleration. However, the contribution of the Coulomb field can also be significant if the fields are to be evaluated inside the intense drive bunch. Here, we specially considered the cases where both the

Coulomb and Cherenkov fields inside the bunch are critical for beam dynamics calculations and the Coulomb contribution has to be taken into account.

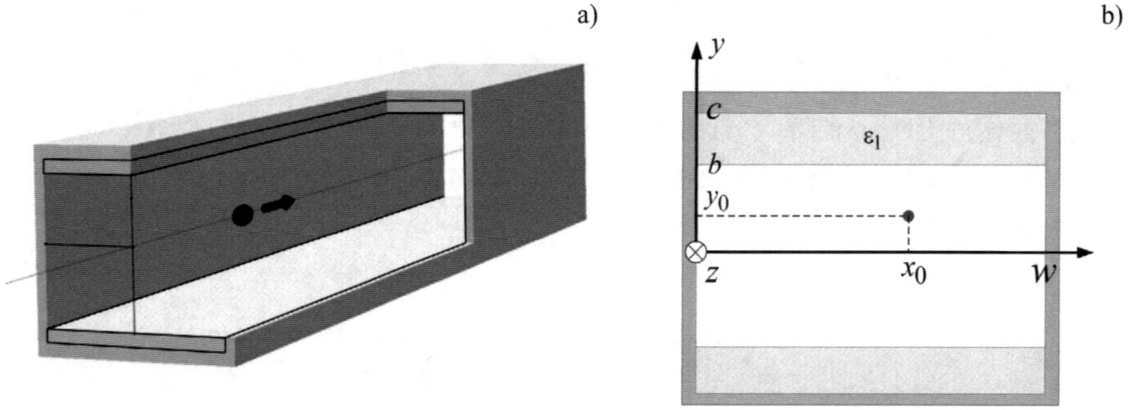

Figure 1. Rectangular dielectric waveguide.

2. Expression for the Green function

We consider a point charge passing through a rectangular dielectric waveguide (figure 1b) along the *z*-axis with the transverse coordinates x_0 and y_0. We use the Maxwell equations in the CGS system to derive an expression for the Green function:

$$\nabla\times\mathbf{E}=-\frac{\partial\mathbf{B}}{c\partial t},\ \nabla\cdot\mathbf{D}=-en,\ \nabla\times\mathbf{H}=\frac{\partial\mathbf{D}}{c\partial t}-\frac{en\mathbf{V}}{c},\ \mathbf{B}=\mu\mathbf{H},\ \mathbf{D}=\varepsilon\mathbf{E},$$

where **E**, **H**, **D**, **B** – electric and magnetic field vectors; **v** – velocity of the electron bunch; *n* - the concentration of electrons in the bunch; *e* – electron charge; *t* – time.

The charge density of a point charge *q*, moving parallel to the longitudinal axis *z* is written as follows:

$$n(x,y,z,t)=\frac{4\pi q}{-e}\delta(z-vt)\delta(x-x_0)\delta(y-y_0).$$

It can be shown for a rectangular dielectric waveguide that the excited modes can be divided into two classes: magnetic modes (*LSM*) with H_y=0, and longitudinal section electric (*LSE*) modes with E_y=0. Other components of the electromagnetic field (E_x, E_z, H_x, H_z) can be expressed in terms of E_y and H_y. We can write the Helmholtz equations for the *LSM* and *LSE*-modes as

$$\begin{aligned}&\frac{\partial^2 E_y}{\partial\zeta^2}+\hat{T}_E E_y=\frac{-e}{\varepsilon_0\left(1-\varepsilon\mu\beta^2\right)}\frac{\partial}{\partial y}\left(\frac{n}{\varepsilon}\right),\\&\frac{\partial^2 H_y}{\partial\zeta^2}+\hat{T}_H H_y=\frac{-ev}{1-\varepsilon\mu\beta^2}\left(\frac{\partial n}{\partial x}\right).\end{aligned}\tag{1}$$

where $\varepsilon=\varepsilon(y)$, $\mu=\mu(y)$ are the permittivity and permeability, and $\zeta=z-vt$.

The transverse operators in expressions (1) are written as

$$\hat{T}_H = \frac{1}{\left(1-\varepsilon\mu\beta^2\right)}\left[\frac{\partial^2}{\partial x^2} + \frac{\partial}{\partial y}\left(\frac{1}{\mu}\frac{\partial}{\partial y}[\mu\cdot]\right)\right],$$

$$\hat{T}_E = \frac{1}{\left(1-\varepsilon\mu\beta^2\right)}\left[\frac{\partial^2}{\partial x^2} + \frac{\partial}{\partial y}\left(\frac{1}{\varepsilon}\frac{\partial}{\partial y}[\varepsilon\cdot]\right)\right].$$

Using the boundary conditions for the components of E_y and H_y, we define the domain of the operators $\hat{T}_H$ and $\hat{T}_E$.The components E_y and H_y can be written as an eigenfunction expansion despite the fact that the operators $\hat{T}_H$ .and $\hat{T}_E$ are not self-adjoint. Using the inverse Fourier transform of the coordinate ζ we get the Green functions:

$$E_y(x,y,\zeta) = 2\pi q \sum_{\lambda_E} \frac{\partial\left[\phi_E^{\lambda_E}(x_0,y_0)\right]^*}{\partial y} \psi_E^{\lambda_E}(x,y)\left[F_1\left(\lambda_E^-,\zeta\right) - F_2\left(\lambda_E^+,\zeta\right)\right], \quad (1)$$

$$H_y(x,y,\xi) = 2\pi q\beta \sum_{\lambda_H} \frac{\partial\left[\phi_H^{\lambda_H}(x_0,y_0)\right]^*}{\partial x} \psi_H^{\lambda_H}(x,y)\left[F_1\left(\lambda_H^-,\zeta\right) - F_2\left(\lambda_H^-,\zeta\right)\right], \quad (2)$$

were

$F_1\left(\lambda,\zeta\right) = \frac{\exp(-\sqrt{|\lambda|}|\zeta|)}{\sqrt{|\lambda|}}$ is the function corresponding to the Coulomb field,

$F_2\left(\lambda,\zeta\right) = 2\frac{\sin(\sqrt{\lambda}\,|\zeta|)\theta(-\zeta)}{\sqrt{\lambda}}$ is the function corresponding to Cerenkov radiation,

λ_E^+, λ_H^+, λ_E^-, λ_H^- are the positive and negative roots of the dispersion relations for the LSM and LSE modes,

$$\psi_E^{\lambda_n}(x,y) = \frac{1}{\sqrt{A_E^{n,n}}}\Psi_E^n(x,y),\quad \phi_E^{\lambda_n}(x,y) = \frac{1}{\sqrt{A_E^{n,n}}}\Phi_E^n(x,y),\quad \Phi_E^n = \hat{J}\Psi_E^n,$$

$$\psi_H^{\lambda_n}(x,y) = \frac{1}{\sqrt{A_H^{n,n}}}\Psi_H^n(x,y),\quad \phi_H^{\lambda_n}(x,y) = \frac{1}{\sqrt{A_H^{n,n}}}\Phi_H^n(x,y),\quad \Phi_H^n = \hat{J}\Psi_H^n,$$

$$\hat{J} \equiv \begin{cases} 1, & y\in[-b,b] \\ -1, & y\in[-c,-b)\cup(b,c] \end{cases}.$$

$A_E^{n,n}$, $A_H^{n,n}$ are normalization factors, $\Psi_E^n(x,y) = X_E^n(x)Y_E^n(y)$, $\Psi_H^n(x,y) = X_H^n(x)Y_H^n(y)$ are eigenfunctions of the transverse operators $\hat{T}_E$, $\hat{T}_H$, and $\theta(\zeta)$ is the Heaviside unit step function.

Since there are two perfect conducting walls at $x = 0$ and $x = w$, the expansion of the field in x is a superposition of standing waves $X_E^n(x)$ and $X_H^n(x)$. The dielectric filling in y is inhomogeneous and accordingly, the functions $Y_E^n(y)$ and $Y_H^n(y)$ may be symmetric or asymmetric relative to the plane $y = 0$. Thus the total field is a superposition of four types of modes: symmetric and asymmetric LSM modes, and symmetric and asymmetric LSE modes.

3. General solution

The resulting expressions for the Green functions (2, 3) are used to calculate the wakefields for BBU-3000 code [8]. The field inside the bunch is fund as convolution of Green function with Gaussian distribution. We compare our results with the CST program that uses a self-consistent wakefield algorithm. We consider as a test problem a Gaussian line charge propagating along the z-axis and offset by the coordinates x_0 and y_0. The parameters of the bunch and waveguide are given in Table 1. The comparison of the results of this test case for the BBU code using the field algorithm derived here to the CST results are shown in figure 2. From figure 2a it is seen that including only Cerenkov radiation leads to a large discrepancy with the CST results. However, if the Coulomb field is taken into account together with Cherenkov radiation, the agreement with CST is much better (figure 2b). It should be noted that the comparison was made for the case of the coordinate of the observation point (x, y) lying outside the electron bunch. The coordinates x_0 and x differ by 0.02 cm (Table 1). The advantage of the BBU code is the possibility of achieving high accuracy of field calculations inside the electron bunch.

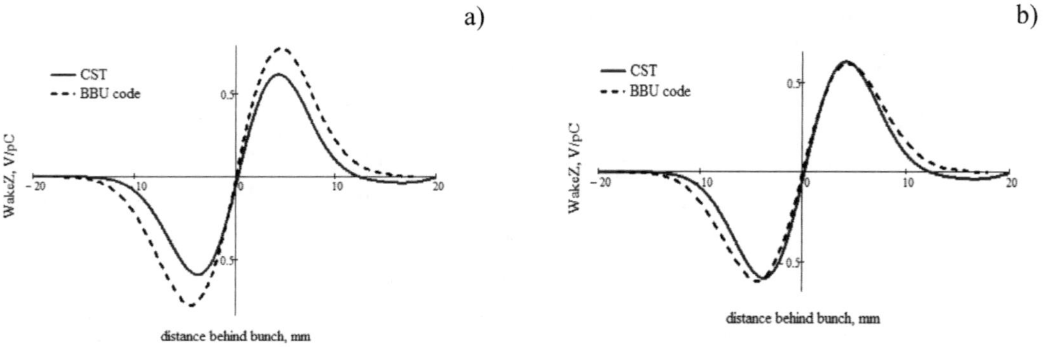

Figure 2. Comparison of test problem results from CST and BBU (using the field expressions derived here) a) Only Cherenkov radiation; b) both Cherenkov and Coulomb fields.

Table 1. Parameters of the rectangular waveguides and Gaussian beam.

b, cm	c, cm	w, cm	ε_1	Q, nC	energy, MeV	Sigma-z, cm	x_0, cm	y_0, cm	x, cm	y, cm
0.3	0.5	2.00	1.2	100	5	0.45	0.99	0	1.01	0

4. Conclusion

- We have obtained an expression for the Green function of the Coulomb field for a point charge using a new approach, the transverse operator method.
- The BBU code originally developed for beam dynamics simulations using Green function based analytical solutions was modified to use our new results.
- It was shown that the Coulomb field contribution at the position of the bunch is comparable with the Cherenkov field magnitude for relatively low beam energies and dielectric constant.

References

[1] Kanareykin A, Antipov S, Jing C *et al* 2013 Phys. Rev. Lett. **111** 134802
[2] Antipov S, Jing C, Fedurin M, Kanareykin A *et al* 2013 Rev. Sci. Inst **84** 022706
[3] Antipov S, Jing C, Kanareykin A, Butler J E *et al* 2012 Appl. Phys. Lett. **100** 132910
[4] Antipov S, Jing C, Fedurin M, Gai W, Kanareykin A *et al* 2012 Phys. Rev. Lett. **108** 144801
[5] Kanareykin A, Altmark A, Antipov S, Gai W *et al* 2010 AIP Conf. Proc. 1 **1299** 359
[6] Baturin S, Sheinman I, Altmark A and KanareykinA 2013 Phys. Rev. STAB **16** 051302
[7] Kiselev V , Linnik A, Marshall T, Onishchenko I *et al* 2006 AIP Conf. Proc. **877** 851
[8] Schoessow P, Jing C., Kanareykin A *et al* 2011 PAC NY **1689**

Journal of Physics: Conference Series **517** (2014) 012006 doi:10.1088/1742-6596/517/1/012006

Intense Cherenkov radiation from a charged particle revolving along a shifted equatorial orbit about a dielectric ball

L Sh Grigoryan[1], H F Khachatryan, M L Grigoryan
Institute of Applied Problems in Physics, 25 Hr. Nersessian Str., 0014 Yerevan, Armenia

E-mail: levonshg@mail.ru

Abstract. The radiation from a charged particle uniformly revolving along a shifted equatorial orbit about a dielectric ball has been investigated. The results of numerical calculations based on exact solutions of Maxwell equations with due regard for dielectric losses of energy within the ball material testify that (a) for definite «resonant» values of the particle revolution frequency relativistic electron may generate microwave Cherenkov radiation at the first harmonic, that is approximately 900-1100 times more intensive than synchrotron radiation generated at the revolution of the same particle in vacuum and (b) in case of large values of permittivity of ball material (strontium titanate) the «resonant» radiation is generated by the revolving electron with energy of the order of 30 KeV.

1. Introduction and background

The operation of a number of devices intended for production of electromagnetic radiation is based on the interaction of relativistic electrons with matter [1–5]. Numerous applications motivate the importance of exploring different mechanisms of amplification and control of produced radiation. Specifically, one may use radiation of different kinds as well as reflecting surfaces for monitoring the flows of radiation.

A research of such a kind was carried out in [6–8], where the emission of radiation from a relativistic particle revolving about a dielectric ball has been investigated (figure 1). In such a geometry in addition to the synchrotron radiation (SR) the particle may generate the Cherenkov radiation (CR), since the field associated with the particle partially penetrates the ball depths and revolves together with the particle. In case of short distances of particle from the ball surface the velocity of its field displacement within the ball may exceed the phase velocity of light in the ball material and, hence, CR would be generated. Here the flow of produced CR may be controlled either by changing the radius of ball or the relative position of ball with respect to particle orbit.

[1] To whom any correspondence should be addressed.

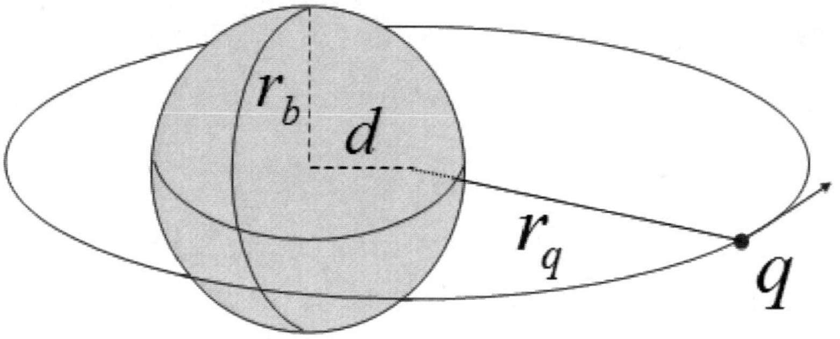

Figure 1. A relativistic charged particle rotating along a shifted equatorial orbit about a dielectric ball. ε_b, r_b are the permittivity of the ball material and its radius and q, r_q are the charge of particle and the radius of its orbit, d is the distance between the centres of the ball and of particle orbit.

As a result of the combination of SR and CR, as well as of the mechanism of radiation flow control (the ball-to-vacuum interface) the revolving particle may, at separate harmonic, generate CR tens of times more intense than CR and SR from the same particle generated in a continuous, infinite and transparent medium having the same permittivity as the ball material. The theoretical substantiation of this effect and its visual explanation are given in [6]. However, the case under consideration in [6] was confined to the simplest configuration of ball located in the centre of particle orbit (figure 1, $d=0$). In reality (electron beam), such a symmetry is difficult to provide.

This work is a continuation of papers [6–8], in particular [7, 8], where it was shown that the electron revolving along a non-equatorial [7] or shifted equatorial [8] orbits abound a dielectric ball can generate a "resonant" CR at the harmonics with the numbers $k \geq 8$, if its energy $E_q \geq 2\,\text{MeV}$.

In the present paper it is shown that a revolving non-relativistic electron ($E_q \sim 30\,\text{KeV}$) also may generate such an intense radiation and not at the 8th, but at the first harmonic ($k=1$). New features of this radiation have been revealed. The possibility of practical applications of this phenomenon is discussed.

2. The basic formula

Now consider the uniform rotation of a charged particle in the magnetic field in vacuum about a dielectric ball in its equatorial plane under the assumption that the centers of ball and of particle orbit are shifted one with respect to the other (figure 1, d is the distance between the centers of the ball and particle orbit).

In spherical coordinates r, ϑ, ϕ with origin in the center of ball, the permittivity $\varepsilon(r)$ of the substance is a step function of radial coordinate:

$$\varepsilon(r) = \varepsilon_b \quad \text{for} \quad r \leq r_b \quad \text{and} \quad \varepsilon(r) = 1 \quad \text{for} \quad r > r_b, \tag{1}$$

where ε_b is the permittivity of ball with radius r_b. The magnetic permeability is taken to be 1. The current density may be written as

$$\vec{j}(\vec{r},t) = \frac{q(\mathrm{v}_r(t)\vec{e}_r + \mathrm{v}_\phi(t)\vec{e}_\phi)}{r_0^2(t)}\delta(r - r_0(t))\delta(\vartheta - \pi/2)\delta(\phi - \phi_q(t)), \tag{2}$$

where

$$r_0(t) = \sqrt{d^2 + r_q^2 + 2dr_q\cos\omega_q t}, \qquad \mathrm{v}_r(t) = -\frac{\omega_q r_q d\sin(\omega_q t)}{r_0(t)},$$

$$\mathrm{v}_\phi(t) = \frac{\omega_q r_q(r_q + d\cos(\omega_q t))}{r_0(t)}, \qquad \phi_q(t) = \mathrm{Re}\left(\frac{1}{\mathrm{i}}\cdot\ln\left(\frac{d + r_q\cdot \mathrm{e}^{\mathrm{i}\omega_q t}}{r_0(t)}\right)\right) \tag{3}$$

q, $v = r_q\omega_q$ and ω_q are the charge, linear velocity and cyclic frequency of particle rotation, and r_q is the radius of particle orbit.

The rotation of particle entails radiation at some discrete frequencies (harmonics) $\omega_k = k\omega_q$ with $k = 1;2;3\dots$. We assume that an exterior force would make up for the braking of particle due to the radiation, by forcing the particle to uniformly rotate about the ball. The total energy losses of particle during one revolution period $T = 2\pi / \omega_q$ may be written as

$$\sum_k (E_k^{(d)} + E_k^{(r)}) = \sum_k W_k, \tag{4}$$

where $E_k^{(d)}$, $E_k^{(r)}$ are the dielectric losses of energy and energy losses due to particle radiation.

It is convenient to introduce a dimensionless quantity

$$E_k^{(r)} / \hbar\omega_k \equiv n_k, \tag{5}$$

where $\hbar\omega_k$ is the energy of corresponding electromagnetic wave quantum. So, n_k is the "number of electromagnetic field quanta" emitted during one revolution period of particle. We have derived the following final formula for n_k:

$$n_k = \frac{c}{\hbar k\omega_q^2}\sum_{l=0}^{\infty}\sum_{m=-l}^{l}\left[\left|a_E^{lm}(k)\right|^2 + \left|a_H^{lm}(k)\right|^2\right] \tag{6}$$

based on exact solutions of Maxwell equations for the case under consideration. Here

$$a_E^{lm} = \frac{4\pi\left(P_{\underline{l-1}}(k) - P_{\underline{l+1}}(k)\right)}{cr_b^2(2l+1)}\left[\sqrt{l(l+1)}\left(u_{1;l-1}^{lm}(k) - u_{1;l+1}^{lm}(k)\right) + (l+1)u_{2;l-1}^{lm}(k) + lu_{2;l+1}^{lm}(k) + \right.$$

$$\left. +\gamma_l\sqrt{l(l+1)}\left(lB_{1;l-1}^{lm}(k) - (l+1)B_{1;l+1}^{lm}(k) + \sqrt{l(l+1)}\left(B_{2;l-1}^{lm}(k) + B_{2;l+1}^{lm}(k)\right)\right)\right],$$

$$a_H^{lm} = \frac{2q\omega_q\cdot(-\mathrm{i})^{l+1}}{cr_b^2\sqrt{l(l+1)}}\frac{\partial}{\partial\theta}Y_{lm}(\pi/2;0)\int_0^T v_\phi(t)b_{\underline{l}}(t)\exp[\mathrm{i}(k\omega_q t - m\phi_q(t))]\mathrm{d}t \tag{7}$$

are amplitudes describing the contributions of electrical (*E*) and magnetic (*H*) type multipoles respectively, $Y_{lm}(\theta,\phi)$ are the spherical harmonics. We have introduced the following notations:

$$u_{1;l_1}^{lm} = \frac{q\omega_q Y_{lm}(\pi/2;0)\cdot(-\mathrm{i})^{l_1+1}}{2\pi}\int_0^T v_r(t)b_{\underline{l_1}}(t)\exp[\mathrm{i}(k\omega_q t - m\phi_q(t))]\mathrm{d}t,$$

$$u_{2;l_1}^{lm} = \frac{-\mathrm{i}mq\omega_q Y_{lm}(\pi/2;0)\cdot(-\mathrm{i})^{l_1+1}}{2\pi\sqrt{l(l+1)}}\int_0^T v_\phi(t)b_{\underline{l_1}}(t)\exp[\mathrm{i}(k\omega_q t - m\phi_q(t))]\mathrm{d}t,$$

$$B_{1;l_1}^{lm} = \frac{q\omega_q}{2\pi r_b}Y_{lm}(\pi/2;0)\cdot u\cdot j_l(u)\int_0^T v_r(t)h_{\underline{l_1}}(u_b r_0(t)/r_b)\exp[\mathrm{i}(k\omega_q t - m\phi_q(t))]\mathrm{d}t.$$

$$B_{2;l_1}^{lm} = -\frac{\mathrm{i}mq\omega_q}{2\pi\, r_b\sqrt{l(l+1)}}Y_{lm}(\pi/2;0)\cdot u\cdot j_l(u)\int_0^T v_\phi(t)h_{\underline{l_1}}(u_b r_0(t)/r_b)\exp[\mathrm{i}(k\omega_q t - m\phi_q(t))]\mathrm{d}t. \tag{8}$$

$$P_l = (-\mathrm{i})^l u_b r_b[\{j_l(u_b);h_l(u)\}j_l(u_b) + \{j_l(u_b);j_l(u)\}h_l(u_b)],$$

$$b_l = -\mathrm{i}\cdot u_b r_b[\{j_l(u_b);h_l(u)\}j_l(u_b r_b\cdot r_0(t)) + \{j_l(u_b);j_l(u)\}h_l(u_b r_b\cdot r_0(t))],$$

$$u = k\frac{r_b v}{r_q c},\qquad u_b = u\sqrt{\varepsilon_b},\qquad \gamma_l = \frac{1-\varepsilon_b}{\varepsilon_b\left(lz_{l-1}^l + (l+1)z_{l+1}^l\right)},\qquad z_v^l \equiv \frac{\varepsilon_b j_v(u_b)uh_l(u) - u_b j_l(u_b)h_v(u)}{j_v(u_b)uh_l(u) - u_b j_l(u_b)h_v(u)}.$$

Here $j_l(\tau)$ and $h_l(\tau)$ are the spherical Bessel and Hankel functions of the first kind respectively. Eventually,

$$\{a_l(u_b);\ g_l(u)\} \equiv u_b a_{l+1}(u_b)g_l(u) - a_l(u_b)ug_{l+1}(u),\qquad f_{\underline{l}}(\tau) \equiv f_l(\tau)/\{j_l(u_b);h_l(u)\}. \tag{9}$$

It is worthwhile to mention that all foregoing expressions are valid in all the cases when a particle rotates uniformly in the equatorial plane of a dielectric ball under the condition that its trajectory doesn't pass through the substance of ball. Two cases are possible here: the particle rotates around the ball ($0 \le d < r_q - r_b$) and outside the ball ($d > r_q + r_b$). The main purpose of this paper is to study the first case.

3. Numerical results

We shall assume that the permittivity of the ball material is equal to $\varepsilon_b = 232 \cdot (1 + 0.0001 \cdot \mathrm{i}) = \varepsilon_b' + \mathrm{i}\varepsilon_b''$ (strontium titanate in $10^8 - 10^{10}\,Hz$ frequency range [9]). We shall also assume $r_b = 1cm$ and confine ourselves to the consideration of electron radiation at a first harmonic: $k = 1$. The radius of the particle orbit $r_q = 1.1$ cm.

First, let us consider the case of $d = 0$. In figure 2 we give the number of electromagnetic field quanta n_1 emitted at the first harmonics versus of the cyclic frequency ω_q. As is seen from the plot, the height of sharp peaks $n_1^{(i)}$ observed at definite values of the cyclic frequency $\omega_q = \omega_q^{(i)}$ («resonant» frequency) is ≈ 1000 times more intensive:

$$n_1^{(i)}(\text{ball}) / n_1(\text{vac}) \cong 925;\ 1180, \qquad i = 1; 2 \tag{10}$$

than SR generated at the revolution of the same particle in vacuum (the dashed line in figure 2, $n_1(\text{vac}) \approx 0.003$). At these peaks the electron energy is approximately $29\,keV$ and nevertheless the Cherenkov condition for the particle speed and the ball material is satisfied:

$$\mathrm{v} > c/\sqrt{\varepsilon_b'}\,. \tag{11}$$

Mentioned peaks witnesses to the strong effect that the ball-vacuum interface has on CR generated by the particle in the substance of ball.

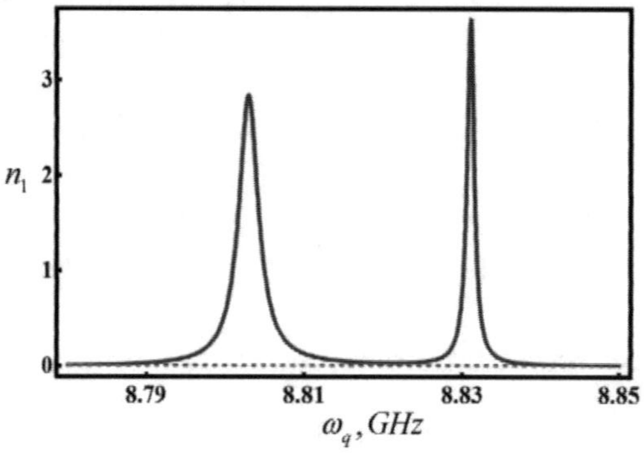

Figure 2. The number of electromagnetic field quanta n_1 emitted at the first harmonic versus of the electron cyclic frequency ω_q, $r_b = 1$ cm, $r_q = 1.1$ cm, $\varepsilon_b = 232 \cdot (1 + 0.0001 \cdot \mathrm{i})$.

As it was mentioned in Introduction in addition to the SR the rotating charged particle may also emit CR inside the ball material. In general, the contribution of CR is small and total radiation differs slightly from SR. However, in case of weak absorption ($\varepsilon_b'' << \varepsilon_b'$) in the ball material and at definite «resonant» values of the cyclic frequency ($\omega_q = \omega_q^{(i)}$ in figure 2) the particle may generate $n_k \geq 1$ quanta of electromagnetic field during one rotation period. Such high power radiation rises due to the fact that electromagnetic oscillations of CR induced by the particle along its entire trajectory are partially locked inside the ball and are superimposed in nondestructive way. More detailed visual explanation of this phenomenon is given in [6].

Now direct our attention to the case when the centers of the ball and of particle orbit do not coincide ($d \neq 0$). In figure 3 the dependence of the number n_1 of electromagnetic field quanta (6) generated per revolution of electron about a dielectric ball on the distance d between the centers of the ball and of particle orbit is shown. The rotation cyclic frequency of electron is equal to the first resonance value $\omega_q^{(1)} = 8.803$ GHz (see figure 2). The corresponding kinetic energy of electron is $\cong 30$ KeV.

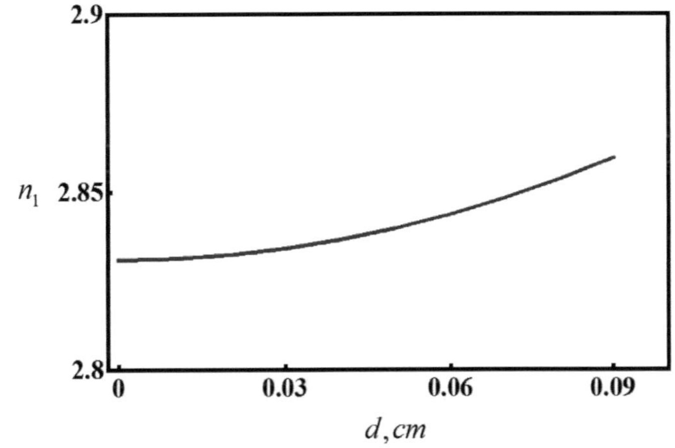

Figure 3. The dependence of n_1 on the distance d between the centres of the ball and particle orbit, $\omega_q = \omega_q^{(1)} = 8.803$ GHz. The remaining parameters are being the same as those in figure 2.

As is seen in figure 3, the radiation stays intensive when d increases from 0 to its maximum value:

$$n_1(d) / n_1(\text{vac}) \approx 930 \qquad \text{at} \qquad 0 < d < (r_q - r_b) = 0.1\,\text{cm}. \qquad (12)$$

This fact simplifies the feasible applications of such a radiation.

At larger values of d ($d > r_q + r_b$) the particle rotates outside the ball in its equatorial plane. In figure 4 the dependence of n_1 on d is shown in this case. Numerical calculations were performed using formulae given in Section 2.

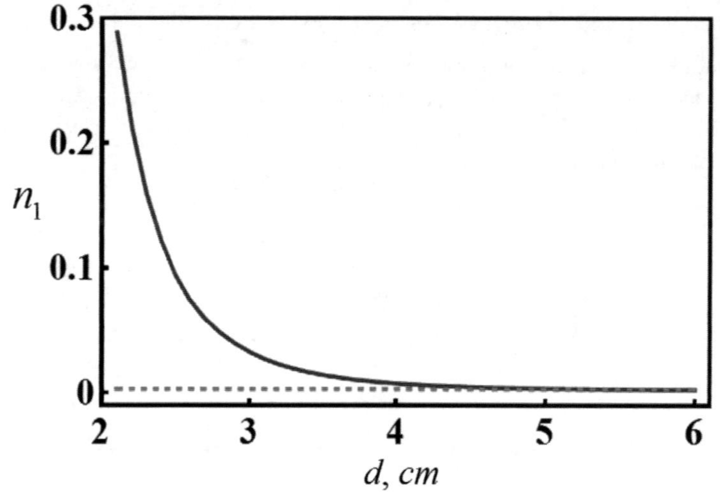

Figure 4. The dependence of n_1 on d, when the particle rotates outside the ball in its equatorial plane: $d > r_q + r_b$. The remaining parameters are being the same as those in figure 3. The dashed line represents SR generated at the revolution of the same particle in vacuum.

As is seen in figure 4 the function $n_1(d)$ decreases rapidly with increasing d and tends to $n_1(\text{vac})$ when $[d-(r_q+r_b)] >> r_{eff}$ ($r_{eff} \sim \gamma r_q / 2 \approx 0.5$ cm is the characteristic transverse dimension of the Coulomb field of a rotating charged particle, and γ is the Lorentz factor of particle). This fact clearly indicates that the sharp peaks $n_1^{(i)}$ observed in figure 2 are due to the radiation generated inside the substance of dielectric ball.

As is shown in [6] the resonance frequencies

$$\omega_q^{(i)} = a^{(i)} c / r_b \sqrt{\varepsilon_b'} \tag{13}$$

are inversely proportional to the radius (r_b) of ball and do not depend on the radius (r_q) of particle orbit. The dimensionless proportionality coefficient in (13) is determined by the order number (i) of resonant frequency, the number (k) of radiated harmonic and the dielectric constant (ε_b) of ball: $a^{(i)} = a^{(i)}(k, \varepsilon_b)$. Numerical results obtained in the framework of this paper indicate that $\omega_q^{(i)}$ are independent not only of r_q but also of d.

4. Conclusions

We have studied the radiation from a charged particle uniformly revolving along a shifted equatorial orbit about a dielectric ball. The results of numerical calculations based on exact solutions of Maxwell equations with due regard for dielectric losses of energy in the ball material testify that:

1) a non-relativistic electron revolving at resonance frequency about a strontium titanate ball generates the radiation at the first harmonic, that is ≈ 1000 times more intensive than synchrotron radiation generated at the revolution of the same particle in vacuum. The intensity of radiation is practically unchanged for notable variations of the distance between the centres of the ball and particle orbit (see (12)).

2) High power radiation rises due to the fact that electromagnetic oscillations of Cherenkov radiation induced by the particle along its entire trajectory are partially locked inside the ball and are superimposed in nondestructive way [6].

The intensity of such radiation is highly sensitive to the values of ω_q and ε_b. This fact may be used for accurate measurement of the permittivity of matter and detection of weak (e.g., ultrasonic) waves.

Acknowledgments
The authors are thankful to the anonymous reviewer, the valuable comments of which helped to improve the statement of results obtained in this paper.

.

References

[1] Rullhusen P, Artru X and Dhez P 1998 *Novel Radiation Sources Using Relativistic Electrons* (Singapore: World Scientific)
[2] Potylitsyn A P 2009 *Radiation from Electrons in Periodic Structures* (Tomsk: NTL Press) (in Russian)
[3] Jelley J V 1958 *Cherenkov Radiation and its Applications* (London: Pergamon Press)
[4] Bolotovskii B M 1961 *Physics-Uspekhi* **75** 295
[5] Zrelov V P 1968 *Vavilov-Cherenkov Radiation (and its Applications in High Energy Physics)*, (Moscow: Atomizdat) (in Russian)
[6] Grigoryan L Sh, Khachatryan H F, Arzumanyan S R and Grigoryan M L 2006 Some features of radiation from a particle rotating about a dielectric ball, Preprint IAPP NAS RA 1-06 (Yerevan) 50pp (in Russian); 2006 *Nucl. Instr. and Meth.* B **252** 50
[7] Grigoryan L Sh, Khachatryan H F, Arzumanyan S R and Grigoryan M L 2006 *Izv. Nats. Akad. Nauk Arm. Fiz. (Engl. Transl.: J. Contemp. Phys.)* **41** 163
[8] Grigoryan L Sh, Khachatryan H F, Arzumanyan S R and Grigoryan M L 2010 Microwave radiation from a particle revolving along a shifted equatorial orbit about a dielectric ball, *presentation at the 35th Int. Conf. IRMMW-THz 2010* (Rome, Italy, 5-10 September, 2010) (arXiv:1007.2776)
[9] Voronkova E M, Grechushnikov B N, Distler G I and Petrov I P 1965 *Optical Materials for Infrared Technology* (Moscow: Nauka, 1965) (in Russian)

Coherent Production of Relativistic Singlet Ps Atoms by High Energy Photons and Electrons in Aligned Crystals – Revisited

Yu P Kunashenko[a,b] **and Yu L Pivovarov**[a]

[a]National Research Tomsk Polytechnic University
[b]Tomsk State Pedagogical University

kunashenko@tpu.ru

Abstract. Cross-sections of coherent production of relativistic (moving with relativistic velocities) singlet Ps atoms in the ultrathin crystal have been calculated and estimations of counting rates are obtained to propose the experiments.

1. Introduction: a short historical survey, key questions, challenges and motivations for future studies

The first papers on production of relativistic (moving with a relativistic velocity) positronium (Ps) atoms by high-energy photons and electrons in the Coulomb field of the separate atom [1–4] were published more than 27 years ago. The further developments of the theory (beyond the Born approximation) was performed in the papers [5–8], again many years ago.

The next step in this story was connected with theory of coherent production of singlet Ps atoms by high-energy photons in a crystal [9]. The next paper [10] was dedicated to coherent production of relativistic Ps (singlet and triplet) by high energy electrons. According [9] and [10], the sharp coherent peaks may appear in photo – and electroproduction cross-sections at definite Ps energies and emission angles.

The investigations in this field have been continued in the later works [11–13]. In particular, more detailed theoretical investigations of Ps coherent type B photo- and electroproduction in the crystals was presented in Ref. [12]. In fact, besides theoretical studies of cross - sections, it was the 1st real proposal of experiment on coherent Ps photoproduction at REFER (Hiroshima University, Japan). Unfortunately, the experiment was not done at that time.

The QED process closely connected to coherent Ps photoproduction in a crystal is the coherent photoproduction of free (non-bound) electron-positron pairs. The idea of the first experiment on coherent type B e+e- pairs photoproduction in a crystal [14] was to approach Ps production kinematics, i.e. to detect created e+e- pairs under hard collimation. The experiment [14] was successful and resulted in observation of brilliant coherent effect in the yield of symmetric (equal energy) electron-positron pairs in the kinematics “approaching Ps kinematics”.

The problems connected with Ps generation and their experimental investigation were the subject of International Workshop "Hadronic atoms and positronium in the standard model" Dubna 1998. [15] and XII International Workshop on Positron and Positronium Physics (Sandbjerg, Denmark 19–21 July 2003) [16]. More than 10 years passed since these Workshops, but predicted coherent effects still have not been observed experimentally.

Why is it interesting to study these effects experimentally?

1. The new coherent effects in a crystal target at a few hundreds MeV photon/electron beams energies can be discovered, while the detection seems the worth challenge for experimentalists.
2. Having the monochromatic beam of Ps atoms, one can consider the possibility the direct measurement of singlet Ps lifetime in a vacuum. All previous measurements were concerned with lifetime measurements of Ps stopped in a matter.
3. Production of Ps atoms by the photons/electrons in the Coulomb field of an atom is one of the QED effects, still not observed experimentally. Using of the crystal target allows increase in Ps yield in comparison with amorphous target.

Here, we reexamine the experimental proposal [14] in connection with programs of current and planned experiments on crystal-assisted processes using high energy photon and electron beams at SAGA-LS (Japan), MAMI (Germany) and BTF (LNF Frascati).

2. Coherent production of relativistic singlet Ps atom by high-energy photons

Differential over emission angles cross - section of singlet Ps production by a photon with the energy ω in the Coulomb field of an atom is [2–3]:

$$\frac{d\sigma_1}{d\Omega}=\frac{Z^2\alpha^6}{8m^2\gamma^4}\frac{\sin^2\Theta}{\left(1-\beta^2-2\beta\cos\Theta\right)^2\left(1-\beta\cos\Theta\right)^2}. \tag{1}$$

Here, α is the fine structure constant, γ and β are the relativistic factor and velocity of created Ps, Θ is Ps emission angle with respect to a photon momentum, m is an electron rest mass, Z is nuclear charge of the target atom (units $\hbar=c=1$ are used). The transferred to an atom momentum squared is:

$$q^2=\left(E^2+p^2-2Ep\cos\Theta\right)=E^2\left(1-\beta^2-2\beta\cos\Theta\right). \tag{2}$$

Here, $E=\omega$ is the Ps energy. If the z axis is parallel to a photon momentum, $\mathbf{k}\parallel OZ$, then the components of the momentum transferred are

$$\mathrm{q}=\begin{cases} q_x=-p\sin\Theta\cos\phi \\ q_y=-p\sin\Theta\sin\phi \\ q_z=\omega-p\cos\Theta=E\left(1-\beta\cos\Theta\right). \end{cases} \tag{3}$$

Here, p is the momentum of created Ps, φ is the angle of Ps momentum in a plane perpendicular to the photon momentum. Let us introduce the transverse component of the momentum transferred, $\mathbf{q}_\perp=\{q_x,q_y\}$, and express the differential cross-section of Ps production by high energy photon in the Coulomb field of an atom in variables of momentum transferred:

$$d\sigma_1=\frac{Z^2\alpha^6E}{8\beta\, m^2\gamma^4}\frac{q_\perp}{q^4q_z^2}\frac{dq_xdq_y}{\sqrt{p^2-q_\perp^2}}. \tag{4}$$

It is well known that in an aligned crystal a cross - section of any coherent process can be written as a sum of coherent and incoherent parts [17–18]:

$$d\sigma_{cr}=d\sigma_{coh}+d\sigma_{incoh},$$

$$d\sigma_{coh}=I(\mathbf{q})\exp\left(-q^2\overline{u^2}\right)d\sigma_1;\qquad d\sigma_{incoh}=N\left[1-\exp\left(-q^2\overline{u^2}\right)\right]d\sigma_1. \tag{6}$$

The exponential $\exp(-q^2\overline{u^2})$ is the Debye-Waller factor, which takes into account the thermal vibrations of the crystal atoms, $\overline{u^2}$ is the mean-squared deviation of the crystal atom from its equilibrium position, $I(\mathbf{q})$ is an interferential multiplier responsible for appearance of the coherent effect and N is the number of atoms. We will consider the case when a photon initial momentum is parallel to a crystallographic axes – so-called type-B coherent photo-production, and the number of atoms in the crystallographic axes is not very large, i.e. the case of a thin crystal. In this case $I(\mathbf{q})$ should be written in a following form:

$$I(\mathbf{q}) = N_x N_y \frac{(2\pi)^2}{d_x d_y} \sum_{n,l} \delta(q_x - g_x n)\delta(q_y - g_y l) \times \frac{\sin^2(N_z q_z d_z/2)}{\sin^2(q_z d_z/2)} |S(\mathbf{q})|^2 \tag{7}$$

In the last equation $N_x >> 1$, $N_y >> 1$, N_z are the numbers of the crystal atoms in x, y, z directions respectively, which contribute to coherent process; $g_x = (2\pi/d_x)n$, $g_y = (2\pi/d_y)l$, $n,l = 1,2,3,...$ are the components of the reciprocal lattice vector $\vec{q}$ in x and y directions, d_x, d_y, d_z are the lattice constant in x, y, z directions respectively, and $S(\mathbf{q})$ is the crystal structure factor.

In the proposed experiment [12] it is planed to measure the Ps yield within small angular cone. To compare our results with [12] (based on 1D model) we integrated $d\sigma_{cr}$ over Ps emission angles. In our case the integration over Ps emission angles is replaced by integration over the transverse transferred momentum $\mathbf{q}_\perp$. After substitution of Eq. (7) into Eq. (5) and integration over $\mathbf{q}_\perp$ we arrive at

$$\begin{aligned}\sigma_{coh} = N_x N_y \frac{(2\pi)^2}{d_x d_y} \sum_{n,l} \sigma(\mathbf{q}_\perp^{nl}, q_z(\mathbf{q}_\perp^{nl})) \left| S(\mathbf{q}_\perp^{nl}, q_z(\mathbf{q}_\perp^{nl}))\right|^2 \times \\ \times \exp\left\{-\left[(\mathbf{q}_\perp^{nl})^2 + q_z(\mathbf{q}_\perp^{nl})^2\right]\overline{u^2}\right\} \times \frac{\sin^2(N_z q_z(\mathbf{q}_\perp^{nl}) d_z/2)}{\sin^2(q_z(\mathbf{q}_\perp^{nl}) d_z/2)}\end{aligned} \tag{8}$$

Here $\mathbf{q}_\perp^{n,l}$ is the transverse transferred momentum with components: $q_x = g_x n$ and $q_y = g_y l$. As it follows from Eq. (3),

$$q_z(\mathbf{q}_\perp^{n,l}) = E\left\{1 - \beta\cos\left[\arcsin\left(\left|\mathbf{q}_\perp^{n,l}\right|/p\right)\right]\right\}.$$

If the atomic potential is the screened Coulomb potential $V(r) = \frac{Ze}{r}\exp\left(-\frac{r}{R}\right)$, than $\sigma(q_\perp, q_z)$ takes a form:

$$\sigma(q_\perp, q_z) = \frac{Z^2\alpha^6\beta E^4}{8m^2\gamma^4} \frac{q_\perp}{(q^2 + R^{-2})^2} \frac{q_z^{-2}}{\sqrt{p^2 - q_\perp^2}},$$

here R is the atomic screening radius.

The summation in Eq. (5) is performed taking into account following condition: $q_x^2 + q_y^2 = q_\perp^2 \le E^2\beta^2 \sin\Theta_{\max}$, here $\Theta_{\max}$ is the maximal emission angle of Ps with respect to the incident photon momentum. The value of $\Theta_{\max}$ is determined by experimental setup. The incoherent part of the Ps photoproduction cross - section in a crystal is defined in a similar way:

$$\sigma_{incoh} = \int_0^{2\pi} d\varphi \int_0^{\Theta_{\max}} \sin\Theta d\Theta \frac{d\sigma_{incoh}}{d\Omega}.$$

3. Coherent singlet Ps production by relativistic electrons in a crystal

The triplet and singlet Ps can be produced also by relativistic electron passing through a crystal [10]. We restrict our consideration only by singlet Ps coherent production. In this case one can derive the cross - section in a framework of the virtual photon method [10, 13]:

$$\frac{d\sigma_E^{cr}}{dE_1} = \sigma_{cr}(\omega)\cdot n(\omega) , \tag{9}$$

where $n(\omega)$ is the virtual photon spectrum associated with incident relativistic electron

$$n(\omega) = \frac{\alpha}{2\pi}\frac{1}{E_1}\left\{\left[1+\left(\frac{E_1-\omega}{\omega}\right)^2\right]\ln\frac{E_1(E_1-\omega)}{\omega^2} - 2\left(\frac{E_1-\omega}{\omega}\right)\right\}. \tag{10}$$

In Eq. (10), E_1 is an initial electron energy, ω is the virtual photon energy. The cross-section $\sigma_{cr}(\omega)$ has been studied in Sec. 2. Therefore, further investigation is straightforward. The cross - sections of coherent of type B production of Ps by relativistic electrons (lover part) and photon (upper part) as a function of created Ps energy are plotted in the figure 1. The energies of the incident electron are E_1 = 855 MeV (figure 1a) and E_1 = 1500 MeV (figure 1b). The target is silicon Si <100>.

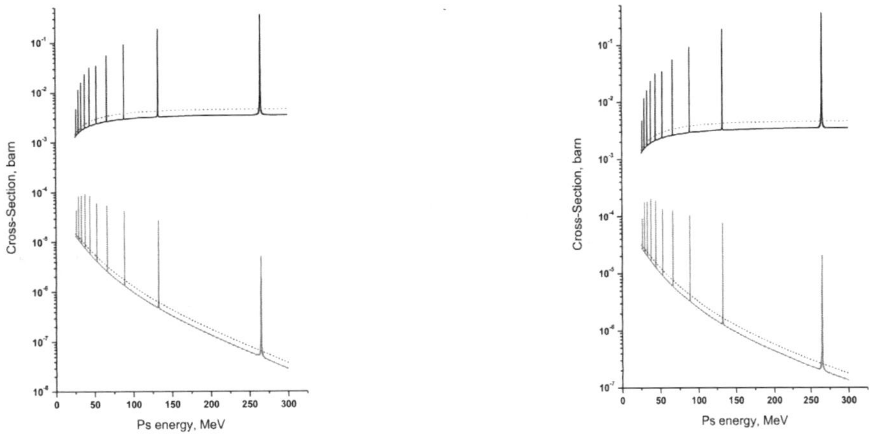

a b

Figure 1. Comparison of Ps cross-sections: coherent photoproduction vs coherent electroproduction in a crystal

The cross-section is integrated over emission angles of Ps assuming that the maximal emission angle of Ps is Θmax = 5 mrad. The number of atoms in a crystal axis is 1000, the crystal temperature T = 300 K. Dotted lines show the cross-sections of Ps production by electrons and photons in an amorphous target of equivalent thickness, with the same number of atoms as in a crystal target.

4. New Experimental Possibilities: SPARC, SAGA-LS & MAMI

Our calculations showed the appearance of brilliant coherent beaks at definite Ps energies (up to several hundreds MeV) in photo and electroproduction of relativistic singlet Ps atoms. The difference of these two processes is as follows: if one uses the tagged photon beam, coherent maxima arise at very definite photon energies, since the photon energy is completely converted to Ps energy. If one uses the relativistic electron beam, there arises very broad energy spectrum of created Ps atoms, with brilliant coherent maxima, since the broad equivalent photon spectrum is converted into broad Ps energy spectrum, with many brilliant coherent maximums. Since the intensity of tagged photon beam is as usual several orders of magnitude less than that of bremsstrahlung (BS), the use of BS beam (standard Schiff spectrum) seems to be more preferable to observe the coherent photoproduction of relativistic Ps atoms. Since the first proposal of experiment [12] at REFER facility (Hiroshima University, Japan), the experiments have not been performed up to day due to some reasons. We may suggest, that the experimental setup could be the same as suggested in [12], with appropriate scaling of the dimensions. In Ref [12] we performed calculations for electron energies E = 150 MeV and corresponding two-photon decay length of singlet Ps atoms.

New possibilities are open up at the modern operating moderate-energy electron accelerators: SPARC LNF Frascati (150 MeV), SAGA-LS Japan (255 MeV) and in Mainz at MAMI B (855 MeV) and MAMI C (1508 MeV), where the various crystal-assistant experiments (channeling radiation, parametric X-radiation, scattering and deflection by crystals) are going on. Why: type B coherent effect appears at photon energies up to several hundreds MeV. In the case of electroproduction, there appear a wide energy spectrum of created Ps atoms, but the heights of coherent peaks slightly increases (as ln E) with increase of electron beam energy. Besides, an increase of the beam energy guarantees the staright – line electron trajectory parallel to a crystal axis (neglecting the channeling effect). The principal is the use the ultrathin crystal targets, since when creating the Ps atom penetrates through the bulk of the crystal and can be broken (the break-up cross-sections are known). The complete theory must take into account the simultaneous processes of Ps generation and break-up (similar to generation of X-Rays by relativistic particles in a solid target).

The important figures for experimentalists are of course the counting rates which defines necessary beam time. Let us consider the case of MAMI C (electron energy up to 1508 MeV). The electron beam has the intensity about 10^{13} electron/s. The intensity of the BS beam can be expected as $N_{BS} = 10^{10}$ photons/s, while the intensity of tagged photon beam is much lower, $N_T = 10^6$ photons/s.

The schematics of possible experimental setups could be the same as suggested in [12] for coherent production of relativistic singlet Ps by BS photon beam and in for coherent Ps production by electron beam, respectively.

Consider the BS spectrum from 1508 MeV electrons, we arrive at the Schiff spectrum of photon beam. The convolution of this BS spectrum with singlet Ps photoproduction cross-section from Section 2 will result in a specific Ps energy spectrum emitted from the crystal, which in fact is similar to Ps spectrum resulting from electroproduction (Section 3). The difference is that it is possible to produce singlet Ps with energies up to 1500 MeV using the BS photon spectrum, while electroproduction of only the soft part of singlet Ps energy spectrum is accessible. In both cases, the Ps energy spectrum has brilliant coherent peaks at the same Ps energies.

Since the maximal decay length of produced singlet Ps is expected at maximal Ps energy close to 1500 MeV, $l_\tau \cong \gamma\beta\tau \approx 51.2$ m, this value may characterize the maximal longitudinal size of experimental setup. For type-B singlet Ps production, as considered here, the crystal axis must be precisely aligned with photon/electron beam direction. There are few possible methods to align the crystal: to use the orientation dependence of channeling radiation (CR), to measure the orientation dependence of Ka-X ray emission, or to measure parametric X-Ray (PXR) yield. In fact, the last method can be used both for alignment and electron beam monitoring. Among two first ones, the CR measurement seems to be more effective, see, e.g. [19].

The expected Ps photoproduction rate in vicinity of strongest coherent peak when the incident BS photon beam of intensity $N_{BS} = 10^{10}$ photons/s will be used, can be estimated using the cross-section given, in figure 1 and the density of atomic strings on the crystal surface, of order of $1/d^2$. By integration between Ps energy E = 263 – 266 MeV, we easily obtain the production rate of order of 350 singlet Ps per hour from Ge crystal target, if the number of atoms which give rise to coherent effect is N_z = 1000. This rate is to be contrasted with that from an amorphous target, which is approximately 16 times smaller.

In the case of electroproduction, using the cross-sections given in figure 1 our calculation gives the production rate of order of 16 ± 20 singlet Ps per hour integrated within energy region E = 260 – 270 MeV, from <100> Si crystal and much less from amorphous (misaligned) target. In addition, in an experiment on Ps coherent electroproduction in a crystal, the long-lived triplet Ps can be produced [7] with smaller cross-section.

5. Conclusions

New (besides scattering, deflection and radiation) physics can be explored in the field of interaction of high-energy photons (> 100 MeV) and relativistic electrons (positrons) with aligned crystals, i.e. new QED effects – production of bound electron-positron pairs – relativistic (moving with relativistic velocity) Ps atoms. Coherent effecs in a crystal lead to enhancement of production cross-sections at definite energies of emitted Ps, i.e. allowing generation of more intense Ps beams for future Ps physics and tests of QED.

6. Acknowledgments

The authors are grateful to Prof. H Backe (MAMI Mainz), Dr. A Mazzolari and L Bandiera (Ferrara University) for their interest to our work.

References

[1] Meledin GV, Serbo V G and Slivkov A K 1971 *JETP Lett.* **13** 68
[2] Olsen H A 1986 *Phys. Rev.* D **33** 2033
[3] Lyuboshits V L 1987 *Sov. J. Nucl. Phys.* **45** 682
[4] Holvik E and Olsen H A 1987 *Phys. Rev.* **D35** 2124
[5] Gevorkyan S, Kuraev E, Schiller A, Serbo V G and Tarasov A 1998 *Phys. Rev.* **A58** 1556
[6] Kotkin G L, Kuraev E A, Schiller A, Serbo V G and Tarasov A 1999 *Phys .Rev.* **C59** 2734
[7] Olsen H A 2001 *Phys. Rev.* **A63** 032101
[8] Olsen H A 1999 *Phys. Rev.* **A60** 1883
[9] Kunashenko Yu P and Yu L Pivovarov Yu L 1990 *Sov. J. Nucl. Phys.* **51** 397
[10] Sandnes G I and Olsen H A 1993 *Phys.Rev.* **A48** 3725
[11] Kunashenko Yu P and Yu L Pivovarov Yu L 1996 *Nucl. Instrum. and Meth.* B **114** 237
[12] Pivovarov Yu L, Kunashenko Yu P, Endo I and Isshiki T 1998 *Nucl. Instrum. and Meth.* B **145** 80
[13] Kunashenko Yu P 2005 *Nucl .Instrum. and Meth.* B **229** 219
[14] Okazaki Y, Andreyashkina M, Chouffania K and *et al* 2001 *Phys. Letters* A **271** 110
[15] Proceedings of the Workshop 1998 *Hadronic atoms and positronium in the standard model* ed. Ivanov M A, and Kuraev E A (E2-98-254, Dubna)
[16] Proceedings of the XII International *Workshop on Positron and Positronium Physics* (Sandbjerg, Denmark 19–21 July 2003) 2004 *Nucl. Instrum. and Meth.* B **221** ed Uggerhoj U I, Knudsen H and Ichioka T
[17] Ter-Mikelian M L 1972 *High Energy Electromagnetic processes in Condensed Media,* (New-York: Wiley Interscience)
[18] Uberall H 1956 *Phys. Rev.* **103** 1055
[19] Abdrashitov S V, Bogdanov O V, Dabagov S B, Pivovarov Yu L and Tukhfatullin T A 2013 *Nucl. Instrum. and Meth.* B **309** 59

Gamma Cherenkov-transition radiation of high energy electrons and methods for the measurement of the refractive index of MeV photons using total internal and external reflections (Invited talk)

K A Ispirian

Alikhanian National Scientific Laboratory (Yerevan Physics Institute)
Br. Alikhanians 2, Yerevan 0036, Armenia

E-mail: karo@mail.yerphi.am

Abstract. It is given a review of the theoretical works showing that gamma ray Cherenkov-transition radiation (GCTR) is produced in some materials the refractive index of which in gamma region is greater than 1 according to the famous results of [1]. Since there are some published doubts [2] on the theoretical results of [1], and taking into account the difficulties of the experiment [1], in order to confirm or decline the results [1] it is proposed to study experimentally GCTR and an experimental method for the measurement of refractive index of MeV photons.

1. Introduction or formulation of the problem

In figure 1 it is shown qualitatively the behaviour of the dependence of the refractive index $n(\hbar\omega)$ of substances on photon energy $\hbar\omega$. As it is seen up to the X-ray region, excluding some narrow regions at anomaly dispersion frequencies, the refraction index is greater than 1, while when $n(\hbar\omega)$ >0.1 keV $n(\hbar\omega)<1$.

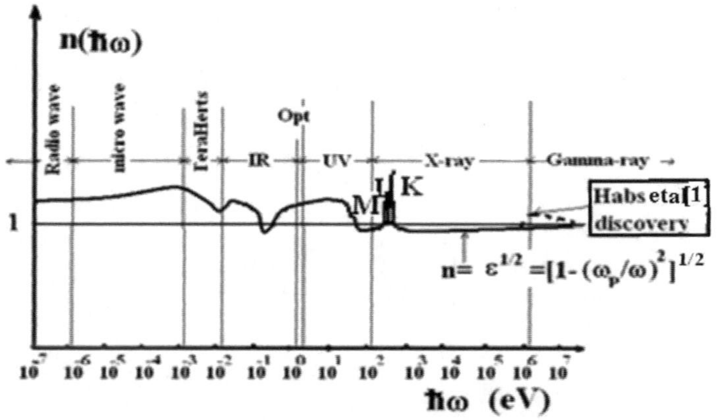

Figure 1. The dependence of n of substances upon ħω.

Usually, one presents the complex values of $\varepsilon(\omega)$ and $n(\omega)$ in the forms: $\varepsilon(\omega) = n^2(\omega) = \varepsilon'(\omega) + i\varepsilon''(\omega) = 1 + \chi'(\omega) + i\chi''(\omega)$ and $n(\omega) = 1 + \delta(\omega) + i\Delta(\omega)$ where χ' and χ'' are the real and imaginary parts of the susceptibility, and $\chi'(\omega)$ and $\delta(\omega)$ are connected with the refraction, while $\chi''(\omega)$ and $\Delta(\omega)$ describe the absorption of photons with absorption length $L_{abs} = 1/\mu(\omega) = c/(\chi''(\omega)\omega)$ where μ is the linear absorption coefficient.

Now according to sensational experimental and theoretical results of [1] for silicon and gold $n(\hbar\omega) > 1$, or $\delta > 0$ for $\hbar\omega \geq 1$ MeV. This has many important consequences and applications (see [3–6]) for nuclear physics, gamma optics, etc. Gold lenses, similar to Snigirev X-ray lenses, have been prepared in order to focus gamma ray beams. However, no new experimental results and answer to [2] has been published. The difficult experiment [1] has been carried out at Institut Laue-Langevin on the facility GAMS5 measuring the nanoradian relative diffraction angle for three lines of Cl^{36} target bombarded by thermal neutrons when a Si wedge is inserted between the two crystals of a double Bragg diffraction spectrometer. The measured relative diffraction angles allowed to determine $\delta = (+1.83\pm1.57)\text{x}10^{-10}$, $(+1.48\pm0.13)\text{x}10^{-9}$ and $(+1.11\pm0.30)\text{x}10^{-9}$, for 786, 1165 and 1951 keV photons, respectively, while the δ measured for lower energies are negative according to the formula $n(\omega) = [1 - (\omega_p/\omega)^2]^{1/2}$.

The authors of [1] using the cross section of various processes, mainly of the QED nonlinear process of Delbruck scattering calculated with the help of a complicated non- perturbative QED theory obtained good agreement between the experimental and theoretical results on δ. However, A comment [2] on the work [1] has been published in which it has been shown that even using the Delbruck scattering cross sections given in [1] the contribution of Delbruck effect into δ is 10^{-6}–10^{-5} times less than it was estimated in [1]. Since there is no answer to [2] and no new experimental and theoretical results one can conclude that there is a problem in XXI century physics which it is better to solve experimentally.

On the other hand, it is well known that when a particle passes through a substance with $n(\omega) > 1$ it can produce Cherenkov radiation (CR) in radio-optical region (see [7, 8]) and X-ray region (XCR) near K-, L- edges (see [9,10]). Using the results of [1] on $n(\hbar\omega) > 1$ for photons with $\hbar\omega \geq 786$ keV in the work [11a] it has been developed the theory of GCTR produced in a plate of material using the more confident experimental data of [1] in the region 786–2000 keV. In [11b] the theory of GCTR produced in a plate of material using the theoretical data of [1] in a wider region 786-10000 keV has been given, while the problems of the formation length and GCTR produced at a single interface and in stack of plate have been considered in [11c]. Looking for other methods of manifestation of the fact that in gamma ray region the refractive index can be greater than 1 in [11d] it is considered a simpler and direct method for the measurement of $n(\hbar\omega)$ using total internal (TIR) and external refection (TER) of MeV photons. In this work it is given a review of the works [11].

2. Theory of GCTR based on Ginzburg-Frank-Garibian Theory of TR

2.2a. GCTR produced at an interface between two media and its formation length

Let us first consider the formation length in medium of GCTR. According to [12, 13] the formation length of XTR is a complex magnitude and is given by the expression

$$Z_{Med}(\theta) = \frac{2\pi V}{\omega(\gamma^{-2} + \theta^2 - \chi)} \quad . \tag{1}$$

From (1) one obtains for the formation length, L_{Med}^{Form}

$$L_{Med}^{Form} = |Z_{Med}| = \frac{2\pi V}{\omega} \frac{1}{\sqrt{(\gamma^{-2} + \theta^2 - \chi')^2 + \chi''^2}} , \tag{2}$$

from which the formation length in vacuum ($\chi'=\chi''=0$) is equal to $L_{Vac}^{Form}=2\pi V/\omega/(\gamma^{-2}+\theta^2)$ with a maximum at $\theta=0$ equal to $L_{GCTRvac}=\lambda\gamma^2$ as for other types of radiation. As it follows from (8) when GCTR takes place and the condition $\gamma^{-2}+\theta^2-\chi_1^{'}=0$ is satisfied $L_{Med}^{Form}=2\pi L_{abs}$.

The spectral-angular distribution of the number of photons of GCTR produced at an interface of two media (Si-vacuum) is derived substituting the values of χ' and χ'' into the formulae (1.59) [13] for XTR. In small angle and γ>>1 approximation one obtains

$$\frac{d^2N_{GCTR}}{d(\hbar\omega)d\theta}=\frac{2\alpha\theta^3}{\pi\hbar\omega}\frac{(\chi'^2+\chi''^2)}{(\theta^2+\gamma^{-2})^2[(\theta^2+\gamma^{-2}-\chi')^2+\chi''^2]}. \quad (3)$$

2.2b. GCTR produced in a plate

Inserting the values of the complex χ into the XTR spectral-angular distribution (formula (2.21) of [13] or formula (3.27) of [9] for a plate one derives the following formula for the spectral-angular distribution of GCTR

$$\frac{d^2N_{GCTR}}{d\hbar\omega d\theta}=\frac{2\alpha\theta^3}{\pi\hbar\omega}\frac{(\chi'^2+\chi''^2)}{(\theta^2+\gamma^{-2})^2[(\theta^2+\gamma^{-2}-\chi')^2+\chi''^2]}\times$$
$$\left\{\left[1-\exp\left(-\frac{L_{rad}}{2L_{abs}}\right)\right]^2+4\exp\left(-\frac{Lrad}{2L_{abs}}\right)\sin^2\frac{\hbar\omega L_{rad}(\theta^2+\gamma^{-2}-\chi')}{4\hbar c}\right\}, \quad (4)$$

where L_{rad} is the thickness of the plate-radiator.

2.2c. GCTR produced in a stack of plates

The necessary expressions obtained with the help of corresponding formulae (3.15) of [13] as well as numerical results are given in [11c] and will not be discussed here. Before proceeding it is important to note the following: In contrast to XTR, due to the facts that for GCTR $L_{rad}<L_{abs}\approx$ a few cm, and the requirement that the particle velocity must be changed insignificantly, therefore, it is necessary to have $L_{rad}\approx 0.1X_0$ (X_0 is the radiation length of the radiator material), GCTR produced in a stack of plates as well as at one interface have no practical values since the GCTR yield from a stack of plates can not exceed the GCTR yield from a plate and it is not reasonable to consider GCTR production at one interface because one can not replace the infinite medium of a single interface with a plate thinner than L_{abs}. Therefore, further it will be considered only GCTR produced in a plate with $L_{rad}\le 1$ cm given by (4) which will be integrated (numerically, because analytically it is very complicated or impossible) over θ, $\hbar\omega$ and over both θ and $\hbar\omega$ to obtain $dN_{GCTR}/d(\hbar\omega)$, $dN_{GCTR}/d\theta$ and N_{GCTR}, respectively.

3. Numerical results on GCTR produced in a plate

The below given numerical results described in details in [11b], have been obtained for electron energy E = 20 GeV if there is no especial remark, and only for Si radiators with thickness $L_{rad}=0.1X_0=1$ cm using the values of $\chi'=2\delta$ (from [1]) and $\chi''=\mu c/\omega$ (from various tables for μ).

3.1. The spectral distribution of the number of GCTR photons

$dN_{GCTR}/d(\hbar\omega)$ is shown in figure 2 (curve "1"). The curve "2" of figure 2 shows the spectral distribution of the background bremsstrahlung calculated taking into account the longitudinal density or Ter-Mikaelian

Journal of Physics: Conference Series **517** (2014) 012008 doi:10.1088/1742-6596/517/1/012008

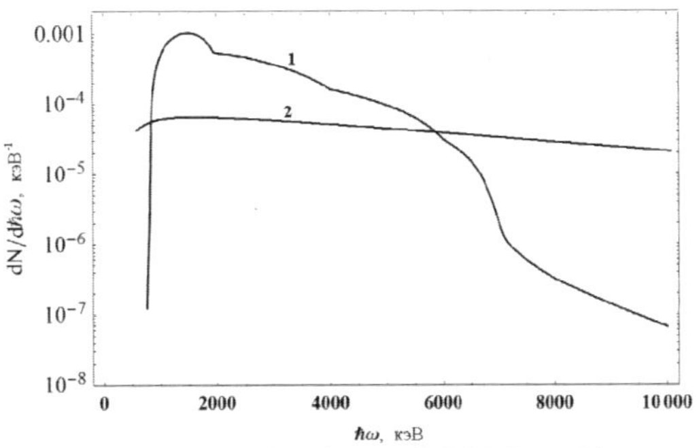

Figure 2. The spectral distributions of GCTR and bremsstrahlung.

effect (see [12]). As it follows from figure 1 GCTR intensity exceeds the background bremsstrahlung intensity in a wide interval from 800 keV up to 6000 keV with S/N>>2. The total number of GCTR and background photons obtained by numerical integration of the curve "1" and "2" of figure 2 over ħω are N_{GCTR}=1.34 and $N_{Br} = 0.1$.

3.2. The angular distribution of GCTR

$dN_{GCTR} / d\theta$ obtained integrating (4) over $\hbar\omega$ in the interval 800–10000 keV is shown in figure 3.

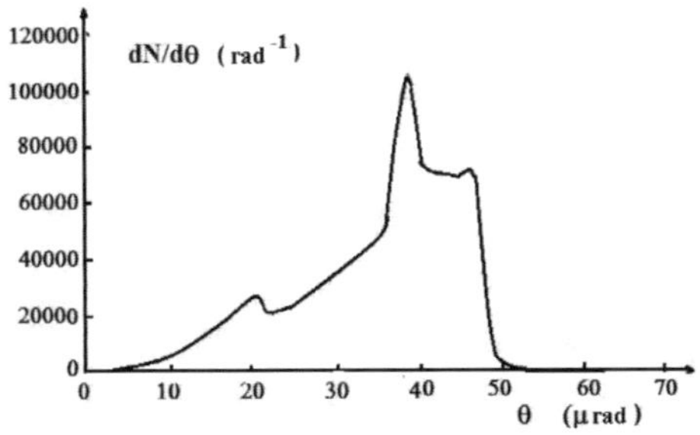

Figure 3. The angular distribution of GCTR.

From figure 3 it follows that GCTR is emitted mainly in the angular interval from 30 μ rad up to 50 μ rad.

3.3. The dependence of GCTR on the electron energy

The energy dependence of the total number of GCTR (solid curve) and bremsstrahlung (dashed curve) photons is shown in figure 4. As it seen in figure 4 the number of the GCTR photons as in case of XTR exceed the background and begins to increase sharply beginning from a "threshold" energy about 12 GeV. Then with the increase of E a "saturation" achieves with $N_{GCTR} \approx 3.4$. As it is has been shown still in [12, 14] the number of the XTR photons in transition radiation detectors (TRD) from a stack of plates without and with taking into account the absorption is equal to N_{XTR} ~11 and ~2 and exceeds the background bremsstrahlung only in a relatively narrow region $\Delta(\hbar\omega) \approx$ (2–30) keV, and this allows to construct TRD detectors, which found wide application for particle identification. Therefore one can use GCTR also for the same purpose.

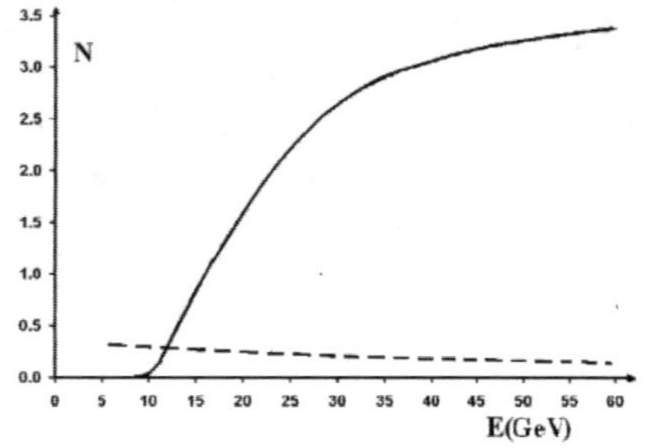

Figure 4. The dependence of the number of the GCTR (solid line) and background (dashed line) photons.

3.4. The dependence of the GCTR yield on the radiator thickness

Figure 5 shows how N_{GCTR} depends on L_{rad}. Around L_{rad}= 1 cm N_{GCTR} is proportional to L_{rad}, and the electron energy losses are still small. Therefore, it is reasonable to take L_{rad}~(2-10) % X_0, and as the calculations (see [11]) show the use of a stack of plates instead of a plate has no advantage.

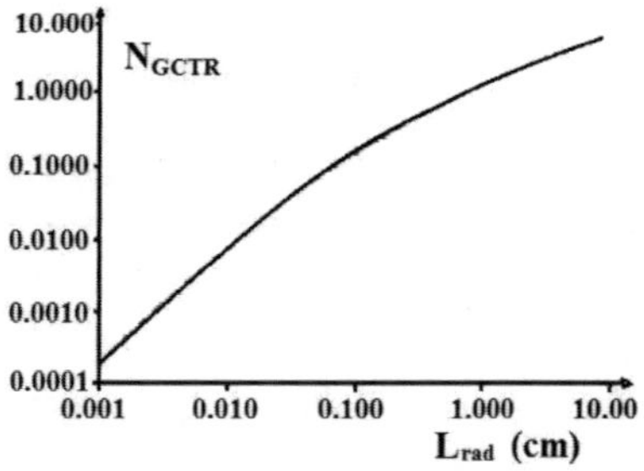

Figure 5. The dependence of N_{GCTR} on L_{rad}.

4. Experimental possibilities

One can use an experimental arrangement similar to those used at SLAC [15] and CERN SPS [16] and [17] for the study of various types of radiation of 1–200 GeV electrons. Without discussing the proposed arrangement let us note that the SLAC beam has the difficulty that it must contain 1 electron per pulse, while the SPS beam being much longer, allows to detect a few GCTR photons per pulse. The analysis of the spectra measured by 25 GeV electrons in $\hbar\omega$= (1–10) MeV region did not reveal any exceeding over the calculated spectra of bremsstrahlung of various theories [15]. Probably, one cannot use this fact to decline the results of [1] may be because no measurements on Si have been performed.

5. Measurement of the reflection index of MeV photons using total internal and external reflection

It is well known that if $n(\omega)>1$ (optical) it can take place TIR at the boundary medium-vacuum with θ_{cr}^{TIR} = arcsin(1/n). Similarly, if $n(\omega)<1$ (X-rays) it can take place TER at the boundary vacuum – medium with $\theta_{cr}^{TER} = (2\delta)^{1/2}$. If δ~ 10^{-9}, θ_{cr}^{TER} = 45 μrad, which is not very small, μrad , as the angles in [1]. In many works of the past (see, [18]) it has been proposed, but not realized, to use TER for various purposes, in particular, for the “monochromatization” of MeV photon beams. However, such MeV experiments become realistic only recently connected with the advance of the method of inverse Compton scattering intense photon sources [19]. Suitable parameters are available at present at the facility HiGS of Duke university, which provides 10^8–10^{11} ph/s in a ~10 % BW, and soon at the ELI facility in Romania, which will provide ~10^{13} ph/s in a ~1 % BW.

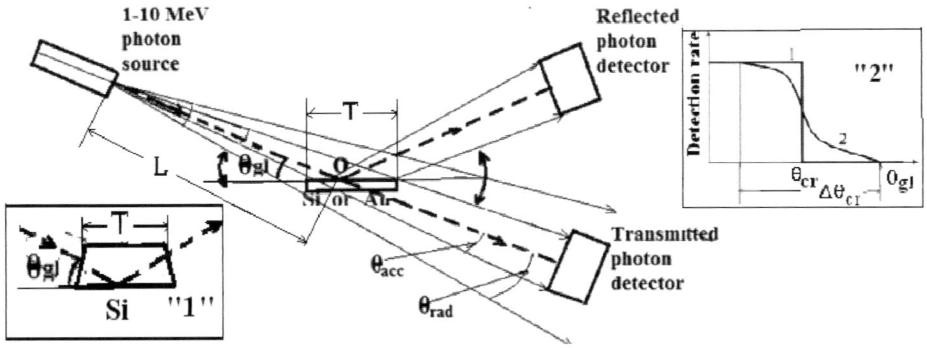

Figure 6. An arrangement for the measurement of $n(\omega)$ of MeV photons.

Figure 6 in which the main Si reflector will be replaced by that of the insert "1" for $n(\omega)<1$ and $n(\omega)>1$, respectively, explains the method of the measurements. First let us assume that $n(\omega)>1$. Increasing the glancing angle θ_{gl} by rotating the Si targets one can expect the well known dependences of the rate of TER (or TIR) photons detection rate of the reflected (transmitted) photon detector upon θ_{gl} (see the insert '2' of figure 6), from which δ is determined from the measured $\theta_{cr}^{TER}=(2\delta)^{1/2}$ with an error $\pm\ \Delta\theta_{gl}/2$. To carry such measurements it is necessary to have (1–10) MeV photon beams with 1) angular spread much less than the above $\theta_{cr}^{TER}=45\,\mu\text{rad}$ and 2) sufficient intensity. The first condition is satisfied by the fact that only a small fraction $F\approx(\theta_{acc}/\theta_{rad})^2$ of the beam of the photon source with angular spread $\theta_{rad}\approx 1/\gamma$ emitted under angles smaller than $\theta_{acc}=T\theta_{gl}/2L$ (T is the sizes of the reflector, L is the distance between the source and reflector), strikes the Si reflector. Taking T=3 cm, L=10 m, one obtains for HiGS ($\gamma=2.4\text{x}10^3$), $F=10^{-8}$.

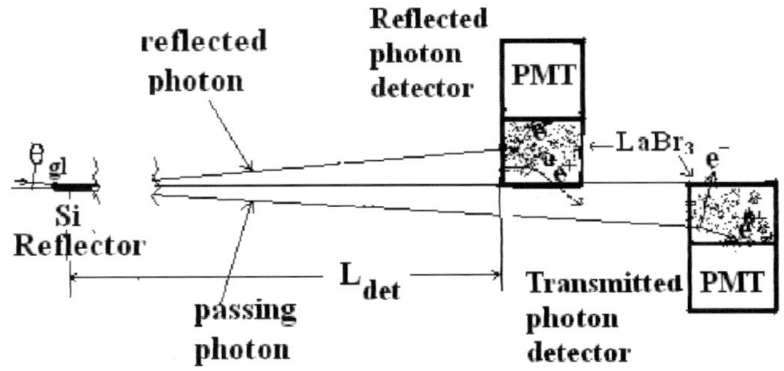

Figure 7. The proposed set-up for the reflected and transmitted photon detectors.

Therefore, for HiGS one expects to detect ~1–10³TER ph/s, and the experiment is feasible. However, in reality one must take into account that the HiGS beam is available after 2 collimators at a distance of $L_{\text{det}}\approx 60$ m from the source as well as the difficulties of photon detection by the reflected and transmitted photon detectors connected with the smallness of the reflection angles (see [11d]). For this purpose the set-up of figure 7 is proposed in which the photons are detected separately at $L_{\text{det}}\approx 60$ m using 2 $LaBr_3$, scintillators. These problems can be solved better with the help of PET liquid xenon time projection chamber (LXeTPC) [20] or proposed LXeTPC on the basis of graphene [21].

6. Conclusion

At present it is possible to study the predicted GCTR using $E \geq 20$ GeV electrons at SLAC or of CERN and to measure directly $n(\omega)$ of various materials in MeV region using TER or TIR. These experiments can confirm or decline the discovery of [1].

References

[1] Habs D, Gunther M M, Jentschel M and Urban W 2012 *Phys. Rev. Lett.* **108** 184802
[2] Donohue J T 2013 *Phys. Rev. Lett.* **110**, 129501
[3] Habs D, Gunther M M, Jentschel M and Thirolf P G 2012 ArXiv:1201.4466
[4] Jentschel M 2013 *Gamma Ray Optics* http://www.inn.jir
[5] Habs D 2013 *Delbruck scattering, non-perturbative QED and possible applications* http://www.eli-np.ro
[6] *CERN Courier* 2012 July 18
[7] Jelley M 1958 *Cherenkov Radiation and Its Applications* (Pergamon Press, London)
[8] Zrelov V P 1968 *Izluchenie Vavilova-Cherenkova i Ego Primenenie v Fizike Visokikh Energii* (Moscow, Atomizdat)
[9] Bazilev V A and Zhevago N K 1987 *Izluchenie Bistrikh Chastits v Veshchestve i vo Vneshnikh Polyakh* (Moscow, Nauka)
[10] Ispirian K A 2005 *X-Ray Cherenkov Radiation*, NATO Sc. Ser. II, Math. Phys Chem. **199** 217
[11] Aginian M A, Ispirian K A and Ispiryan M, To be published in a) 2013 *European Physics Letters*; b) 2014 *Izvestya of NAS of Armenia;* c) 2014 Proc of this Conf.; d) 2014 Proc of this Conf.
[12] Ter-Mikaelian M L, *High Energy Electromagnetic Processes in Cond. Media,* Wiley Intersc., N Y, 1972
[13] Garibian G M, Yan Shi 1983 *Rentgenovskoe Perekhodnoe Izluchenie,* Publishing home of Academy of Sciences of Armenia (Yerevan, Armenia)
[14] Ispirian K A 1965 Dissertation, Yerevan State University
[15] Klein S 1999 *Rev. Mod. Phys.* **71**, 1501
[16] Hansen H D, Uggerhoy U I, Biino C *et al* 2004 *Phys. Rev. D* **69** 032001
[17] Mazzolari A *et al* 2014 To be published in Proc of this Conf.
[18] Kumakhov M A 1986 *Radiation of Channeled Particles in Crystals*, Energoatomizdat, Moscow,
[19] Wu Y K, *Overview of present and future Compton photon sources,* PAC 2012, TUXB03
[20] Lewellen T K 2008 *Phys. Med. Biol* **53** R287-R317
[21] Ispirian K A and Ispiryan R K 2011 *Nuovo Cimento* **34** 37

Journal of Physics: Conference Series **517** (2014) 012009 doi:10.1088/1742-6596/517/1/012009

Investigation of the characteristics of EUV backward transition radiation generated by 5.7 MeV electrons in mono- and multilayer targets

S R Uglov[1], V V Kaplin[1], A P Potylitsyn[1], L G Sukhikh[1], A V Vukolov[1] and G Kube[2]

[1]Tomsk Polytechnic University, Lenin avenue, 30, Tomsk, 634050, Russia
[2]Deutsches Elektronen–Synchrotron (DESY), Notkestrasse 85, 22607, Hamburg, Germany

E-mail: uglov@tpu.ru

Abstract. This paper presents experimental results of a study of radiation in the vacuum ultraviolet and extreme ultraviolet spectral region, generated by electrons with an energy of 5.7 MeV in a multilayer X-ray mirror. The angular distribution of the radiation generated in the rear hemisphere from the input surface of the target was measured. The contribution to the radiation yield caused by the periodic structure of the multilayer target was experimentally investigated.

1. Introduction

Imaging of transverse beam profiles based on backward transition radiation (BTR) in the visible region of the spectrum is widely used in modern linear accelerators. This kind of beam diagnostics is usually termed optical transition radiation (OTR) monitor. However, the latest results from new generation VUV and X-ray free-electron lasers, such as LCLS (USA), FLASH (Germany), and SACLA (Japan), have demonstrated that there are serious limitations in using such a beam diagnostics because of instabilities arising in individual bunches. This microbunching causes the appearance of coherent effects (quadratic dependence of the radiation power on the microbunch population), which makes it impossible to use standard imaging techniques based on OTR. To overcome this problem it is possible to measure at smaller photon wavelengths, and in Ref. [1] it was proposed to use BTR in the spectral region of the extreme vacuum ultraviolet (EUV). In this case, the main problem is connected with the quite small spectral density of the radiation yield from conventional radiators, consisting of a homogeneous substance (monolayer targets). This requirement of a high spectral density is driven by the fact that for beam diagnostics in modern linear accelerators it is necessary to resolve beam profiles on the level of individual bunches, which typically contain $10^9 - 10^{10}$ electrons.

The radiation yield of BTR in the EUV region can be increased by using a large target tilt angle of about $60 - 70$ degrees (with the angle defined between the normal of the target surface and the beam axis), but this results in a significant beam image distortion due to the depth of field effect. On the other hand, OTR monitors are usually implemented with smaller target tilt angles of 22.5 or 45 degrees. In order to achieve a comparable tilt angle and sufficient radiation yield in the EUV region, the use of a periodic multilayer structure as radiator is considered to be more promising in comparison to a conventional monolayer target. Such a periodic multilayer structure is known as a multilayer X-ray mirror. Considering the mechanism of BTR generation as reflection of the relativistic electron field (so called pseudo-photons) at the target structure, it is to expect that the use of a multilayer interference structure with an appropriate period will result in an increase of the target reflectivity and the radiation yield, respectively. The energy of the additional radiation photons will be distributed in a narrow spectral range which is determined by the Bragg diffraction from a periodic structure. Indeed, if the mirror is placed under Bragg condition, X-rays can be scattered out and two processes, diffracted transition radiation (DTR) and parametric X-ray radiation (PXR) can occur. There are several theoretical and experimental investigations [2, 3, 4, 5, 6, 7] connected with these types of radiation. The mechanism of radiation generation in multilayer X-ray mirrors (MXM) are similar to the mechanisms of PXR and DTR generation from periodic crystal structures. Experimental studies [4, 5, 6, 7] have been carried out for MXM generated in the X-ray region with photon energies of $E_\gamma = 6 - 15\,\mathrm{keV}$. Recently, experimental investigations of the MXM mechanism in the EUV region started at Tomsk Polytechnic University using a 5.7 MeV electron beam [8, 9]. This paper presents the results of a measurement of the angular distributions of EUV radiation from a multilayer radiator consisting of 50 Mo/Si bilayers placed onto a silicon substrate. The results are compared with the ones obtained from a monolayer radiator consisting of a silicon wafer coated with a layer of molybdenum.

2. Experimental setup

The experiment was carried out at the external electron beam of the the microtron M-5 of Tomsk Polytechnic University with a total electron energy of $E_e = 5.7\,\mathrm{MeV}$. The pulse repetition frequency amounted to 25 Hz, and the pulse duration of the extracted beam was $0.4\,\mu\mathrm{s}$. The scheme of the accelerator complex together with the experimental setup are shown in figures 1 and 2. The studies were carried out at the experimental chamber (16) of beam line II, cf. figure 1. The electron beam extracted from the microtron passed a bending magnet (3), a collimator system (4,6,7), and a second bending magnet (9). Using the second bending magnet it was possible to reduce the background generated by bremsstrahlung in the collimators. The transverse beam shape at the target (14) in the center of the the experimental chamber was close to a circle with a diameter of 2 − 3 mm, and the beam charge amounted to $q \approx 4\,\mathrm{pC}$ per pulse of beam ejection. The monitoring of the beam current of accelerated electrons was carried out by a NaI (Tl) detector (17) located at a distance of 1200 mm behind the target.

The radiation yield in the EUV range was measured by the secondary electron multiplier SEM-6 (13). The detector was located in the median plane of the accelerator at a distance of 140 mm from the target onto a rotator that allowed to scan the angular distribution of the radiation emitted from the target in the range of $\theta_D = 20°$ up to $\theta_D = 160°$ with respect to

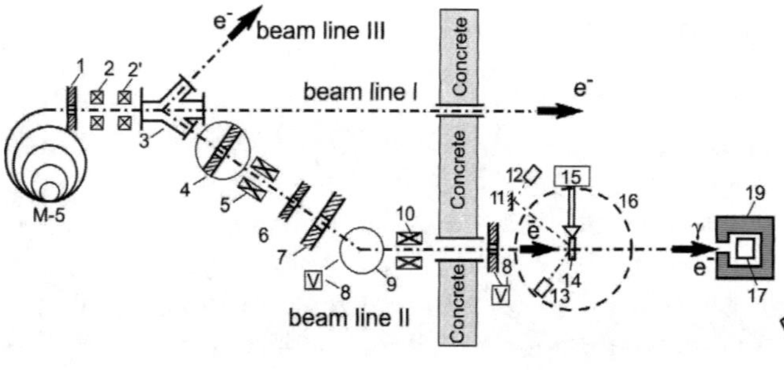

Figure 1. Scheme of the accelerator complex.

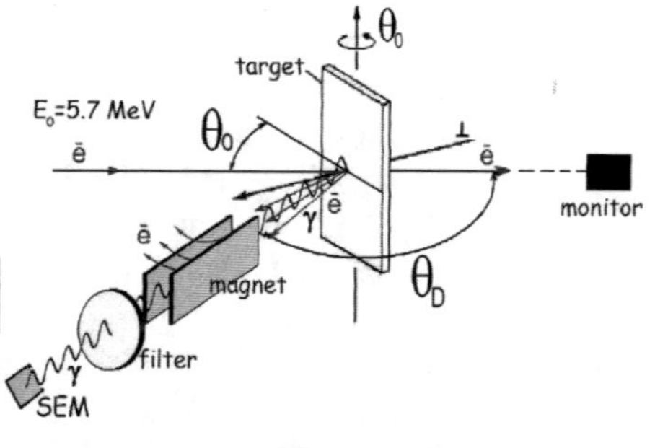

Figure 2. Experimental scheme.

the electron beam axis. Permanent magnets and diaphragms were mounted between target and detector along the direction of radiation propagation for cleaning the photon beam from electrons scattered at the target. To eliminate the influence of stray magnetic fields on the electron beam, rare-earth permanent magnets were placed inside an iron box with two reach-through holes. The detection and data acquisition system for the EUV radiation consisted of the SEM-6 secondary electron multiplier, a preamplifier U3-33, an amplifier BUS2-97, a CAMAC shaper and a counter. The SEM-6 was operated in counting mode. To prevent pile-up effects, the detector load was controlled every 100 pulses of the accelerator and, as a rule, did not exceed 50 counts per 100 pulses. The spectral range of the registered radiation was defined by the spectral sensitivity of the SEM-6 [10] and the absorption filter set directly in front of the entrance window of the detector. The entrance window of the SEM-6 had a diameter of $D_0 = 10\,\mathrm{mm}$. In the experiment however, the size of this window was limited by diaphragms, either with a diameter of $D_1 = 8\,\mathrm{mm}$ in the measurements with filter, or with a diameter of $D_2 = 2.5\,\mathrm{mm}$ in the measurement without. The detector was surrounded by a lead shielding with 10 mm thickness.

As periodic target, a multilayer X-ray mirror consisting of 50 Mo/Si bilayers with a period of $d = 113.2\,\text{Å}$ was used. The multilayer was sitting onto a silicon wafer with dimensions of $40 \times 40 \times 0.53\,\mathrm{mm}^3$. The thicknesses of the alternating Si and Mo layers were $a = 79.2\,\text{Å}$ and $b = 34\,\text{Å}$ $(a + b = d)$, respectively. The substrate was a Si (100) wafer of p-type (boron doping), the cut plane was tilted by $1.5 - 2$ degrees with respect to the (100) plane. Figure 3 shows the spectral reflectivity curve for a photon incidence angle of 22.5°.

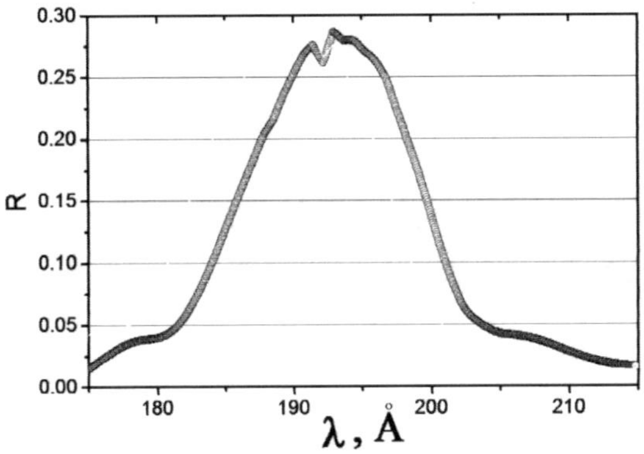

Figure 3. Spectral reflectivity curve of the Mo/Si mirror at an angle of incidence of 22.5°.

The second target consisted of a silicon wafer with dimensions of $40 \times 10 \times 0.68\,\mathrm{mm}^3$ and a 50 nm thick Mo layer at the entrance surface. The parameters of this target correspond to the ones from the target described in Ref. [11]. Both targets were mounted onto a motorized goniometer located in the center of the experimental chamber. The target was preliminarily aligned with respect to the electron beam axis by means of a laser level. In the experiment, the electron beam was steered onto the target with the corrector (10) and observed with a camcorder (12) and two phosphor screens located below and at the side of the target. The angular distributions were scanned in the plane formed by electron beam axis and target surface normal.

3. Experimental results

For a preliminary test of the experimental setup, angular distributions of BTR generated from a polished Si wafer with 250 μm thickness and a polished Al foil with 1 μm thickness were measured. The results are shown in figures 4 and 5. The solid lines indicate the results of a theoretical calculation of the BTR angular distribution according to the theory [12]. Both calculations

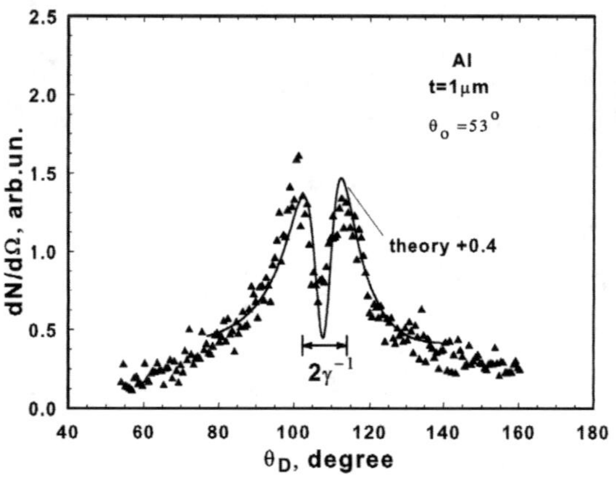

Figure 4. Experimental (triangles) and theoretical (solid curve shifted upwards on 0.4) angular distributions of BTR for the 1 μm thick Al foil and $\theta_0 = 53°$.

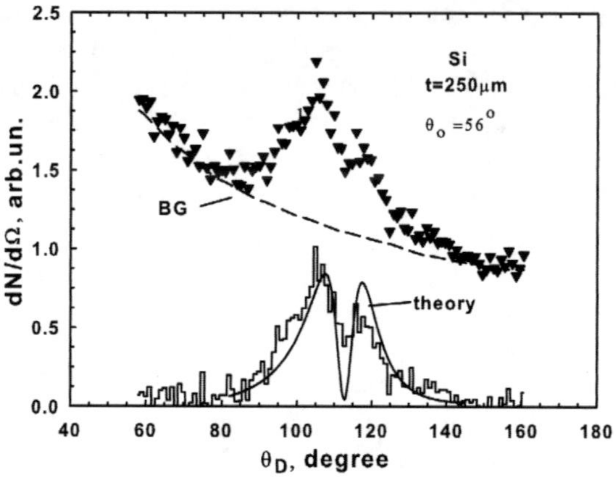

Figure 5. Experimental and theoretical angular distributions of BTR for the 250 μm thick Si wafer and $\theta_0 = 56°$; triangles – experimental data, dashed curve – background, histogram – experimental data without background, solid curve – theoretical distribution.

show a good agreement compared to the experimental data, the expected angular distance $\Delta\theta_D = 2\gamma^{-1}$ between the two lobes of the angular distribution is clearly visible. From the figures it can be seen that the angular distribution for the thick Si target has a considerable background contribution (BG) in comparison to the thin Al target. The nature of this background radiation is bremsstrahlung and characteristic radiation generated by electrons into the bulk of the target.

The main goal of the experiment was to demonstrate that the radiation yield in the EUV region emitted from the multilayer target is larger than the one from the monolayer target because of the additional MXM contribution. According to [2, 3, 4], the expected MXM excess should be observed at the photon energy E_γ defined by:

$$E_\gamma = \hbar\omega = \frac{2\pi\hbar c}{d}\frac{\sin\theta_0}{\beta^{-1} - \sqrt{\varepsilon(\omega)}\cos\theta_D}, \tag{1}$$

with $\beta = v/c$ the reduced electron velocity, d the multilayer structure period, $\varepsilon(\omega) = 1 - [a(1-\varepsilon(\omega)_{Mo}) + b(1-\varepsilon(\omega)_{Si})]/d$ the dielectric constant averaged over a structure period, and $\varepsilon(\omega)_{Mo}$, $\varepsilon(\omega)_{Si}$ the dielectric constants of pure molybdenum and silicon. θ_0 is the Bragg angle and θ_D the angle of observation.

According to equation (1), the detection of the radiation component arising from the periodic mirror structure would in principle be possible by exploiting the radiation spectrum using a detector with sufficient energy resolution. However, the photon counting scheme applied in this experiment gave no access to the radiation spectrum. Therefore, the detection of the radiation component of interest was done based on the analysis of angular distributions. Unfortunately, the MXM component is emitted in the same direction as ordinary BTR.

Measurements of angular distributions were carried out for a Bragg angle $\theta_0 = 67.5°$ by scanning the detector angle θ_D around the angle $2\,\theta_0 = 135°$. For these parameters and assuming $d = 113.2$ Å and $\theta_D = 135°$, radiation with a photon energy of $E_\gamma = 60$ eV should be generated according to equation (1).

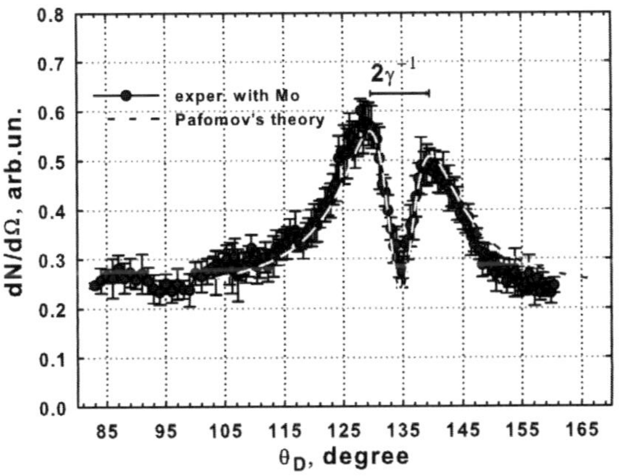

Figure 6. Experimental and theoretical angular distributions of radiation for the Si wafer coated with molybdenum and $\theta_0 = 67.5°$; points – experimetal data, dashed curve shifted upwards on 0.25 – theoretical distribution calculated according to [12].

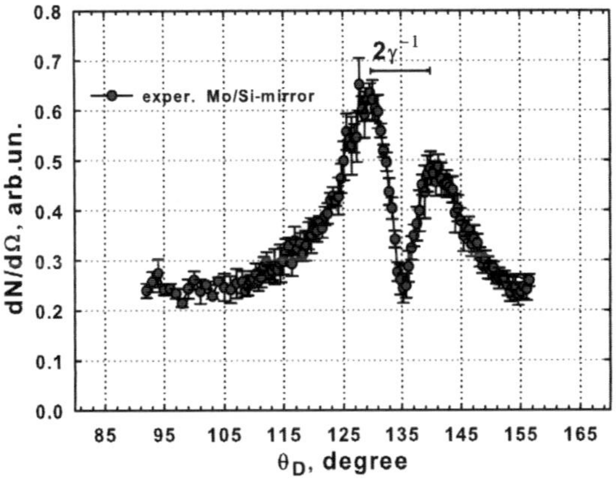

Figure 7. Experimental angular distribution of radiation from the multilayer Mo/Si structure and the Bragg angle $\theta_0 = 67.5°$.

Figures 6 and 7 show measured angular distributions for the multilayer Mo/Si target and the Mo-coated silicon wafer target. For these measurements, the diameter of the diaphragm placed in front of the SEM-6 detector amounted to $D_2 = 2.5$ mm. In this case, the detection efficiency for EUV radiation was solely determined by the detection efficiency of the SEM-6 [10].

Comparing the measured angular distributions in figures 6 and 7, the intensities and the distribution shapes do not show a significant difference because of the rather intense component of conventional BTR in the EUV region which is generated at the entrance surfaces of both targets. Unfortunately, the comparison of these curves gives no direct hint for the observation of an additional MXM contribution. It is necessary to note that the relative intensity values plotted in both figures were obtained by normalizing the SEM-6 count rate with the beam current monitor readings from the NaI (Tl) detector located behind the target, cf. figure 1. As consequence, the quite different target thicknesses have to be considered.

In order to overcome the thickness dependence, in the subsequent measurements both targets were mounted together onto the goniometer close to each other. The radiation under investigation was generated at the first target, the second one was used to compensate the total thickness, and both targets could easily interchange their places.

For a discussion of an additional reduction of the EUV BTR intensity, figure 8 shows BTR spectra for Mo (curve 1), Si (curve 2) and SiO_2 (curve 3), calculated in the range from 6 eV up to 100 eV. They were computed at the maximum of the horizontal profile of the angular distribution. Note that the shape of the spectral distribution of BTR is almost independent on the direction of the radiation inside the cone. Curves 4 and 5 show the transmission coefficients of the Al filter with a thickness of 1 μm. Curve 5 demonstrates the additional influence of two oxide layers (Al_2O_3) of equal thickness at both sides of the filter surface, assuming an overall layer thickness of $2\,t_{Al_2O_3} = 40$ nm. The calculations were based on the database "EPDL-97" [13] for optical constants.

As can be seen from figure 8, the main yield of BTR (curves – 1,2,3) is expected to be in the range of photon energies $E_\gamma < 40$ eV, whereas the energy of the MXM photons generated by the periodic target structure is in the range of 54 – 70 eV. Therefore, by using an Al filter

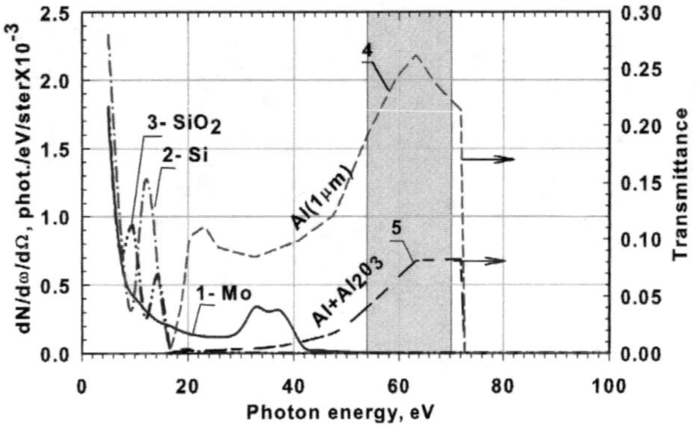

Figure 8. Spectral angular density of BTR at the maximum of the angular distribution for three materials. Mo: curve 1, Si: curve 2, and SiO_2: curve 3. Curves 4 and 5 show the photon transmittance through a $1\,\mu m$ thick Al foil without and with additional Al_2O_3 layers ($2t_{Al_2O_3} = 40\,\mathrm{nm}$). The gray region indicates the expected MXM region.

(curve 5) it is possible to suppress sufficiently the main component of the EUV BTR without significant suppression of the radiation in the energy range of MXM photons.

Figures 9 and 10 show the measurements of the angular distributions using a $1\,\mu$m thick Al filter. In these measurements, the diaphragm diameter in front of the SEM-6 was increased up to $D_1 = 8\,\mathrm{mm}$. Figure 9 (data set 1) shows the angular distribution for the radiation from a silicon target with molybdenum coating. Data sets 2 and 3 correspond to angular distributions measured with homogeneous Si targets. Data set 2 is measured for a target with a polished surface, data set 3 was measured for a target with a rough surface (the reverse side of the multilayer mirror substrate). As can be seen from this figure, no distinctive features of BTR can be observed, indicating that the photons of BTR were absorbed by the Al filter. At the same time, in the angular distribution of radiation from a multilayer target (data set 1) it is possible to observe a double-humped structure with a downward excursion at $\theta_D = 135°$ and a characteristic distance between the peaks of $\Delta\theta_D \simeq 10°$, see figure 10. For a better comparison, data set 1 from figure 9 is plotted additionally and named data set 2 in figure 10.

Thus, according to the results of the angular distribution measurements, an additional component of radiation is present in the radiation generated at a multilayer target. The origin of this component can be explained by the generation mechanisms of PXR and/or DTR in a

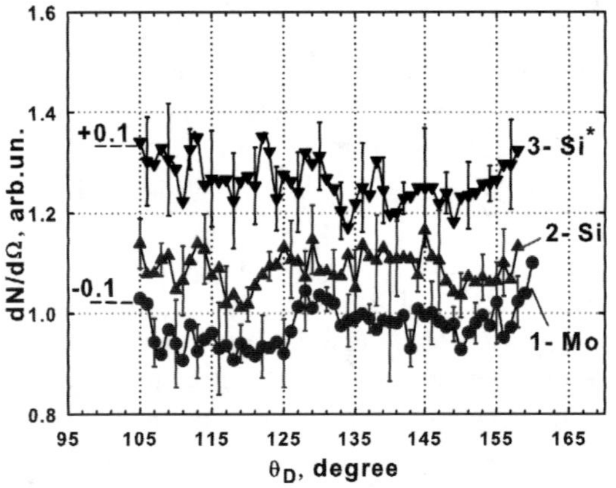

Figure 9. Angular distribution of radiation from uniform targets using the $1\,\mu$m thick Al filter. 1: Mo target, 2: polished Si plate, 3: unpolished Si plate.

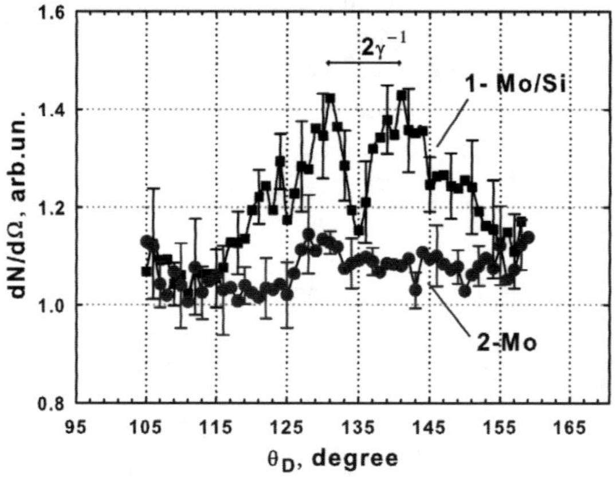

Figure 10. The angular distribution of radiation using the $1\,\mu$m thick Al filter. 1: radiation from multilayer Mo/Si mirror, 2: radiation from uniform Mo target.

periodic structure of a multilayer X-ray mirror. Taking into account the data concerning the detector efficiency from [10] together with the calculated transmittance of the 1 μm thick Al foil, a rough estimate of $dN/d\Omega \approx 5.6 \times 10^{-4}$ ph/$\bar{e}$/sr for the angular density of MXM at the peak position can be stated.

4. Conclusion

The experimental investigation of EUV BTR angular distributions generated by 5.7 MeV electrons in multilayer structures and targets with homogeneous composition indicates that the periodic structure of the target makes an additional contribution to the radiation intensity. The observed angular distribution of the additional contribution in the form of a double-humped distribution with a minimum at $\theta_D = 135°$ and peaks at $\theta_D = 130°$ and $\theta_D = 140°$ can be explained by the mechanisms of PXR and/or DTR generation from the periodic structure of a multilayer X-ray mirror [2, 3, 4, 5, 6]

From the experimental data, it can be seen that there is a significant background contribution from characteristic radiation and bremsstrahlung of the target material. This background can be reduced to a large extent by reducing the thickness of the multilayer structure substrate down to several microns. Measurements carried out with a thin Al target (figure 4) illustrate the possibility of obtaining angular distributions with a high signal-to-noise ratio.

The estimation of an absolute yield of the detected MXM is difficult because of two reasons: the EUV photon efficiency of the SEM-6 detector is not precisely known, and experimental data concerning the transmittance of the Al filter used in the experiment are missing, especially taking into account the contribution of an additional Al_2O_3 oxide layer. However, as rough estimate a value of $dN/d\Omega \approx 5.6 \times 10^{-4}$ ph/$\bar{e}$/sr for the angular density of MXM at the peak position can be stated. This result is close to the value $dN/d\Omega \approx 4.6 \times 10^{-4}$ ph/$\bar{e}$/sr calculated according to the Nasonov's theory [5].

Acknowledgments

We would like to thank I.A. Artyukov for his help with the manufacturing of the Mo/Si targets. This work was partially supported by the Russian Ministry of Education and Science program "Nauka", Grant 2456, and by the Russian Foundation for Basic Research, Grant 14-02-01032.

References

[1] Sukhikh L G, Bajt S, Kube G, Popov Yu A, Potylitsyn A P and Lauth W 2012 *Proc. International Particle Accelerator Conference IPAC'12 (New Orleans, Louisiana, USA)*, p 819, MOPPR019
[2] Andre J-M, Barchewitz R, Bonnelle C and Pardo B 1993 *J. Opt. Paris* **24** 31.
[3] Andre J-M, Pardo B and Bonnelle1 C 1999 *Phys. Rev.* E **99** 968
[4] Kaplin V V, Uglov S R, Zabaev V N, Piestrup M A, Gary C K, Nasonov N N and Fuller M K 2000 *Appl. Phys. Lett.* **76** 3647
[5] Nasonov N N, Kaplin V V, Uglov S, Piestrup M and Gary C 2003 *Phys. Rev.* E **68** 036504
[6] Nasonov N N, Kaplin V V, Uglov S R, Zabaev V N, Piestrup M and Gary C 2005 *Nucl. Instrum. and Meth.* B **227** 41
[7] Kaplin V V, Uglov S R, Sohoreva V V, Bulaev O F, Voronin A A, Piestrup M, Gary C and Fuller M 2009 *Nucl. Instrum. and Meth.* **267** 777
[8] Uglov S R, Zabaev V N, Kaplin V V and Kuznetsov S I 2012 *Jour. of Phys.: Conf. Ser.* bf 357 012012
[9] Uglov S R, Zabaev V N and Kaplin V V 2013 *Nucl. Instrum. and Meth.* B **309** 79
[10] Einbund M R and Polenov B V 1981 *Secondary electron multipliers and their application* (Moscow: Energoatomizdat) (in Russian)
[11] Sukhikh L G, Krambrich D, Kube G, Lauth W, Popov Yu A and Potylitsyn A P 2011 *Proc. 10th European Workshop on Beam Diagnostics and Instrumentation for Particle Accelerators DIPAC'11 (Hamburg, Germany)* p 544, WEOA02
[12] Pafomov V E, 1971 *Proceedings P.N. Lebedev Physics Institute, ed. by D.V. Skolbel'tsyn* **44** (New York: Consultants Bureau) 25-157
[13] www.esrf.fr/computing/expg/subgroups/theory/DABAX/dabax.html

Journal of Physics: Conference Series **517** (2014) 012010 doi:10.1088/1742-6596/517/1/012010

Radiation from a varying velocity charge in flight through a plate

A R Mkrtchyan[a,b], L Sh Grigoryan[a,1], H F Khachatryan[a]

[a] Institute of Applied Problems in Physics, 25 Hr. Nersessian str., 0014 Yerevan, Armenia
[b] National Research Tomsk Polytechnic University, 30 Lenin Avenue, 634050 Tomsk, Russian Federation

E-mail: levonshg@mail.ru

Abstract. The radiation from a particle traversing with varying velocity through a plate in straight motion (along the normal to its surface) has been investigated. It is assumed that the particle coordinate (inside the plate) is an arbitrary monotone function of time. The calculations are based on appropriate exact solutions of the Maxwell equations. It was shown that as a result of variation of particle velocity inside the plate (e.g., due to its slowing-down) the spectral-angular distribution of Cherenkov radiation generated by the particle inside the plate may be essentially affected. Possible practical application of this phenomenon is discussed.

1. Introduction

In the presence of substance the uniform and straight-line motion of a charged particle may go with radiation of electromagnetic waves (e.g., Cherenkov Radiation (CR), transition radiation) that are widely used in practice. In this regard, in investigations on the influence of matter on the charged particle radiation it is usually assumed that the particle motion is uniform and the calculations in this approximation well agree with experiment. In the present work our aim is to study the cases when this approximation is not correct and the nonuniformity of particle motion has to be taken into consideration.

General expressions concerning the electromagnetic field of a charged particle at its arbitrary motion in solid uniform medium have been derived earlier [1]. Analogous problem at the presence of two plane interfaces between media having different permittivities has been solved (see, e.g., [2, 3]). In these works, however, the influence of nonuniformity of particle motion on CR generated by the particle has not been investigated. In this work the effect of the nonuniformity of particle motion on CR generated by particle inside the plate has been studied and possible uses of this effect have been discussed.

In [5-7] the influence of particle slowing down on the angular distribution of CR generated by the particle has been studied. It was assumed that (i) the particle moves in an unlimited and transparent dielectric and (ii) CR (with the wavelength λ_m) is accumulated from a part of particle trajectory of $l >> \lambda_m$ length. It was predicted that the structure and the width of angular distribution of CR depends (a) on the energy loss of particle in the dielectric, the initial energy of particle (heavy ion) [4-7] and (b)

[1] To whom any correspondence should be addressed.

on the ion mass (isotope effect [6,7]). A simple formula (see (17)) has been proposed in [7] allowing one to estimate the width of angular distribution of CR in case of $l >> \lambda_m$. In this paper the influence of nonuniformity of particle motion (slowing down or acceleration stimulated by an external field) on the spectral-angular distribution of CR from the particle passing through a dielectric plate of finite thickness l was investigated. The absorption of radiation in the plate material is also taken into account.

2. Formulation of the problem

Now let us consider the motion of a charge q along OZ axis from $-\infty < z < -l$ range (the range of $\alpha = -$, see figure 1) that passes into a plate at some instant of time t_- ($-l < z < 0$ range), travels in it with varying velocity $v_0(t)$ and then, at some instant of time t_+, leaves the plate and continues its motion in $0 < z < +\infty$ range ($\alpha = +$ range). In what follows it is assumed that inside the plate the particle coordinate $z_q = z_0(t)$ is an arbitrary monotone function of time (such that there is $t_q = t_0(z)$ function that is inverse for it for all $-l < z < 0$) and that the plate is made of radiation absorbing uniform and isotropic substance.

Without loss of generality it is also assumed that the space on both sides of the plate is filled with transparent uniform and isotropic substances (gases for instance) with dielectric permittivity and magnetic permeability ε_α and μ_α respectively. To simplify the calculations we shall assume that the motion of particle to the left and to the right of the plate is uniform (with velocities v_- and v_+ respectively).

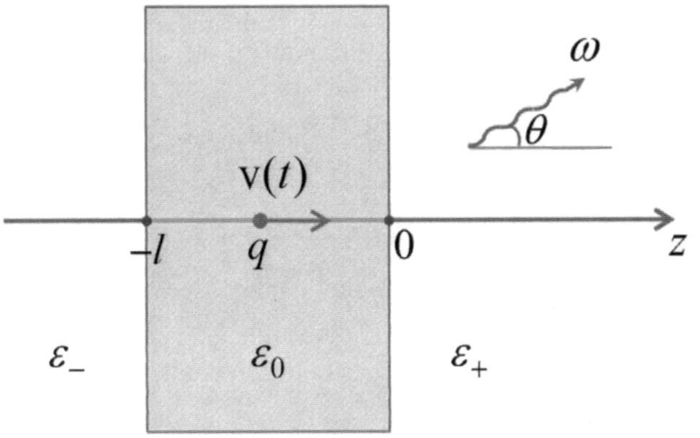

Figure 1. Charged particle traversing the plate with varying velocity.

As a result, we can represent the dielectric permittivity ε and magnetic permeability μ of the medium, the velocity v_q and coordinate z_q of the particle by means of following expressions:

$$\psi = \begin{cases} \psi_- & z < -l \\ \psi_0 \quad \text{at} & -l < z < 0 \\ \psi_+ & z > 0 \end{cases} \quad \text{and} \quad v_q = \begin{cases} v_- \\ v_0(t) \\ v_+ \end{cases}, \quad z_q = \begin{cases} -l + (t - t_-)v_- \\ z_0(t) \\ (t - t_+)v_+ \end{cases} \quad \text{at} \quad \begin{matrix} t < t_- \\ t_- < t < t_+ \\ t > t_+ \end{matrix} \tag{1}$$

where $\psi = \varepsilon, \mu$ and $v_0 = \mathrm{d}z_0/\mathrm{d}t$. The parameters of laminated medium ε и μ will be assumed below to be independent of transverse coordinates $(x, y) \equiv \vec{\chi}$.

Now introduce the radiation energy

$$W_\alpha = \int I_\alpha(\omega, \theta)\mathrm{d}\omega\mathrm{d}\theta, \tag{2}$$

penetrating the plane $z = \text{const}$ distant from the plate during all the time of particle motion. It may be both to the left ($\alpha = -$ range), and to the right ($\alpha = +$ range) of the plate. In (2) $I_\alpha(\omega, \theta)$ is the spectral-angular distribution of the radiation energy, propagating either against ($\alpha = -$) or along ($\alpha = +$) the direction of OZ axis. We shall focus on the investigation of some features of this function.

3. Stages of analytical calculations and finite expressions

Making allowance for the azimuthal symmetry of problem and performing the Fourier transform

$$f_{u\omega}(z) = \frac{1}{(2\pi)^3}\int f(\vec{r},t)\exp[\mathrm{i}(\omega t - \vec{u}\vec{\chi})]\mathrm{d}\vec{\chi}\mathrm{d}t \tag{3}$$

($f \equiv \vec{E},\vec{H}$), the solution of the set of Maxwell equations is reduced to solution of a single equation [8]

$$[\varepsilon\frac{\partial}{\partial z}\left(\frac{1}{\varepsilon}\frac{\partial}{\partial z}\right) + \frac{\omega^2}{c^2}\varepsilon\mu - u^2]\varepsilon E_{u\omega} = 4\pi[\varepsilon\frac{\partial}{\partial z}\left(\frac{\rho_{u\omega}}{\varepsilon}\right) - \mathrm{i}\frac{\omega}{c^2}\varepsilon\mu\, j_{u\omega}] \tag{4}$$

(the basic equation of the problem under consideration). In this equation $\varepsilon \equiv \varepsilon_\omega$ and $\mu \equiv \mu_\omega$ are Fourier transforms of the dielectric permittivity and magnetic permeability (the allowance for the frequency dispersion is made, the contribution of spatial dispersion being considered negligibly small). Here we assume that $\varepsilon_0 = \varepsilon'_{0\omega} + \mathrm{i}\varepsilon''_{0\omega}$, $\mu_0 = \mu'_{0\omega} + \mathrm{i}\mu''_{0\omega}$ are complex quantities (with due regard for the absorption of radiation in the plate material). Based on solution $E_{u\omega}$ of equation (4), one can calculate the strengths of electric and magnetic field using the following formulae

$$\vec{E}_{u\omega} = E_{u\omega}\vec{n}_z + \frac{\mathrm{i}\vec{u}}{\varepsilon u^2}\left(\frac{\partial}{\partial z}\varepsilon E_{u\omega} - 4\pi\rho_{u\omega}\right), \qquad \vec{H}_{u\omega} = \frac{\omega}{cu^2}(\varepsilon E_{u\omega} + 4\pi\frac{\mathrm{i}}{\omega}j_{u\omega})\vec{u}\times\vec{n}_z. \tag{5}$$

Here

$$\rho(\vec{r},t) = q\delta(x)\delta(y)\delta[z - z_q(t)] \quad \text{and} \quad \mathrm{v}_q(z)\rho_{u\omega}(z) = \frac{q}{(2\pi)^3}\exp[\mathrm{i}\omega t_q(z)] = j_{u\omega}(z) \tag{6}$$

is the current density associated with the charge ($\vec{n}_z$ is the unit vector showing the direction of OZ axis).

In regions $0 < z < +\infty$ ($\alpha = +$) and $-\infty < z < -l$ ($\alpha = -$) equation (4) must have the following solutions:

$$E_{u\omega\pm} = E^q_{u\omega\pm} + \frac{\mathrm{i}q}{2\pi^2\omega}a_\pm\exp[\pm\mathrm{i}\omega\sigma_\pm(z - z_\pm)/c], \quad \sigma_\pm = \sqrt{\varepsilon_\pm\mu_\pm - u^2c^2/\omega^2}, \tag{7}$$

where the 1-st summand is the known field of a charge uniformly moving in a homogeneous and solid medium with parameters $\varepsilon_\pm$, $\mu_\pm$, and the 2-nd one describes the free field (radiation field); $z_+ = 0$, $z_- = -l$ [8-10]. Far from the plate ($z \to \pm\infty$) the spectral-angular distribution of radiation energy is determined by an expression [8,10]

$$I_\alpha(\omega,\theta) = \frac{2q^2}{\pi c}\frac{\cos^2\theta}{\sin\theta}\left|a_\alpha(\omega,u)\right|^2, \qquad \text{where} \qquad u^2 = \frac{\omega^2}{c^2}\varepsilon_\alpha\mu_\alpha\sin^2\theta, \tag{8}$$

and θ is the angle giving the direction of radiation with respect to $z = +\infty$ and $z = -\infty$ for the forward ($\alpha = +$) and backward radiations ($\alpha = -$) respectively. For calculation of $a_\pm$ it would be adequate to equate (7) to the complete solution of equation (4) that is valid for all $-\infty < z < \infty$. One can determine this solution, e.g., by using the Green function method.

Dropping the intermediate calculations, we give the final expression

$$a_+ = \frac{\phi_+\varepsilon_0}{2\varepsilon_+\sigma_0\eta}\left\{\left[(\eta + \eta_-)\frac{c}{\varepsilon_+\mathrm{v}_+} - (\frac{\eta}{1 - \sigma_+\mathrm{v}_+/c} + \frac{\eta_-}{1 + \sigma_+\mathrm{v}_+/c})\chi_+\right]\frac{1}{k_+} + \right.$$
$$\left. + (\frac{\chi_-}{1 - \varepsilon_-\mathrm{v}_-/c} - \frac{c}{\varepsilon_-\mathrm{v}_-})\exp[\mathrm{i}\omega(t_- - t_+)] - \frac{\mathrm{i}\omega}{2c}\int_{-l}^{0}\xi_+(z)\exp[\mathrm{i}\omega(t_0(z) - t_+)]\mathrm{d}z\right\}. \tag{9}$$

The expression for a_- is obtained from (9) first by means of replacing v_α by $-v_\alpha$, and second, $+\leftrightarrow -$ in all the indices. The following notations are introduced in (9):

$$4\eta = (1+k_-)(1+k_+)\exp[-i\omega\sigma_0 l/c] + (1-k_-)(1-k_+)\exp[i\omega\sigma_0 l/c], \quad \sigma_i = \sqrt{\varepsilon_i\mu_i - u^2c^2/\omega^2}, \quad i = 0,\pm$$

$$4\eta_\pm = (1-k_\pm)(1+k_\mp)\exp[i\omega\sigma_0 l/c] + (1+k_\pm)(1-k_\mp)\exp[-i\omega\sigma_0 l/c], \quad k_\pm = \frac{\varepsilon_0\sigma_\pm}{\varepsilon_\pm\sigma_0}, \quad \chi_\pm = \frac{c}{\varepsilon_\pm v_\pm} - \frac{\mu_\pm v_\pm}{c}$$

$$\xi_+ = (1+k_-)(\mu_0 - \frac{\sigma_0 c}{\varepsilon_0 v_0})\exp[-i\omega\sigma_0(z+l)/c] + (1-k_-)(\mu_0 + \frac{\sigma_0 c}{\varepsilon_0 v_0})\exp[i\omega\sigma_0(z+l)/c] \tag{10}$$

$$\xi_- = (1+k_+)(\mu_0 + \frac{\sigma_0 c}{\varepsilon_0 v_0})\exp[i\omega\sigma_0 z/c] + (1-k_+)(\mu_0 - \frac{\sigma_0 c}{\varepsilon_0 v_0})\exp[-i\omega\sigma_0 z/c]$$

and, at last, $\phi_\pm = -\exp(i\omega t_\pm)$ is an inessential phase factor. In the particular case of $v_- = v_+ = v_0(t) = \text{const}$ and $\varepsilon_\pm = \mu_\pm = 1$, the well known [8-10] expression for CR and transition radiation from a particle leaving the plate for vacuum is obtained from (9).

4. Numerical results and their analysis

Below we shall confine ourselves to the consideration of the simplest case when the force

$$F = \frac{d}{dt}\frac{mv_0}{\sqrt{1 - v_0^2/c^2}} \tag{11}$$

with which the plate material slows down the particle motion is constant: $F(t) = \text{const}$. In this case [11] the velocity v_0 and coordinate $z = z_0(t)$ of the particle are determined from equalities

$$v_0 = \sqrt{1 - 1/\gamma^2} \qquad \gamma(z) = \gamma_- + (z+l)(\gamma_+ - \gamma_-)/l \tag{12}$$

and the corresponding instant of time $t = t_0(z)$ is described by the function

$$t_0(z) = t_- + (\sqrt{\gamma^2 - 1} - \sqrt{\gamma_-^2 - 1})l/c(\gamma_+ - \gamma_-). \tag{13}$$

We shall also assume that (a) the plate is made of molten quartz with

$$\varepsilon_0 = 3.78\cdot(1 + 0.0001\cdot i) \quad \text{and} \quad \mu_0 = 1 \tag{14}$$

in the frequency interval $\sim 10^{10}\,GHz$, (b) the Lorentz factor γ of particle (electron) at incidence on the plate is $\gamma_- = 2$ and that (c) the thickness of plate is such that at an escape from plate the Lorentz factor of particle is reduced (owing to the slowing down) to the value of $\gamma_+ = 1.5$. As for such a choice of the values of parameters inside the plate the Cherenkov condition is observed

$$c/v_0(t)\sqrt{\varepsilon_0'} < 1 \tag{15}$$

and the Cherenkov radiation is generated during all the time period of particle flight inside the plate.

The spectral-angular distribution of the energy radiated in the forward direction by electron ($\alpha = +$) as shown in figure 2 is added up of CR and transition radiation generated at an incidence on and escape of particle from the plate.

For simplification of analysis we additionally show in figure 3 the spectral (at the left) and the angular (right) distributions of $I_+(\omega,\theta)$ function. The oscillations in the radiation spectrum (in figure 3, at the left) are due to the superposition and interference of electromagnetic oscillations generated by the particle at the flight through the plate of finite thickness. The first maximum is at $\omega_{max} l/c = 5$. So, with known value of ω_{max} one can determine the plate thickness l from this equality.

If the particle were not slowed down in the plate, its energy would stay constant: $\gamma_- = \gamma(z) = \text{const}$. In figure 3 the upper curve *b* on the right corresponds to this case. As is seen from the comparison of *a* and *b* curves in figure 3, *the slowing down of particle notably reduces the radiation energy.*

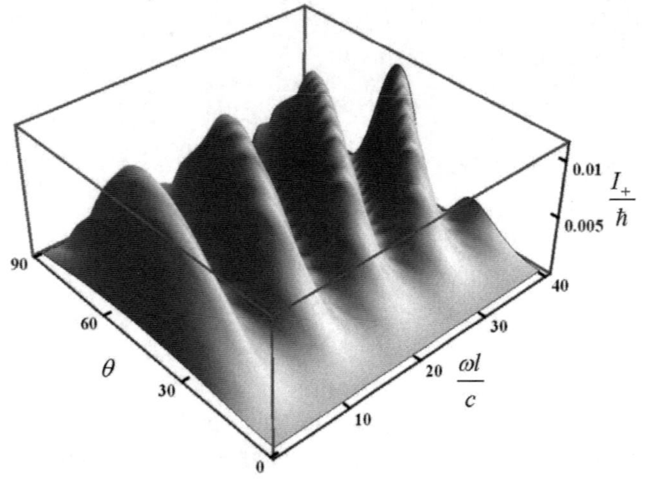

Figure 2. Spectral-angular distribution of the energy of total radiation in the forward direction $I_+(\omega,\theta)$ at the flight of relativistic electron through the plate (molten quartz). In calculations, the allowance for slowing down of particle inside the plate is made under the assumption that the force, with which the plate material slows down the particle motion is constant. The electron energy at the incidence is $\mathrm{E}=2m_e c^2$ and is $\mathrm{E}=1.5m_e c^2$ at the escape from the plate.

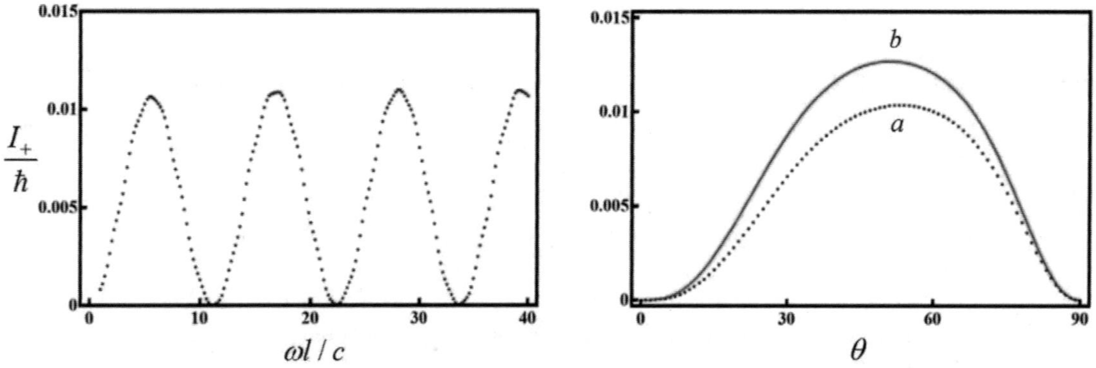

Figure 3. Leftwards is the spectral distribution of the same radiation as that in figure 2 in $\theta=50^{\circ}$ direction. On the right is the angular distribution of the same radiation as that in figure 2 at the intended frequency $\omega l/c=5$ (curve *a*). Curve *b* (on the right) is the angular distribution of $I_+(\omega,\theta)$ function for $\omega l/c=5$ and uniform motion of the electron inside the plate with $\gamma_-=\gamma(\mathrm{z})=\mathrm{const}$.

Based on an analysis of radiation spectrum some effective value E_{eff} of particle energy $\mathrm{E}=\gamma mc^2$ is determined, such that the spectrum of CR for uniform motion with this energy would best approximate the "actual" radiation spectrum (in which the allowance for slowing down of particle is made). The results of our numerical calculations ascertain that this effective value of particle energy is determined by means of the following equality:

$$\mathrm{E}_{eff} \cong (\mathrm{E}_- + \mathrm{E}_+)/2. \qquad (16)$$

The energy losses of particle $\mathrm{E}_- - \mathrm{E}_+$ due to its slowing in plate may be determined from (16) using the value of E_{eff} and, e.g., E_-. As a result, the force *F* slowing down the particle motion inside the plate, is also determined.

Formula

$$\Delta\theta_S = arc\cos(c/\mathrm{v}_-\sqrt{\varepsilon_0'}) - arc\cos(c/\mathrm{v}_+\sqrt{\varepsilon_0'}) \qquad (17)$$

allows to estimate the portion of the width of angular distribution of CR, which is due to the slowing down of particle in the plate. According to [7] formula (17) describes good enough the width of the angular distribution of CR (with wavelength λ_m) in the case when the length of the particle trajectory

$l >> \lambda_m$ ($\lambda_m = 2\pi c / (\omega\sqrt{\varepsilon_0'})$ is the wavelength of CR in an unlimited transparent medium with permittivity ε_0'). As seen from the data shown on the right-hand side in figure 3 the half-width $\Delta\theta \cong 50^o$ of the angular distribution of CR is greater than the value $\Delta\theta_S \cong 7.2^o$ calculated by formula (17). This is understandable, since plotted in the right-hand side in figure 3 is the case $l \approx \lambda_m$. In this case, CR is only formed when the particle is flying out of the plate so one cannot neglect the contribution of transition radiation.

It is worthwhile to add that CR from a relativistic particle travelling a limited trajectory of length l in a gas has been investigated theoretically and experimentally in [12,13]. In these papers the alternative criterion $\mathrm{v}_0(\sqrt{\varepsilon_0'} - \lambda_m / l) > c$ for CR generation has been proposed (cf. with (15)). We rewrite this criterion as $l > \lambda_m / (\sqrt{\varepsilon_0'} - c / \mathrm{v}_0)$. For the right-hand side case in figure 3 it leads to $l > 1.4\lambda_m$ where $\lambda_m = 2\pi c / (\omega_{\max}\sqrt{\varepsilon_0'})$. This inequality indicates directly that CR just begins to form when the particle left the plate of $l \approx \lambda_m$ length.

5. Conclusions

The electromagnetic field of a charged particle in flight through a plate along the normal to its surface in straight-line motion at varying velocity with no restrictions on the variation law of this velocity (save the fact that its projection on the direction of motion should be a constant-sign function of time) has been investigated based on appropriate exact solutions of the Maxwell equations. The obtained analytical expressions are essentially simpler than those to be derived from earlier reported general formulae (e.g., [3]). The influence of irregularities in particle motion on CR generated by this particle inside the plate was not dealt with in [3]. This deficiency is filled up in the present work.

It was shown that the variation of particle velocity inside the plate (e.g., due to the slowing down of particle) may essentially influence the spectral-angular distribution of Cherenkov radiation of particle. This fact may be used for practical purposes, particularly for simultaneous measurement of three quantities, - the force that slows down the particle motion in plate, the thickness of the plate and the refractive index of the plate material.

In this paper we do not take into account the effects due to the multiple scattering of charged particle and ignore fluctuations of energy losses in plate substance. Numerical calculations were carried out on the assumption that the particle decelerating force F is not changed when the particle moves in the plate. In reality F depends on the velocity of particle and therefore it should vary during the slowing down. This fact is taken into consideration, for example, in [7]. A detailed study of this problem for a particle traversing through a plate requires a separate consideration.

Acknowledgments

The authors are thankful to the anonymous reviewer, the valuable comments of which helped to improve the statement of problem.

References

[1] Tsytovich V N 1951 *Vestnic MGU* **11** 27 (in Russian)
[2] Gharibian G M and Magomedov M R 1963 *Reports NAS RA* **36** 77 (in Russian)
[3] Pafomov V E 1969 *Trudy FIAN* **44** 28 (in Russian)
[4] Kuzmin E S and Tarasov A V 1993 *JINR Rapid Commun.* **4** 64
[5] Altapova V R, Bogdanov O V and Pivovarov Yu L 2009 *Nuclear Instrum. Methods* B **267** 896
[6] Bogdanov O V and Pivovarov Yu L 2011 *Nuovo Cimento* **34** 103
[7] Bogdanov O V, Fiks E I and Pivovarov Yu L 2012 *JETP* **142** 442 (in Russian)
[8] Ginzburg V L and Tsytovich V N 1990 *Transition Radiation and Transition Scattering* (Bristol: Adam Hilger)
[9] Ter-Mikaelian M L 1972 *High Energy Electromagnetic Processes in Condensed Matter* (New

York: Wiley)
[10] Garibian G M and Yang C 1983 *X-Ray Transition Radiation* (Yerevan: AN Arm. SSR Press) (in Russian)
[11] Landau L D and Lifshitz E M 1980 *The Classical Theory of Field* (Oxford: Pergamon Press)
[12] Zrelov V P, Klimanova M, Lupiltsev V P and Ruzicka J 1983 *Nuclear Instrum. Methods* **215** 141
[13] Takahashi T, Kanai T, Shibata Y *et al* 1994 *Phys. Rev.* E **50** 4041

Sub-micrometer transverse beam size diagnostics using optical transition radiation

K Kruchinin[1], A Aryshev[2], P Karataev[1], B Bolzon[3,4,5], T Lefevre[3], S Mazzoni[3], M Shevelev[2], S T Boogert[1], L J Nevay[1], N Terunuma[2] and J Urakawa[2]

[1]John Adams Institute at Royal Holloway, Egham, Surrey, TW20 0EX, UK
[2]KEK, 1-1 Oho, Tsukuba, Ibaraki 305-0801, Japan
[3]CERN: European Organization for Nuclear Research, CH-1211 Geneva 23, Switzerland
[4]Cockcroft Institute, Warrington, Cheshire, UK
[5]The University of Liverpool, Liverpool, UK

E-mail: `Konstantin.Kruchinin.2012@live.rhul.ac.uk`

Abstract. Optical transition radiation (OTR) arising when a relativistic charged particle crosses a boundary between two media with different optical properties is widely used as a tool for diagnostics of particle beams in modern accelerator facilities. The resolution of the beam profile monitors based on OTR depends on different effects of the optical system such as spherical and chromatic aberrations and diffraction. In this paper we present a systematic study of the different optical effects influencing the OTR beam profile monitor resolution. Obtained results have shown that such monitors can be used for sub-micrometer beam profile diagnostics. Further improvements and studies of the monitor are discussed.

1. Introduction

Transverse electron beam diagnostics are crucial for stable and reliable operation of the future electron-positron linear colliders such as CLIC [1] or ILC [2]. Sub-micrometer resolution beam profile monitors are required to measure high-energy particle beam emittance. The-state-of-the-art in transverse beam diagnostics is based on the laserwire technology [3, 4], i.e. when a high power laser is focused in the interaction region down to a micrometer dimension and is scanned across the electron (positron) beam generating intense inverse Compton scattered photon beam. The photon intensity is proportional to the number of particles in the beam and to the laser power and position with respect to the beam center. By measuring the Compton photon yield versus laser position one can directly measure the transverse beam profile. However, the use of high power laser significantly increases the cost of the laserwire system. In linear colliders over 70 such stations are required. Therefore, a simpler and relatively inexpensive method is required. A beam profile monitor based on Optical Transition Radiation (OTR) is very promising. Although such a monitor destroys the beam or the beam can destroy the monitor itself, it still can be used as an addition to the laserwire for diagnostics of low current "pilot" beams.

The resolution of a conventional OTR monitor is defined by the root-mean-square of the so-called Point Spread Function (PSF) [5]. The PSF can be defined as the source distribution generated by a single electron and projected by an optical system onto a detector. In the optical wavelength region the resolution is limited and defined by diffraction and aberration effects of the optical system. Such effects lead to the broadening of the PSF and to a lower resolution. The best resolution achieved by conventional OTR monitors is about a few micrometers [6]. However, in [7] we demonstrated that the OTR PSF differs from a conventional PSF of an optical system. The vertical polarization component of the OTR PSF has a two-lobe structure, the visibility of which can be used to monitor vertical beam size with sub-micrometer resolution. On the other hand if the beam is flat, which is true for linear colliders, the horizontal projection of the distribution represents a direct measurement of the horizontal beam size that is much larger than the vertical one. It gives an opportunity to diagnose an electron beam size in two directions in a single shot.

2. Experimental setup

The measurements were performed at the Accelerator Test Facility 2 (ATF2) in KEK (Tsukuba, Japan). The accelerator consists of 1.3 GeV linac followed by damping ring to produce a very low emittance beam and extraction line [8]. The ATF2 is an upgrade of the ATF that was performed in 2008 that extends the extraction line and aims to focus the electron beam to a vertical beam size of about 37 nm [9].

In our previous works [10,11] the experimental setup of the OTR monitor was well described. In this section we briefly overview the major components of the setup. In recent operation the setup was slightly modified. The schematic setup and photograph of the OTR monitor is presented in figure 1.

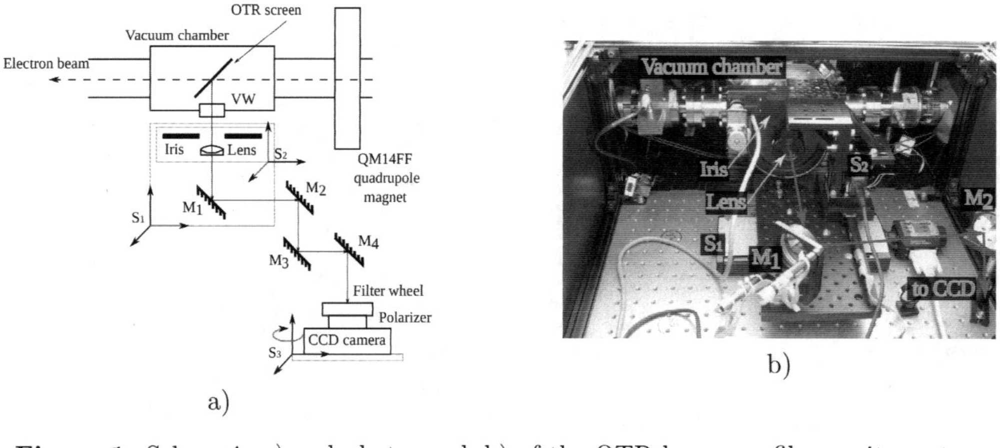

Figure 1. Schemaic a) and photograph b) of the OTR beam profile monitor setup.

The setup consists of 30×30×0.3 mm aluminized silicon target tilted at 45 degrees with respect to the electron beam line. The target was mounted on a 4D vacuum manipulator that allows remote control of its vertical and angular position, and manual adjustment of the transverse position. From the target, the OTR passes through an optical system to a CCD camera, which consists of a motorized iris, lens holder with a set of focusing lenses ("CVI Laser Optics" cemented achromat with f=120 mm; "Sigma Koki" achromatic doublets with f=120.1 mm and f=100.7 mm and "Sigma Koki" spherical plano-convex lens with f=100 mm), four aluminum coated mirrors forming a periscope, set of optical filters with bandwidth from 500±20 to 600±20 nm placed in a remotely controlled filter wheel and a linear polarizer placed in front of a CCD camera mounted on a remotely controlled rotation stage. Both the iris and the lens holder were mounted on a same board to keep them centered. The lens holder mounted on the board has a possibility of 3D adjustment of the lens position. The whole setup was mounted on an optical bread board surrounded by a light-tight enclosure. To record the OTR images a high resolution (~8 megapixel), cooled CCD camera (SBIG-ST8300M) based on Kodak KAF-8300 (monochrome) sensor with 5.4 μm pixel size and ~50% quantum efficiency was used.

3. Experimental results

Initially the accelerator was carefully tuned to minimize the background level and provide a stable working regime during the data taking. The longitudinal position of the focusing lens was adjusted to provide a clear OTR image with a minimum distribution width. After that, the monitor was ready to perform beam size measurements. All measurements were performed in single-bunch mode, with a bunch repetition rate of 3.25 Hz and ~0.3 nC bunch charge.

A typical CCD image of the OTR spot taken with linear polarizer and 550±20 nm optical filter is presented in figure 2a. Since the horizontal beam size is much bigger than the vertical size, it can be directly extracted from the simple gaussian fit applied to the horizontal projection (figure 2c) of the OTR image. The vertical projection of the OTR image has a two-lobe structure (figure 2b) and the vertical beam size can not be directly obtained from the fit function applied to the vertical projection. The idea of the vertical beam size extraction is based on fact that the contrast ratio (minimum to maximum ratio) of the vertical projection strongly depends on the real beam size. To calculate the contrast ratio a special empirically found fit function has been proposed:

$$f(x) = a_0 + \frac{a_1\left(a_4 + (x-a_3)^2\right)}{1 + \left(a_2\left(x-a_3\right)\right)^4}, \tag{1}$$

where fit parameters are: a_0 the vertical offset of the distribution with respect to zero; a_1 the amplitude of the distribution; a_2 the smoothing parameter; a_3 the horizontal offset of the distribution with respect to zero and a_4 the distribution width. In order to calculate the vertical beam size from the contrast ratio, a special self-calibration procedure was introduced [12, 13].

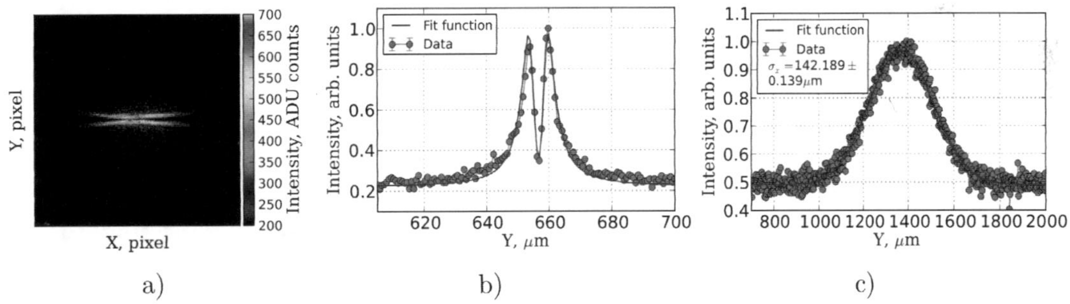

Figure 2. a) CCD image of the OTR spot taken with linear polarizer, 550±40 nm optical filter and DLB lens with f = 120 mm; b) vertical projection; c) horizontal projection.

Recent results of the vertical beam size measurements are presented in figure 3. The horizontal and vertical beam size could be controlled by changing the current in the nearest quadrupole magnet. We performed the measurements using four lenses with different optical properties in order to understand how the optical effects such as chromatic and spherical aberrations influence the measured beam size. The use of achromatic doublets (see figure 3a, b) allowed to achieve a sub-micrometer resolution, while using a simple singlet lens without aberration corrections (see figure 3c) leads to lower resolution. The smallest beam size measured in the experiment is consistent with the SAD [14] prediction within uncertainty.

The form of the PSF, and hence the resolution of the OTR monitor, strongly depends on the diffraction and aberration effects of the optical system, which lead to broadening of the PSF and a lower resolution. The broadening can be easily measured as the distance between peaks in the vertical projection of the image.

In order to study such effects a series of measurements with different optical elements was performed. Firstly, to study chromatic aberration effects, we performed a several focus-scans (measuring a distance between peaks of the distribution at different longitudinal lens positions) with different optical filters in a wavelength range 500±20 - 600±20 nm. The resultant dependencies are presented in figure 4a. These measurements were performed using "Sigma Koki" achromatic doublet with a focal length of 120.1 mm and 30 mm diameter. As one can see from the picture, the focus positions are different for different optical filters even for the lens designed to minimize the chromatic aberrations. Therefore for any lens its position must be carefully adjusted to maximize the resolution of the instrument.

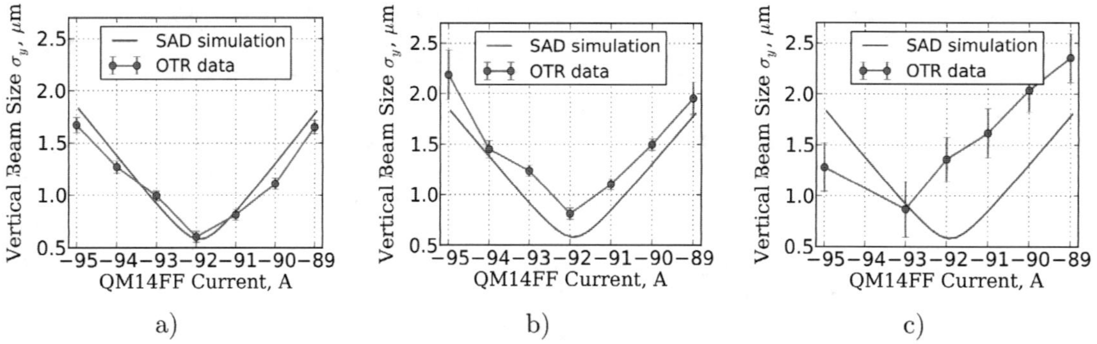

Figure 3. Vertical beam size as a function of the QM14FF quadrupole magnet current for different lenses: "CVI Laser Optics" cemented achromat with 120 mm focal length a), "Sigma Koki" achromatic doublet with focal length of 100.7 mm b), "Sigma Koki" plano-convex lens with 100 mm focal length c).

Another optical effect that decreases the resolution is the diffraction of the OTR tails by the hard aperture of the optical elements. To study this effect, an iris was mounted in front of the lens (see figure 1). Similar to aberrations, diffraction effects lead to broadening of the distribution and, consequently, to a decrease in resolution. In order to estimate the influence of the diffraction on distribution width the distance between peaks as a function of the iris aperture diameter was measured (see figure 4b). It is clear that diffraction effect becomes noticeable at the iris diameter of ∼15 mm (∼360 γ^{-1}). This means that the diameter of the focusing lens can be reduced down to 15 mm, thereby the lens can be manufactured thinner which leads to reduction of the spherical aberrations.

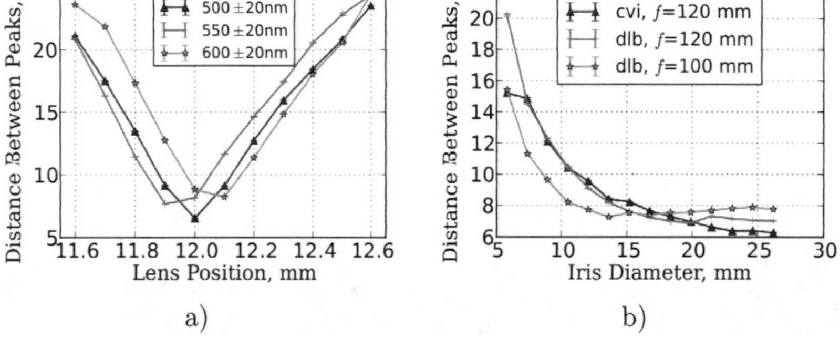

Figure 4. a) Distance between peaks as a function of the lens position for different optical filters and b) distance between peaks as a function of the iris diameter for different lenses.

4. Aberration test bench

In order to study chromatic and spherical aberration effects more deeply a special test stand was developed. The schematic picture and photograph of the stand is presented in figure 5, 6.

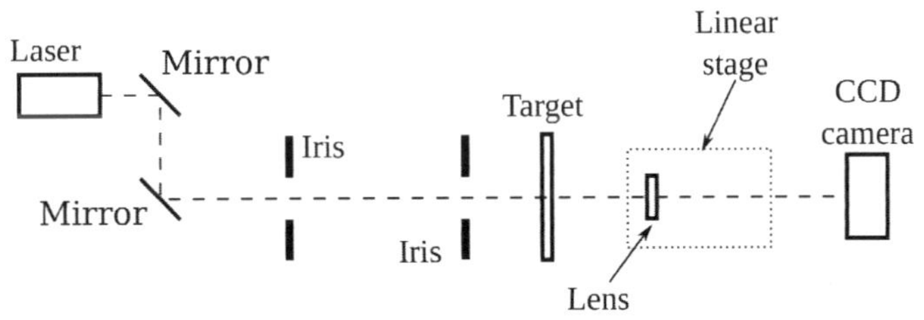

Figure 5. Schematic of the test bench.

The whole setup was assembled on an optical table. Three continuous wave (CW) lasers with 532 nm, 400 nm and 632 nm wavelengths were used as sources of light. The lasers were placed on a stage that allows to change the lasers without changing the alignment. Alignment of the laser beam was performed using two aluminum coated mirrors with diameter of 50 mm and two irises. To imitate the OTR screen a "ThorLabs" R3L3S1P - positive USAF target was used. A lens was mounted on the remotely controlled linear stage. The image of the pattern was detected by the CCD camera and recorded by the camera software. Images were recorded as a function of the lens position. Focusing properties of the lens were measured by analyzing the target images.

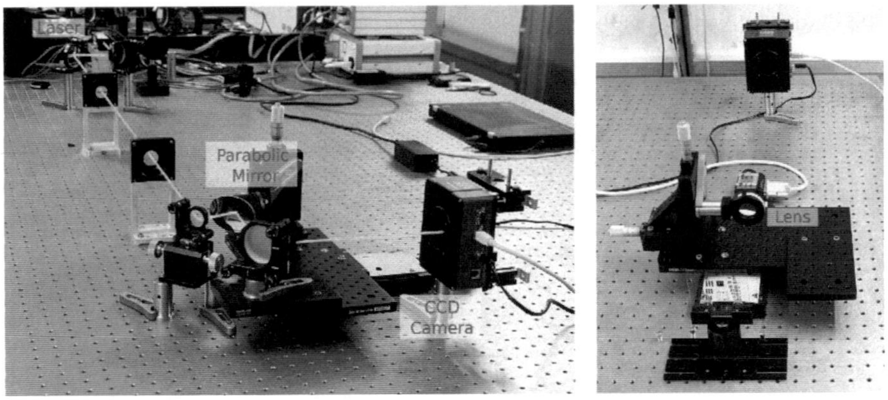

Figure 6. Photograph of the test stand setup with parabolic mirror (left) and with lens (right).

One way to reduce chromatic aberrations is to use reflective optics instead of refractive one. In order to compare these two approaches we performed the measurements of the focusing properties using both an achromatic lens and an off-axis parabolic mirror. In figure 7, a typical focused CCD image of the target taken with the parabolic mirror (figure 7a) and the lens (figure 7b) is presented. Figure 7c represents the vertical projection of the strip edge. As one can see, the image taken with the lens is well focused and has small distortion caused by geometrical aberrations. Whereas the image taken with the parabolic mirror is focused only in a small region and significantly distorted because of the aberration effects. Such distortion of the image taken using the parabolic mirror is caused mainly by imperfections in the mirror alignment. A parabolic mirror has to be perfectly aligned both vertically and horizontally because even small deviations of the beam from center leads to significant distortion of the image. Moreover, an off-axis parabolic mirror surface does not form an ideal image due to coma-like geometrical aberrations. For that purpose an elliptical mirror might be used. However design and manufacturing of such mirror with an optical quality might take time.

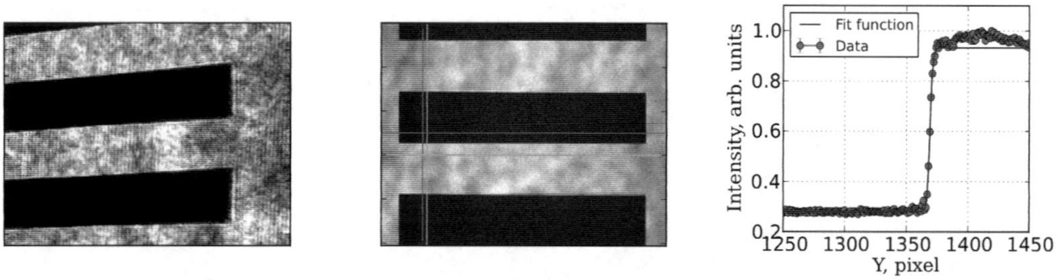

a) b) c)

Figure 7. a) CCD image of the target taken with "Edmund Optics" 90° off-axis parabolic mirror with parent focal length of 50.8 mm and ø=50.8 mm and b) "CVI Laser Optics" lens with focal length of 120 mm; c) vertical projection of the strip edge.

To study the chromatic aberrations of the lenses three lasers with different wavelengths were used. Every lens was gradually moved along the laser beam through the focus. At each lens position the image of the target was recorded by a CCD. A part of the image containing a strip edge was selected (see figure 7b, c). Then vertical projection of this selected area was fitted using the following fit function:

$$f(x) = a_0 + \frac{a_1}{1 + \exp\left(\dfrac{x - a_2}{a_3}\right)}. \tag{2}$$

Here, a_0 is the vertical offset of the distribution with respect to zero; a_1 is the amplitude of the distribution; a_2 represents position of the strip edge. Parameter a_3 of the fit function strongly depends on the image contrast and can be used as a measure of the focusing (e.g. if $a_3 \to 0$ the image is perfectly focused). The final dependencies of the focusing parameter versus the lens position are presented in figure 8.

As one can see, achromatic doublets (figure 8a, b) have small chromatic aberrations because the position of the best focus (i.e. the minimal a_3 parameter) is nearly the same. While singlet lens shows a strong chromatic aberration (figure 8c) for presented wavelength. These results demonstrate a good correlation with beam size measurements described in the previous section. Beside chromatic aberrations spherical aberrations of the lenses can be investigated using these data. However, the data are still under analysis, and the method of extracting the degree of spherical aberration needs to be further developed.

5. Conclusion

The results presented demonstrate progress in the optimization of an OTR beam profile monitor system based on analysis of the PSF structure. Changes in the experimental setup as well as improvement of the analysis allowed a better understanding of the monitor to be achieved. It was demonstrated that the use of a specially developed optics, such as achromatic doublets and optical filters, leads to decreasing of aberration effects. The presence of aberrations (spherical or chromatic) as well as diffraction effects lead to significant degradation of the potential resolution.

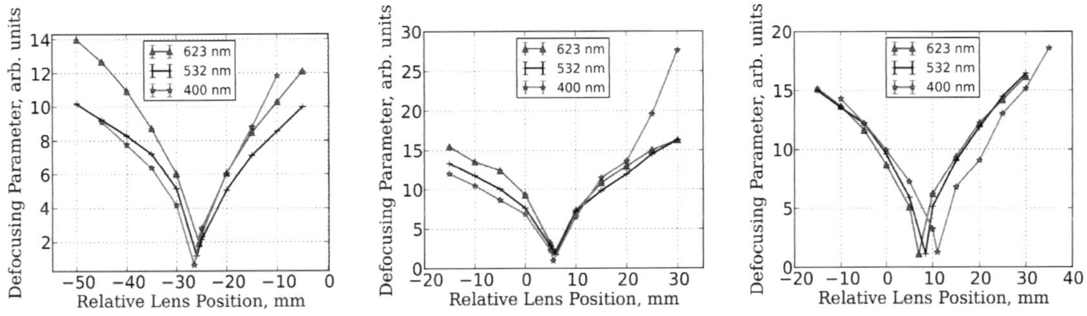

a) b) c)

Figure 8. Chromatic aberration study for different lenses: "CVI Laser Optics" cemented achromat with 120 mm focal length a); "Sigma Koki" achromatic doublet with focal length of 100.7 mm b); "Sigma Koki" plano-convex lens with 100 mm focal length c).

In order to further improve the resolution of the monitor, effects (mainly aberrations of the imaging system) significantly influencing the PSF width, need to be reduced. One way is to use a special simulation tool for optical calculations (such as ZEMAX [15]) in order to better optimize the optical system. It was shown in [16] that the use of such tools gives good agreement with the experimental data. In the next step we are planning to apply either a multi-element optics or a reflective optics to reduce the resolution of the monitor even further. For example, the use of an elliptical mirror can reduce the aberration effects. However, using such mirror in the real experiment becomes challenging because of the mirror alignment. Even small deflection of an incidence angle leads to significant image distortion due to coma-like geometrical aberrations.

Acknowledgments

We would like to acknowledge the Laserwire team, ATF II collaboration and, in particular, T. Okugi for their support. This research was partially supported by the oPAC: FP7 Marie Curie Initial Training Network under contract PITN-GA-2011-289485, Quantum Beam Technology Program of the Japanese Ministry of Education, Culture, Sports, Science, and Technology (MEXT), JSPS KAK- ENHI Grant No. 23226020.

References

[1] A Multi-TeV linear collider based on CLIC technology: CLIC Conceptual Design Report edited by Aicheler M, Burrows P, Draper M *et al* CERN-2012-007
[2] ILC Technical Design Report, URL: http://www.linearcollider.org/ILC/Publications/Technical-Design-Report
[3] Agapov I, Blair G and Woodley M 2007 *Phys. Rev. ST Accel. Beams* **10** 112801
[4] Boogert S, Blair G, Boorman G *et al* 2010 *Phys. Rev. ST Accel. Beams* **13** 122801
[5] Castellano M and Verzilov V 1998 *Phys. Rev. ST Accel. Beams* **1** 062801
[6] Ross M, Anderson S, Frisch J *et al* 2002 *SLAC-PUB* **9280**
[7] Karataev P, Aryshev A, Boogert S *et al* 2011 Phys. Rev. Lett. **107** 174801
[8] ATF 2 Proposal 2005 SLAC-R-771
[9] Bambade P, Alabau Pons M, Amann J *et al* 2010 *Phys. Rev. ST Accel. Beams* **13** 042801
[10] Aryshev A, Boogert S, Howell D *et al* 2010 *J. Phys.: Conf. Series* **236** 012008
[11] Kruchinin K, Boogert S, Karataev P *et al Proc. of IBIC'13* (Oxford) WEAL2
[12] Aryshev A, Terunuma N, Urakawa J *et al* 2010 *Proc. of IPAC'10* (Kyoto) MOPEA052 193
[13] Aryshev A, Terunuma N, Urakawa J *et al* 2011 *Proc. of IPAC2011* (San Sebastian) WEOBB01 1964
[14] SAD, Strategic Accelerator Design, URL: http://acc-physics.kek.jp/SAD/
[15] "ZEMAX", software for optical design, URL: http://www.zemax.com
[16] Bolzon B, Lefevre T, Welsch C *et al* 2013 *Proc. of IBIC'13* (Oxford) MOPF04

Electromagnetic field of a bunch intersecting a dielectric plate in a waveguide

Tatiana Yu Alekhina and Andrey V Tyukhtin

Radiophysics Department, Physical Faculty of St. Petersburg State University
1 Ulyanovskaya str., St. Petersburg, 198504, Russia

E-mail: tanya@niirf.spbu.ru, tyukhtin@bk.ru

Abstract. The electromagnetic field (EMF) of a bunch moving uniformly and traversing a dielectric plate located in a waveguide is investigated. The main attention is focused on the case when Cherenkov radiation is generated in the plate. Analysis of the field components of the mode is performed with methods of the complex variable function theory. An algorithm of computation using the exact expressions for the EMF is also presented. Consideration of the EMF structure for different time moments is given. It is shown that Cherenkov-transition radiation (CTR) is generated in the vacuum area after the plate under certain conditions. Results obtained might be of interest for development of new methods of generation of electromagnetic radiation.

1. Introduction

Investigation of electromagnetic fields of charged particles in waveguides loaded with dielectrics is of interest for development of new methods of generation of electromagnetic radiation and acceleration of particles [1–3]. Radiation of a charged particle (or a small bunch) crossing a boundary between two homogenous media in a metal cylindrical waveguide has been analyzed in detail in our previous papers [4–6]. Such an effect as Cherenkov-transition radiation (CTR) at the dielectric-vacuum boundary [6] has been investigated. It has been shown that the CTR can be dominant in the vacuum area. It should be noted that, recently, the idea of self-amplified Cherenkov radiation (CR) in a waveguide filled with a periodic dielectric structure was discussed [7]. Therefore, it is of significance to consider an intermediate case when a beam intersects two boundaries in the waveguide. The case under investigation is one of the key problems that are important for discussion of main peculiarities of the CTR effect.

It should be noticed that the case of a charge intersecting a dielectric plate situated in a waveguide was analyzed earlier [8] (as well as the case when a charge intersects a waveguide with dielectric [9]). However, the main attention was paid to consideration of the energetic characteristics, and the EMF structure was not analyzed. Present paper is exactly devoted to a study of the total electromagnetic field for the problem under consideration.

Formulation of the problem is the following: a bunch of charge particles moves uniformly with a velocity $\boldsymbol{V} = c\beta \boldsymbol{e}_z$ in a circular waveguide of radius a along its axis through a dielectric plate with permittivity ε located in a vacuum. The length of the plate is d. The bunch is characterized by a Gaussian distribution with the charge density $\rho = q\delta(x)\delta(y)\exp\left(-\zeta^2/\left(2\sigma^2\right)\right)$ where $\zeta = z - \beta ct$ and σ is a half-length of the beam. The middle of the bunch intersects the first boundary at the moment $t = 0$. We mark the vacuum areas before and after the plate as media 1 and 3 accordingly

($\varepsilon_1 = \varepsilon_3 = 1$), and medium 2 (the dielectric plate) has a permittivity $\varepsilon_2 = \varepsilon$. All of the media are nonmagnetic and have no dispersion.

The analytical solution to this problem is found traditionally as an expansion into a series of eigenfunctions of the transversal operator [8]. The electromagnetic field components in each media have two summands:

$$\boldsymbol{E}_j = \boldsymbol{E}_j^q + \boldsymbol{E}_j^b, \quad j = 1,2,3. \tag{1}$$

The first one $\left(\boldsymbol{E}_j^q\right)$, called by V.L. Ginzburg [10] the "forced" field, is the field in a regular waveguide. It contains CR if the bunch velocity exceeds the Cherenkov threshold. The second summand $\left(\boldsymbol{E}_j^b\right)$ is the "free" field connected with the influence of the boundary. It includes transition radiation (TR).

The forced field $\left(\boldsymbol{E}_j^q\right)$ has been well studied [11] and, mainly, analysis of the free field $\left(\boldsymbol{E}_j^b\right)$ presents some difficulties. Here we give expressions for longitudinal components of the free field in each of the media:

$$E_{zj}^b = \frac{2qc^2}{\pi a^4 \varepsilon_j} \sum_{n=1}^{\infty} \frac{\chi_n^2 J_0(\chi_n r/a)}{J_1^2(\chi_n)} I_{zj}^b, \quad j = 1,2,3, \tag{2}$$

$$I_{zj}^b = \frac{i\beta(\varepsilon - 1)}{\left(1 - \beta^2\right)\left(1 - \beta^2\varepsilon\right)} \int_{-\infty}^{+\infty} \left(A_{nj} e^{ik_{zj}z} + B_{nj} e^{-ik_{zj}z}\right) e^{-i\omega t} e^{-\omega^2\sigma^2/\left(2\beta^2c^2\right)} d\omega, \tag{3}$$

where

$$A_{n1} = 0,\ B_{n1} = \frac{1}{q_n \Delta_n}\left(g_n^+ r_n^+ e^{-ik_{z2}d} + g_n^- r_n^- e^{ik_{z2}d} - 2\sqrt{\omega^2\varepsilon - \omega_n^2}\, p_n^+ e^{i\omega d/\beta c}\right), \tag{4}$$

$$A_{n2} = \frac{1}{q_n \Delta_n}\left(-g_n^- p_n^+ e^{i\omega d/\beta c} + g_n^+ p_n^- e^{-ik_{z2}d}\right),\ B_{n2} = \frac{1}{q_n \Delta_n}\left(-g_n^+ p_n^+ e^{i\omega d/\beta c} + g_n^- p_n^- e^{ik_{z2}d}\right) \tag{5}$$

$$B_{n3} = 0,\ A_{n3} = \frac{1}{q_n \Delta_n}\left(\begin{array}{l} -g_n^+ r_n^- e^{-ik_{z2}d + i\omega d/\beta c} - g_n^- r_n^+ e^{ik_{z2}d + i\omega d/\beta c} + \\ +2\sqrt{\omega^2\varepsilon - \omega_n^2}\, p_n^- \end{array}\right) e^{-ik_{z1}d}, \tag{6}$$

$$g_n^{\pm} = \sqrt{\omega^2\varepsilon - \omega_n^2} \pm \varepsilon\sqrt{\omega^2 - \omega_n^2}, \quad \Delta_n = \left(g_n^+\right)^2 e^{-ik_{z2}d} - \left(g_n^-\right)^2 e^{ik_{z2}d}, \tag{7}$$

$$q_n = \left(\omega^2 + \beta^2\omega_n^2\left(1 - \beta^2\right)^{-1}\right)\left(\omega^2 + \beta^2\omega_n^2\left(1 - \beta^2\varepsilon\right)^{-1}\right), \tag{8}$$

$\alpha_n = \omega^2\left(1 - \beta^2 - \beta^2\varepsilon\right) + \beta^2\omega_n^2$, $p_n^{\pm} = \alpha_n \pm \omega\beta^3\varepsilon\sqrt{\omega^2 - \omega_n^2}$, $r_n^{\pm} = \alpha_n \pm \omega\beta^3\sqrt{\omega^2\varepsilon - \omega_n^2}$.

$k_{z2} = \sqrt{\omega^2\varepsilon - \omega_n^2}/c$, $k_{z1,3} = \sqrt{\omega^2 - \omega_n^2}/c$, $\omega_n = \chi_n c/a$, χ_n is the nth zero of the Bessel function ($J_0(\chi_n) = 0$), ($\operatorname{Im} k_{z1,2,3} > 0$).

We investigate the exact solution (2) – (8) both analytically and numerically. Analytical research is carried out with methods of the complex variable function theory. Computations are based on an algorithm using some separation of the integration path and the singularities of the integrands by taking into account the small absorption in the media. Next, we focus the main attention upon the case when the condition for CR in the dielectric plate is satisfied $\beta^2\varepsilon > 1$.

2. Analytical research

In the analytical method, the singularities of integrands (3) are studied in a complex plane of (ω). It can be seen, there are the following singularities:

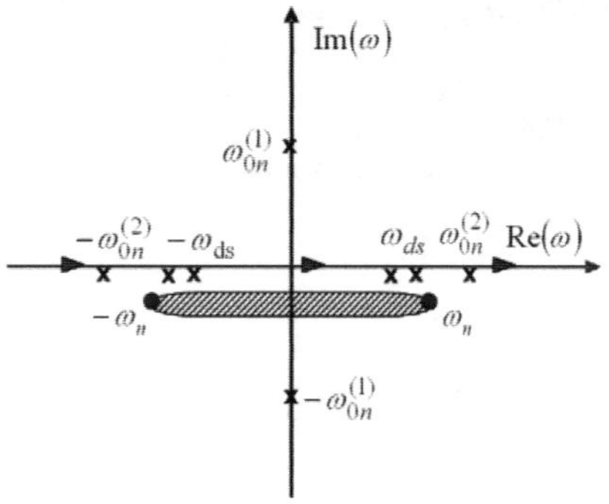

Figure 1. Disposition of the singularities of the integrands, branch cut and integration path in a complex plane of ω for the free field components in vacuum if $\beta > \varepsilon^{-1/2}$.

- two branch points $\pm\tilde{\omega}_n^{(1)} = \pm\omega_n - \mathrm{i}\delta_1$, which are shown at figure 1 with corresponding branch cut defined by the equation $\mathrm{Re}\sqrt{\omega^2 - \omega_n^2} = 0$;
- two poles $\pm\omega_{0n}^{(1)} = \pm\mathrm{i}\beta\omega_n\left(1-\beta^2\right)^{-1/2}$ which is always situated on an imaginary axis;
- two poles on the imaginary axis $\pm\omega_{0n}^{(2)} = \pm\mathrm{i}\beta\omega_n\left(1-\varepsilon\beta^2\right)^{-1/2}$ if $\beta < \varepsilon^{-1/2}$ or on the real axis $\pm\omega_{0n}^{(2)} = \pm\beta\omega_n\left(\varepsilon\beta^2-1\right)^{-1/2} - \mathrm{i}\delta_2$ if $\beta > \varepsilon^{-1/2}$;
- a finite number of pairs of poles $\pm\omega_{ds} - i\delta_3$.

Here $\delta_1, \delta_2, \delta_3$ are positive infinitesimal quantities, as we take into account the very small losses in the media. Therefore, all of the singularities are located below a real axis (figure 1). The poles $\pm\omega_{0n}^{(1)}$ and $\pm\omega_{0n}^{(2)}$ are the zeros of function q_n (8) in the denominators of the integrands. The zeros of function Δ_n (7) give a finite number of poles $\pm\omega_{ds}$. These poles are connected with the finite length of the dielectric plate. They can be found by solution of transcendental equation

$$\Delta_n = \mathrm{tg}(x) - \mathrm{f}(x,\mathrm{d})\big|_{\omega_\mathrm{dj}} = 0, \tag{9}$$

where $$\mathrm{f}(x,\mathrm{d}) = \frac{2\sqrt{\varepsilon}x\sqrt{(\lambda_n d)^2(\varepsilon-1) - x^2}}{x^2(\varepsilon+1) - (\lambda_n d)^2\varepsilon(\varepsilon-1)},\quad k_{z2}d = d\sqrt{\omega^2\varepsilon - \omega_n^2}\big/c \equiv x,\quad \chi_\mathrm{n} d/a \equiv \lambda_\mathrm{n} d.$$

This solution is represented graphically at figure 2. There is an upper branch of roots and also a lower branch of roots. Inequalities

$$\pi(\mathrm{k}-1) < \lambda_n d\sqrt{\frac{(\varepsilon-1)}{(\varepsilon+1)}}\left(\sqrt{\varepsilon+1} - \sqrt{\varepsilon}\right) < \pi\mathrm{k},\quad \pi(2\mathrm{l}-1)/2 < \lambda_n d\sqrt{\frac{\varepsilon(\varepsilon-1)}{(\varepsilon+1)}} < \pi(2\mathrm{l}+1)/2, \tag{10}$$

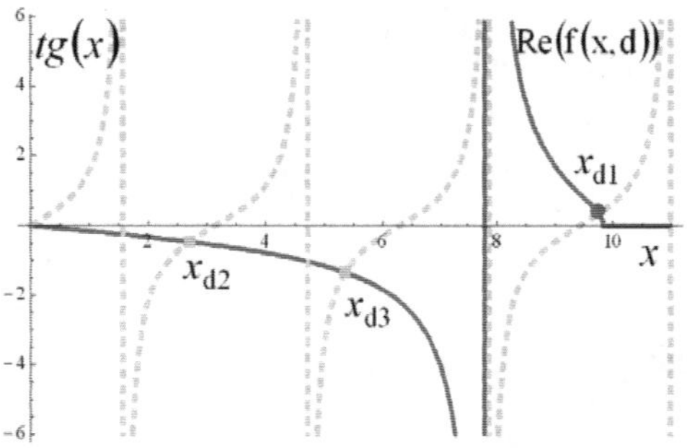

Figure 2. Solution of transcendental equation (9) for the following parameters of the problem: $n = 1$, $d/a = 5$, $\varepsilon = 1.67$, $\lambda_\mathrm{n} d = 12.02$. Red continuous line represents $\mathrm{Re}(\mathrm{f}(x,d))$, and green dashed line is for $\mathrm{tg}(x)$.

indicate that there is always a finite number of such roots $\mathrm{s} = \mathrm{k}+1$ which gives the finite number of pairs of poles $\pm\omega_{ds}$.

It can be obtained that all these poles are under inequality

$$\omega_\mathrm{n}\varepsilon^{-1/2} \le |\omega_\mathrm{dj}| < \omega_\mathrm{n},\quad j = 1,\ldots,s. \tag{11}$$

Thus the contributions from the poles $\pm\omega_{ds}$ to the free field give stationary waves inside the dielectric plate, and, outside the plate, they give the field exponentially decreasing with the distance from the boundaries.

If CR is generated in the dielectric plate, the contributions from the poles $\pm\omega_{0n}^{(2)}$ to the free field give so-called Cherenkov-transition radiation (CTR) which is the reflected and transmitted waves of CR. They can be evaluated by the residue theorem. It can be obtained that the CTR can exist at $\beta_C < \beta < \beta_{CT}$, where the lower threshold $\beta_C = \varepsilon^{-1/2}$ is connected with the condition of the Cherenkov radiation generation and the upper threshold $\beta_{CT} = (\varepsilon - 1)^{-1/2}$ is explained by total internal reflection of Cherenkov waves from the boundaries.

We obtain the expression for the CTR in medium 3 (after the plate) in the following form:

$$E_{z3}^{CTR} = -\frac{4q}{a^2}\sum_{n=1}^{\infty}\frac{J_0\left(\frac{\chi_n r}{a}\right)}{J_1^2(\chi_n)}\exp\left[-\frac{\sigma^2\kappa_n^2}{2}\right]\mathrm{Re}\left[\frac{(1+\varepsilon S)\exp\left[\mathrm{i}\kappa_n\left((z-d)S-\beta ct\right)\right]}{2\varepsilon S\cos(\kappa_n d)-\mathrm{i}\sin(\kappa_n d)\left(1-\varepsilon^2 S^2\right)}\right], \tag{12}$$

where $S = \sqrt{1-\beta^2(\varepsilon-1)}$, $\kappa_n = \lambda_n\left(\varepsilon\beta^2-1\right)^{-1/2} = \chi_n a^{-1}\left(\varepsilon\beta^2-1\right)^{-1/2}$. As distinct from the case of the dielectric-vacuum boundary [6], the CTR from the plate has the finite length which can be estimated with the results of our previous works. A rising edge of the CTR train can be described by formula:

$$z_1(t) = d + \mathrm{v}_{g1}(t - d/\beta c), \quad t > d/\beta c, \tag{13}$$

where $\mathrm{v}_{g1} = c\sqrt{1-\beta^2(\varepsilon-1)}\beta^{-1}$ is a group velocity of the CTR in the vacuum area [6]. For a falling edge of the CTR we have

$$z_3(t) = d + \mathrm{v}_{g1}(t - d/\mathrm{v}_{g2}), \quad t > d/\mathrm{v}_{g2}, \tag{14}$$

where $\mathrm{v}_{g2} = c(\beta\varepsilon)^{-1}$ is a group velocity of the CTR at the dielectric area [5]. So, the CTR train in area 3 has finite length that does not depend on time and the mode number but only on the parameters of the problem:

$$\Delta z^{CTR} = z_1(t) - z_3(t) = d\beta^{-2}\left(\beta^2\varepsilon - 1\right)\sqrt{1-\beta^2(\varepsilon-1)}. \tag{15}$$

The CTR train has maximum length for some optimal parameters which are $\beta_0 = \left(\sqrt{1+8\varepsilon(\varepsilon-1)^{-1}} - 1\right)^{1/2}(2\varepsilon)^{-1/2}$ if permittivity ε is fixed and $\varepsilon_0 = \left(3+2\beta^2\right)\left(3\beta^2\right)^{-1}$ for the fixed velocity of the bunch motion. Therefore, for the case of the ultra-relativistic particle ($\beta \approx 1$) we find that $\varepsilon_0\big|_{\beta\to 1} \approx 1.67$. Note that even for these optimal parameters the maximum length of the CTR is less than the length of the dielectric plate: $\max \Delta z^{CTR}\big|_{\beta\to 1} \approx 0.39d$.

3. Numerical results and discussion

For calculations, we use the exact formulas (3). The integrands behave rather abruptly along the integration contour because it crosses the poles $\omega_{0n}^{(2)}$ and ω_{ds}. To overcome these difficulties, an algorithm of computations is applied. As distinct from our previous works [4-6] we do not transform the initial integration contour (a real axis) but displace the singularities from the integration path (figure 1) by taking into account the small losses in the media. It should be underlined that the results can be obtained with a very good accuracy by optimizing the integration parameters. The exponential factor in (3) appearing because of the finite length σ of the bunch results in decrease of significance of modes with large numbers.

The behavior of the first mode of the longitudinal component E_z of the total field at different time moments ct/a for the plates with different lengths of the plate d/a is presented at figures 3 and 4 for the optimal parameters of the problem discussed above. The CTR determined by formulae (12)-(14) is also shown at pictures.

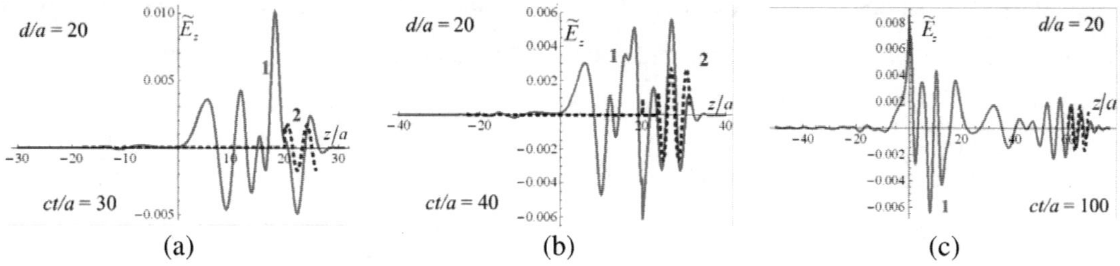

Figure 3. Dependence of the normalized longitudinal component $\tilde{E}_z = E_z a^2 / q$ of the first mode of the total field (continuous line 1) and the CTR (dashed line 2) on distance z/a for different dimensionless time moments *ct/a* for the plate with length *d/a* = 20; $\beta = 0.99$, $\varepsilon = 1.67 + i\,0{,}0001$, $\sigma/a = 1$, $\omega_0 = 2\pi \cdot 10\text{GHz}$, $a = 5\,\text{mm}$. The observation point is on the waveguide axis $r = 0$.

For the plate with length *d* = 20*a* (figure 3) the situation in all the media is presented at different time moments. As one can see almost all the radiation is directed forward, and there is significant field not only inside the plate but also after it in the vacuum area. The moment when the CTR starts to form out of the plate is shown at figure 3,a. When the CTR train is entirely formed (figure 3,b), there is a coincidence between the exact solution and the analytical approximation in frequency but not in amplitude. At figure 3,c, the field has mainly a type of TR than the CTR.

The field in medium 3 after the plate is presented at figure 4 for relatively big plates. For the plates with length *d* =50*a* (figure 4, a, b, c) the CTR approximation is quite valid. At greater time moments (figure 4, b, c), the following wave trains are formed as well. These trains are connected with reflection and re-reflection of the CTR inside the plate. However, the approximation (12) describes only the first CTR train.

For the plates with bigger length *d* =100*a* (figure 4, d, e, f) the CTR train is longer, and the CTR approximation matches rather well with the exact solution but only in a certain limited area.

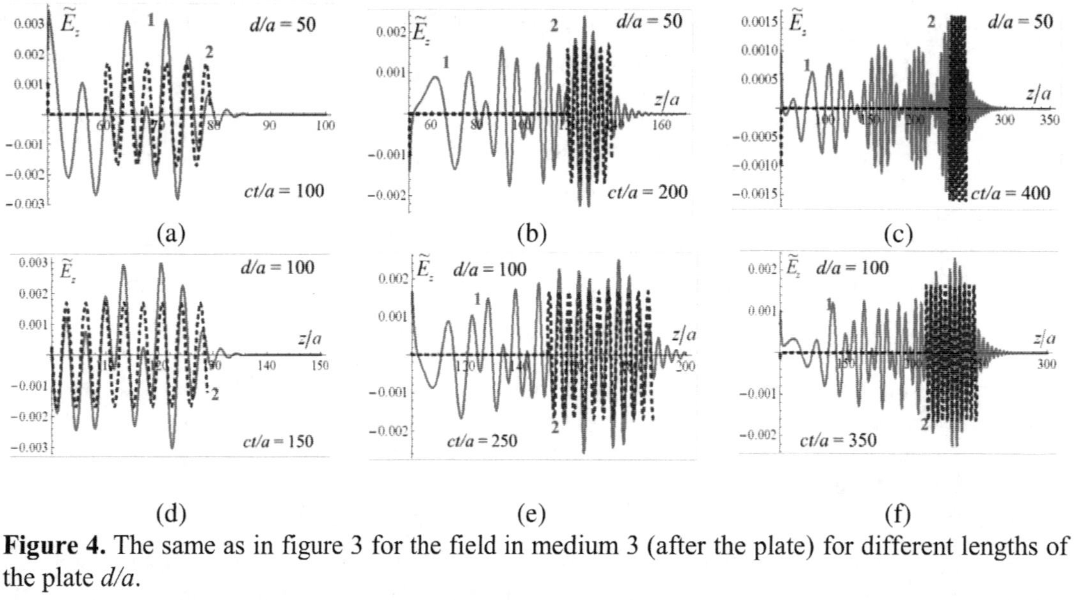

Figure 4. The same as in figure 3 for the field in medium 3 (after the plate) for different lengths of the plate *d/a*.

4. Conclusion

The total electromagnetic field of a bunch moving through a dielectric plate located in a waveguide has been investigated numerically by using exact formulas. Analytical study has been made with methods of the complex variable function theory. The main attention has been paid to the case when the CR is generated in the dielectric plate.

Physical estimation shows that the CTR is generated in the vacuum area after the plate at $\beta_C < \beta < \beta_{CT}$ where the upper limit is explained by the total reflection of the CR from the boundary. The CTR train has finite length which does not depend on the time and the mode number but only on the parameters of the problem: the length of the plate *d*, permittivity ε and velocity of the bunch motion β. The optimal parameters can be found in order that the length of the CTR train can be a maximum. It is shown that this length is always less than the length of the dielectric plate. So, for the CTR effect occurs, the length of the plate should be quite large in order that CR is formed inside the plate.

A complex structure of the field is revealed at the total field computations performed with using of the exact expressions. For thin plates, the CTR does not essentially signify. For thick enough layers, the CTR describes approximately the total field, but only in a certain limited spatial area (at fixed time moment). For sufficiently large time moments, the following CTR trains connected with reflection and re-reflection of radiation inside the plate are generated in addition to the first CTR train. Results obtained here might be of interest for development of new methods of generation of electromagnetic radiation.

References

[1] Gai W 2009 AIP Conf. Proc. **1086** 3
[2] Cook A M, Tikhoplav R, Tochitsky S Y, Travish G, Williams O B, and Rosenzweig J B 2009 Phys. Rev. Lett. **103** 095003
[3] Kanareykin A 2010 J. Phys.: Conf. Series **236** 012032
[4] Alekhina T Yu and Tyukhtin A V 2011 *Phys. Rev. E.* **83** 066401
[5] Alekhina T Yu and Tyukhtin A V 2012 *J. Phys.: Conf. Series* **357** 012010
[6] Alekhina T Yu and Tyukhtin A V 2012 *Phys. Rev. ST – AB* **15** 091302
[7] Grigoryan L Sh, Mkrtchyan A R, Khachatryan H F, Arzumanyan S R, and Wagner W 2012 *J. Phys.: Conf. Series* **357** 012004
[8] Barsukov K A 1960 *Zh. Tekh. Fiz.* **30**, 9 1337
[9] Barsukov K A, Gazazian E D, and Laziev E M 1972 *Izvestiya vuzov - Radiofizika* **15** 191
[10] Ginzburg V L and Tsytovich V N 1990 *Transition Radiation and Transition Scattering* (London: Hilger) p. 445
[11] Bolotovskii B M 1962 *Sov. Phys. Usp.* **4** 781

Overview of experimental investigations of X-ray radiation generated by 500-800 MeV electrons in multicrystal and multilayer radiators carried out at Tomsk synchrotron

V V Kaplin, S I Kuznetsov, S R Uglov, V N Zabaev

National Research Tomsk Polytechnic University, P.O. Box 25, Lenin Ave. 2-a, 634050 Tomsk, Russia

E-mail: vnz@tpu.ru

Abstract. The results of studies on X-ray radiation generated by ultrarelativistic electrons in multicrystal and multilayer radiators are shortly described. The experiments with mentioned radiators have shown that these types of new radiators could generate more intensive and narrower directed beams of x-rays than that generated in single crystals.

1. Introduction

Studying of X-ray generation with fast electrons in single crystals and periodic stacks of amorphous foils performed at the Tomsk synchrotron led to the investigations of artificial radiators such as periodic multicrystal stacks and multilayer X-ray mirrors. The investigation with the multicrystal radiator has been carried out in the collaboration with University of Tokyo and Hiroshima University [1, 2]. The multicrystal stacks consisted of 3, 10 and 100 thin crystals were created by Toshiba Corporation. A triple-crystal radiator was used at the Tomsk synchrotron for the detailed measurement of the characteristics of X-ray radiation generated by 800 MeV electrons in each crystal. The crystals were slightly disoriented relatively each other in the radiator. In principle, it is a disadvantage of any multicrystal radiator which is composed from a set of thin single crystals. But, such disorientation of the crystals of this radiator allowed to observe the separate angular distributions of X-rays emitted from each crystal. Later, in order to exclude the mutual disorientation of the crystals, the multicrystal structure was created on the surface of crystalline plate in Tomsk [3, 4].

Experiments with multicrystals led to the investigations with the multilayer radiators at the Tomsk synchrotron about 15 years ago in the collaboration with Adelphi Technology Inc (USA) [5]. Multilayer X-ray mirrors are widely used in X-ray optics for producing the X-ray beams with required parameters under Bragg reflection of real photons. X-rays are emitted in a narrow cone in the Bragg direction due to the diffraction of pseudophotons of the fast electron field in layers of X-ray mirror.

This paper presents a short review of the results of above mentioned investigations which have shown that both types of new radiators excited by relativistic electrons can generate more intensive and narrow beams of X-rays than that generated in single crystal with equivalent thickness.

2. Experimental installation

A general scheme of installation used for the experiments with new sources of radiation is represented in figure 1. A beam of accelerated electrons of the Tomsk synchrotron with the energy of 500 or 800

MeV was directed to the internal radiator of one or another type (figure 1a). Pulse duration of the electron beam was 10 ms, pulse-repetition frequency was 5 Hz.

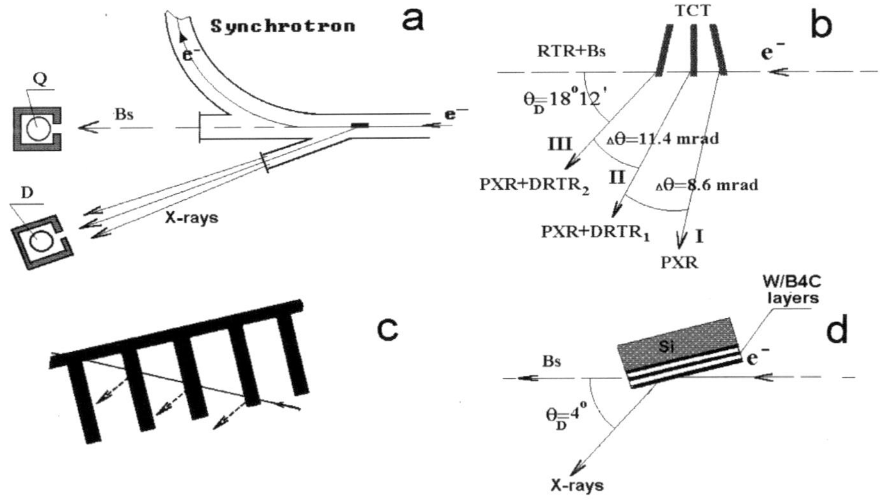

Figure 1. **a** – Scheme of installation: Q – quantometer, D – detector; **b** - Geometry of X-rays generation in triple-crystal target (TCT); I, II, III – X-rays emission from the 1st, 2nd and 3rd crystals, respectively; **c** - Multicrystal radiator on crystal surface; **d** - Multilayer radiator (mirror).

Bremsstrahlung (Bs) generated in a target was recorded by a Gauss-quantometer (Q). Readings of the quantometer were used to normalize the measurement results. Coherent radiation emitted from the radiators at an angle θ_D came out through a 200 μm beryllium window of a synchrotron chamber and got into a detector (D).

3. Triple-crystal radiator

Triple-crystal radiator consisted of three Si crystals with thickness of 16 microns separated by gaps of 147 microns [6]. Crystal wafers were cut for obtaining crystallographic planes (220) oriented perpendicular to their surfaces. Outside left and right crystals were disoriented relative to the middle crystal on 4.3 and 5.7 mrad in different directions. With a narrow electron beam transmitted through such radiator at some angle to planes (220) in Laue geometry, one can obtain three well-separated reflections of X–ray radiation in Bragg directions (figure 1b). Difference in angles is a doubled angle value of mutual disorientations of the crystals.

The first reflex of radiation (I) emitted from the first crystal is formed by parametric x-ray radiation (PXR) and the diffraction of transition radiation (TR). Transition radiation is generated on the input surface of this crystal. TR contribution would be small in comparison with PXR in case of first reflex. The second reflex of radiation (II) is formed by PXR from the second crystal and diffraction of resonance transition radiation (RTR) generated by the surfaces of the first crystal. The third reflex (III), respectively, is formed by PXR from the third crystal and diffracted RTR (DRTR) generated on the surfaces of the previous two crystals. Having measured characteristics of reflexes, it is possible to judge about the formation of coherent X-ray radiation of a relativistic electron passing through a complex crystal structure.

Single Si crystal with equivalent thickness (48 micron) was used to measure "pure" PXR without the contribution of DRTR. Both radiators were fixed on a goniometric head. Moving the goniometer head vertically allows placing the radiators on the electron beam by turns.

In this experiment a 2 mm NaJ(Tl) spectrometer was used. Energy resolution was about 35% for ^{57}Co (line 6.4 keV) and 16% for ^{241}Am (line 59.6 keV). Detection threshold was set at 7 keV. The input beryllium window of the detector has a diameter of 40 mm and a thickness of 200 microns. In front of the detector there was a vertical slit collimator with width of 2 mm. The distance between the target and detector was 458 cm, including 243 cm of air.

There are spectral peaks of the first and second orders in the experimental spectrum of PXR obtained on the equivalent target in Bragg orientation. The ratio of yields of PXR photons into the energy intervals (29 – 44) and (10 – 29) keV corresponding to the second and first spectral peaks is 0.13. Orientation dependencies (OD) of yields of PXR photons in these energy intervals were also measured for the equivalent target. Figure 2 represents the experimental OD (points) of the yield of (10 – 29) keV photons of PXR + DRTR emitted from a triple-crystal radiator. OD has three maxima (I, II and III) appearing with successive passing of crystals through Bragg positions of orientation. Three dashed curves show partial PXR contributions formed in each crystal for the total OD. The shape of these curves was obtained by the processing of OD of radiation, measured on the equivalent target. Figure 2 shows that the peaks of OD of DRTR from second and third crystals are considerably narrower, than OD of PXR yield. Widths of these peaks are 1.25 and 4.75 mrad, respectively. Thus, DRTR is a more narrowly directed X-ray source than PXR. Besides, it follows from the figure 2 that DRTR yield in the maximum of OD increases with the increasing of crystal number, so that the angular density of DRTR from the third crystal exceeds the density of PXR by approximately 1.7 times.

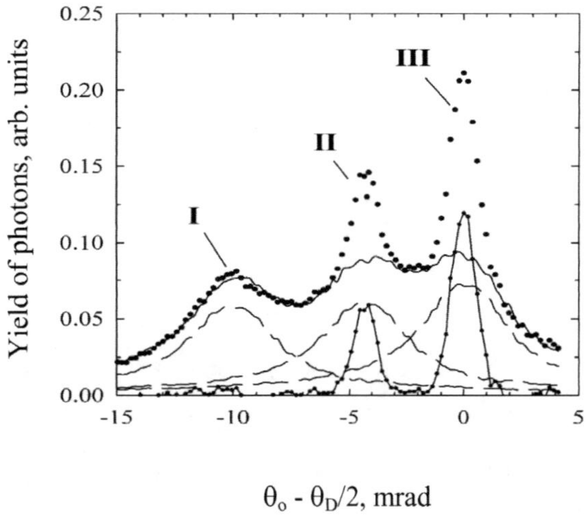

Figure 2. Measured OD (points) of yield of X-rays from triple-crystal radiator; dashed curves – ODs of PXR from three crystals; continuous curve – a sum of these ODs; left and right peaks with points – ODs of DRTR from second and third crystals.

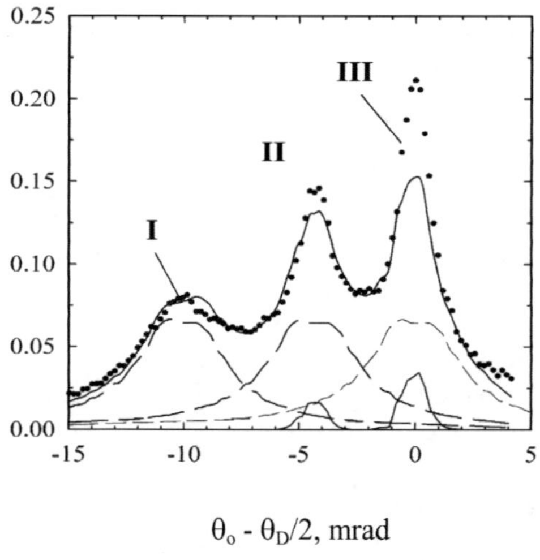

Figure 3. Calculated ODs of PXR from three crystals (dashed curves), ODs of DRTR from the second and third crystals (two peaks below) and ODs of PXR + DRTR (solid curve) with its interference.

Figure 3 shows the comparison of the calculated ODs of PXR and DRTR yields with the experimental one. The calculations were performed in the kinematic approximation. In the calculations the theoretical models for PXR [7] and DRTR [8] were used taking into account the experimental conditions. The dashed curves show the partial yields of PXR from three crystals, two thin continuous curves show ODs of DRTR yields from the second and third crystals. Comparing the data presented in figures 2 and 3, it can be concluded that the calculated ratio of DRTR and PXR yields does not correspond to the ratio, obtained in the experiment, and their sum does not describe a form of experimental orientation dependence. Abnormally high radiation yield in the experimental maxima II and III can be connected with interference of DRTR and PXR. Considering of interference was carried out in model approach when the square of sum of PXR and DRTR amplitudes for a whole cone was calculated. And after that the detector geometry was considered. The dependence, shown in figure 3 by a solid curve, is in better agreement with the experiment, though the difference is still large, especially for peak III. Probably, additional study is required to understand the observed effect.

The experiment showed that the making of radiator as a number of crystals leads to an effective increase of the X-ray yield in comparison with the case of a monocrystal with a thickness which is equivalent of the total thickness of the assembly. At the same time the angular distribution of radiation is much narrower than distribution of PXR from a monocrystal due to DRTR.

4. Multicrystal radiator on crystal surface

Further, at the Tomsk synchrotron a multicrystal radiator has been investigated. This radiator has no disadvantages which are inherent to radiators assembled from thin crystals [3, 4]. It was created on the surface of GaAs wafer by the method of microlithography with plasma chemical etching, and represents a flat, periodic, multistrip structure. The scheme of the radiator cross-section is shown in figure 1c. The 10 μm crystalline elements of the radiator are accurately oriented with respect to each other. This radiator is more compact than multicrystal stack and is easier to manufacture. Electrons cross the system of crystalline strips at an angle θ_0 to its atomic planes, and PXR is emitted at the Bragg angle to them. X-ray photons of RTR generated on the surfaces of strips are emitted near the direction of electron beam and then diffract in the Bragg direction. Such radiator is equivalent to a compound multicrystal radiator with a number of crystals $N = H / L \cot \theta_s$, where H is height of a strip, L is period of structure, θ_s is the angle between surface of GaAs wafer and electron beam. The asymmetric geometry of X-ray generation was used in order to reach a larger value of N. The angle between the crystallographic planes (100) and surface of GaAs wafer was $\theta_a = 2.7°$. The angle θ_s was equal to $\theta_D/2 - \theta_a = 6.8°$ when the angle of radiation detection was $\theta_D = 19°$. Accordingly, the effective number of crystals N = 20 with the parameters of the radiator created for this experiment H = 100 μm, L = 43 μm . Multicrystal radiator was fixed to a goniometer head and was able to move vertically. It was possible to change the position of the electron beam on the radiator and to obtain radiation from a multistrip structure or from a smooth part of the crystal. The photons emitted from multicrystal radiator were detected by a CdTe detector positioned at an angle $\theta_D = 19°$ to the beam of 500 MeV electrons and at a distance of 245 cm from the radiator. A radiation spectrum from a continuous part of a crystal wafer and a radiation spectrum from a structure were measured at a symmetric orientation of radiator $\theta_0 = \theta_D/2 = 9.5°$. These spectra have bright peaks on the photon energy 27 keV and weak peaks of characteristic radiation. The yield of 27 keV photons from a multicrystal structure is much higher than that from a continuous crystal. The increase of radiation yield is created by diffracted RTR (DRTR). RTR is generated on the side surfaces of crystalline strips and then it is diffracted on the crystallographic planes (100) of a set of next strips of multicrystal structure of the radiator.

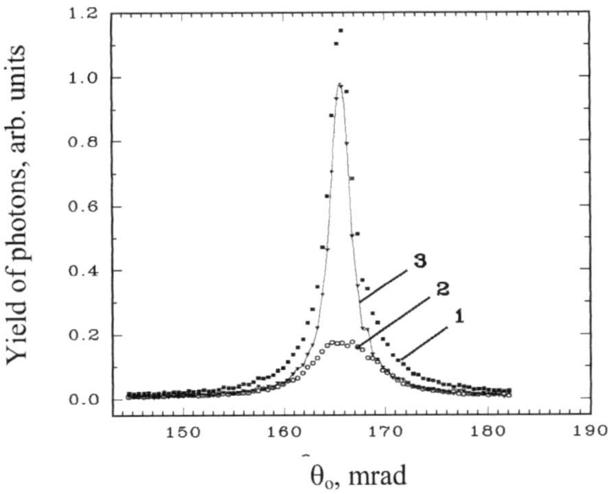

Figure 4. Curve 1 (points) – the measured orientation dependence of radiation yield from multicrystal periodic structure of a radiator; Curve 2 (circles) – contribution of PXR; Curve 3 – contribution of DRTR.

Figure 4 shows the results of measurements of OD of the yield of radiation generated by 500 MeV electrons in the multicrystal radiator. Energy range of detected photons was (25 – 28) keV. Curve 1 represents the measured orientation dependence of the radiation yield generated in a periodic structure. The radiator was oriented at certain angles to the beam of electrons to obtain this dependence and the photons yield was determined in this photon energy range. Curve 2 shows OD of PXR generated in a periodic multicrystal structure. It was obtained from OD of PXR measured from continuous part of radiator, reduced by coefficient k = 2.38. Coefficient k is a relation of a path length in a continuous part of radiator to a path length in a substance of periodic structure. Orientation dependence of radiation from a multicrystal structure is approximately 6 times higher and 3 times narrower than the orientation dependence of the "pure" PXR. This means that the cone of radiation from multicrystal periodic structure is also 3 times narrower than that of PXR from a monocrystal part of the radiator. Curve 3 was obtained by subtracting the data for PXR from the data for radiation from the structure. It represents the orientation dependence of DRTR and the possible contribution of the interference of PXR and DRTR. It can be seen that for a given multicrystal structure and conditions the mechanism of DRTR is about 5 times more effective than that of PXR.

5. X-ray mirror as a radiator of X-ray radiation

Another type of new source of coherent X-ray radiation has been studied at the Tomsk synchrotron [5] using the multilayer interference structures, known as X-ray mirrors. The generation of both PXR and DRTR is also possible when relativistic electrons are passing through such a radiator. Two X-ray mirrors (product of OSMIC Inc.) consisted of 300 W/B4C bilayers with the repetition period d = 1.236 and 1.8 nm were used as multilayer radiators. For example, in the case of the first X-ray mirror, vertical and horizontal dimensions of the multilayer radiator were about 10 and 29 mm, respectively. Thickness of its Si substrate was about 100 microns. The CdTe detector was placed at an angle θ_D = 66.3 mrad to the direction of 500 MeV electron beam and at a distance of 443 cm from the radiator.

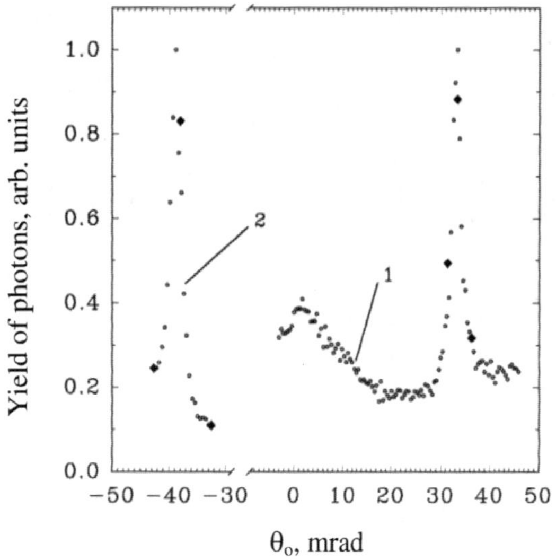

Figure 5. Orientation dependences of X-rays yield in the energy ranges of (9.24 –20.17) and (92–268) keV (curve 1 and 2, respectively).

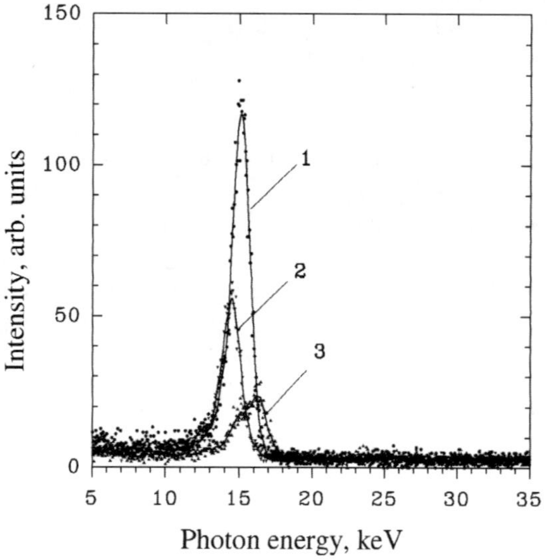

Figure 6. X-rays spectra at X-ray mirror orientation angle θ_o = 33.2; 31.1 and 36.1 mrad (curves 1, 2 and 3, respectively).

In figure 5 the measured OD of radiation yield in the ranges of photons energy of (9.24 – 20.17) and (92 – 268) keV. One can see in figure 5 the effect of coherent radiation generation in the X-ray mirror and possible PXR from its Si substrate. Both dependences demonstrate the bright peaks at the radiator orientation angles of 33.2 and –39 mrad. The first orientation angle corresponds to symmetrical position of multilayer structure of X- ray mirror relative to the direction of electron beam and the direction from the X-ray mirror to a detector. The second orientation angle corresponds to the symmetrical position of (113) Si substrate.

Radiation spectra measured at the X-ray mirror with d = 1.236 nm and at its orientation angle of 31.1; 33.2 and 36.1 mrad are depicted in figure 6. The spectral peak changes its position and amplitude when tilting radiator and it is maximal at the symmetric orientation. The dependence of PXR spectrum on the orientation of atomic planes of Si substrate was also measured. PXR spectral peak from the substrate also changed its position and amplitude when tilting radiator and reached its maximum at photons energy of 142 keV. It must be taken into account, that the (113) crystallographic planes of Si substrate, which are responsible for generation of this radiation, are disoriented relative to the layers of X-ray mirror at an angle of 71.3 mrad. Taking into account that electrons path in multilayer structure is approximately 300 times smaller than that in its substrate, it can be concluded that the efficiency of multilayer radiator is considerably higher, than the efficiency of crystalline radiator.

6. Conclusion

The experiments carried out at the Tomsk synchrotron have shown that new radiators based on multicrystal and multilayer structures can be effective X-rays generators when they are excited by 500–800 MeV electrons. Later, when the operation of the Tomsk synchrotron had been discontinued, the study of multicrystal radiators didn't get continuation, while the experiments with the X-ray mirrors were continued using betatron B–35 [9]. The experiments were carried out to investigate characteristics of radiation generated by the 15–33 MeV electrons in the X-ray multilayer mirrors which were installed inside the betatron chamber. The generation of X-rays by fast electrons in more complex multilayer structure as a triple-layer X-ray waveguide [10] was recently investigated at the betatron [11]. This structure consists of two multilayer substructures separated by waveguide channel. Currently, in the X-ray optics such structures are referred to as non-uniform X-photonic crystals or as Bragg waveguides. The use of these structures as new narrow-focus X-ray source provides the combination of the two effects: parametric X-ray radiation in the channel walls (multilayer structures) and the waveguide effect for X-rays generated in the wall structures and then penetrating the radiator in the waveguide modes similar to the Borrmann effect with low absorption. Recently, the investigation of extreme ultraviolet radiation generated by 5.7 MeV electrons in multilayer radiators was also begun at the Tomsk microtron M–5 [12].

References

[1] Andreyashkin M Yu, Zabaev V N, Yoshida K *et al* 1995 *Russian Pisma v JETP* **62** 770
[2] Andreyashkin M Yu, Kaplin V V, Potylitsin A P *et al* 1996 *Nucl. Instr. Meth.* B **119** 108
[3] Kaplin V V, Uglov S R, Zabaev V N *et al* 2000 *Nucl. Instrum. Meth.* A **448** 66
[4] Kaplin V V, Kuznetsov S I, Timchenko N A *et al* 2001 *Nucl. Instrum. Meth.* B **173** 238
[5] Kaplin V V, Uglov S R, Zabaev V N *et al* 2000 *Appl. Phys. Lett.* **76** 3647
[6] Andreyashkin M Yu, Zabaev V N, Kaplin V V *et al* 1997 *Russian Pisma v JETP* **65** 594
[7] Feranchuk I D and Ivashin A V 1985 *J. Physique* **46** 1981
[8] Garibiyn G M, Yan-Shi *X-Ray Transition Radiation* 1983 (Yerevan: Publishing House of Academy of Science of Armenia)
[9] Kaplin V V, Sohoreva V V, Uglov S R *et al* 2009 *Nucl. Instr. Meth. Phys. Res. B* **267** 777
[10] Kaplin V V, Sohoreva V V, Uglov S R *et al* 2011 *Nucl. Instr. Meth. Phys. Res.* B **269** 1518
[11] Kaplin V V, Uglov S R 2012 *Proceedings of the 7th International Forum on Strategic Technology (IFOST2012) IEEE cat. # CRP12786-PRT,* **2** 422
[12] Uglov S R, Zabaev V N, Kaplin V V *et al* 2012 *Journal of Physics: Conf. Series* **357** 012012

Gamma Cherenkov-transition radiation produced by charged particles at an interface and in a stack of plates

M A Aginian[1], K A Ispirian[1*], M Ispiryan[1], M I Ivanyan[2]
[1]Alikhanian National Scientific Laboratory, Yerevan Physics Institute, Br. Alikhanians 2, Yerevan, 0036, Armenia, [2] CANDLE Synchrotron Research Institute, Acharyan 31, 0040 Yerevan, Armenia

E-mail: karo@mail.yerphi.am

Abstract. The spectral and angular distributions as well as the total number of photons of gamma-ray Cherenkov-transition radiation (GCTR) in the photon energy region (1÷10) MeV produced by charged particles passing through an interface between two media and a stack of plates are calculated using the formulae of X-ray transition radiation. The concept of formation length for GCTR is discussed.

1. Introduction

It is well known [1] that one can present the dielectric constant and the refraction index of materials in the form $\varepsilon(\omega) = n^2(\omega) = \varepsilon'(\omega) + i\varepsilon''(\omega) = 1 + \chi'(\omega) + i\chi''(\omega)$ and $n(\omega) = 1 + \delta(\omega) + i\beta(\omega)$ where χ' and χ'' are the real and imaginary parts of the susceptibility. In this expressions $\chi'(\omega)$ and $\delta(\omega)$ are connected with the refraction, while $\chi''(\omega)$ and $\Delta(\omega)$ describe the absorption of photons with absorption length $L_{abs} = 1/\mu(\omega) = c/(\omega\chi''(\omega))$ where μ is the linear absorption coefficient. On the other hand, it is known that when a charged particle with energy exceeding some threshold value passes through a medium Cherenkov radiation is produced in the optical region (see [2]) as well as in X-ray region (see [3, 4]) if $n(\omega) > 1$.

Recently in a difficult experiment it has been shown [5] that the refractive index of silicon in the gamma ray region ~ (1–2) MeV is also greater than 1. Using QED calculations on Delbruck scattering the authors of [5] explained this fact and showed that for some other materials also $n(\omega) > 1$ in the larger photon energy interval from 1 MeV up to ~10 MeV. This fact opens new applications in gamma optics, nuclear physics, etc. Taking that these results are true it has been shown [6, 7] that Cherenkov radiation will exist also in the gamma ray region. It is more reasonable to call this last radiation as gamma ray Cherenkov-transition radiation, GCTR, since more correct results on GCTR are obtained using the theory of transition radiation [6, 7]. As it is shown in [6, 7] GCTR also can find wide applications, in particular, GCTR can be used in a much easier way for finding new materials with $n(\omega) > 1$ in gamma region.

Let us note that for the first time the energy losses, in particular, the so called parametric Cherenkov radiation of the particles passing through stratified medium has been considered in the work [8]. Further the parametric Cherenkov radiation has been studied in the long wave region in the works [9] and [10]. Then it has been shown theoretically (see [3, 11, 12]) that the transition radiation (TR) formulae include also the CR, and in certain conditions it is difficult to separate CR from TR. It has been shown that the radiation produced by relativistic particles at an interface between two media and in stratified media is emitted in X-ray region and is called X-ray transition radiation (XTR).

This work is devoted to the study of GCTR produced in a stack of plates. However, before developing the theory of the GCTR produced by particles passing stratified media it is necessary to consider GCTR produced at an interface of two media and the concept of formation length, L_{form}, for GCTR, which were not studied in [6, 7].

2. Formation lengths of GCTR

Let us first consider the formation length of GCTR in medium. According to [12] the formation length of XTR is a complex magnitude and for relativistic energies is given by the expression

$$Z_{Med}(\theta) = \frac{2\pi V}{\omega(\gamma^{-2} + \theta^2 - \chi)}, \tag{1}$$

in which θ is the radiation angle, V and $\gamma = E/mc^2 = 1/\sqrt{1-(V/c)^2}$ are the velocity and relativistic factor of the particle. Taking the module of (1) one obtains for the formation length of GCTR in medium, L_{Med}^{Form}, the length at which the GCTR photon is separated from the charged particle

$$L_{Med}^{Form} = |Z_{Med}| = \frac{2\pi V}{\omega} \frac{1}{\sqrt{(\gamma^{-2} + \theta^2 - \chi')^2 + \chi''^2}}. \tag{2}$$

Therefore, the formation length in vacuum ($\chi' = \chi'' = 0$) is equal to

$$L_{Vac}^{Form} = \frac{2\pi V}{\omega(\gamma^{-2} + \theta^2)}, \tag{3}$$

which has a maximum at $\theta = 0$ equal to $L_{Vac}^{Form} = \lambda\gamma^2$ ($\lambda = 2\pi c/\omega$ is the wavelength of the radiation photon) just as the formation length in vacuum of other types of radiation.

As it follows from (2) when GCTR takes place in medium and the condition for CR production $\gamma^{-2} + \theta^2 - \chi' = 0$ is satisfied, the formation length of GCTR in medium $L_{Med}^{Form} = 2\pi L_{abs}$. For GCTR predicted and considered in [6, 7] when the GCTR photons have $\lambda \leq 5\cdot10^{-11}$ cm and $\gamma \approx 4\cdot10^4$, $L_{Vac}^{Form} \approx 0.08$ cm, which is less than the absorption length of GCTR in Si equal to a few cm [6, 7], i.e. $L_{Med}^{Form} > L_{abs} >> L_{Vac}^{Form}$. On the other hand all the total thickness of any GCTR radiator must be less than the radiation length X_0, i.e. $L_{rad} << X_0$. Otherwise, the electron energy will be degraded due to radiation losses. For Si $X_0^{Si} = 9.36$cm. Therefore, it is reasonable to take the thickness of the single radiator and the total thickness of the stack of plate equal to, say, $L_{rad} \approx 0.1X_0 \approx 1$ cm. Therefore, for the above parameters for production of GCTR one has $L_{rad} < L_{abs} < L_{Med}^{Form}$. These facts result in the following two interesting consequences:

a) It is well known that the formula for XTR from one interface produced between a semi-infinite medium and vacuum it is not obligatory to have a medium with infinite thickness: it is enough to have a medium thicker than a few formation length in the medium, and the validity of the XTR formula for an interface is justified because in contrast to GCTR for XTR $L_{abs}^{XTR} >> L_{Med}^{XTRForm}$. For $h\omega >> (1-10)$ keV and $\gamma \approx 4\cdot10^4$ L_{abs}^{XTR} is less then a few tens and $L_{Med}^{XTRForm}$ is about a few microns, see, for instance [12]). On the other hand, for the validity of GCTR formula for an interface (see below) requiring to have instead of semi-infinite medium a radiator thicker than a few formation length one has to take radiator thicker than X_0 in which the velocity and energy will be changed essentially due to radiation losses. Therefore, the validity of below formula for GCTR from one interface is under question.

b) It is well known that the formula for XTR (or resonance transition radiation, RTR, see, for instance [12]) produced in a stack of M plates with plate thickness l_1^{XTR} greater than the formation length of XTR in medium, $L_{Med}^{XTRForm}$, and distances l_2^{XTR} between the plates greater than the XTR formation length $L_{Vac}^{XTRForm}$ gives RTR yield equal to the sum of XTR from M plates. Now using the above said one can show that for a stack of plates the maximal GCTR yield can be equal to the sum of the GCTR yield from one plate, if the total thickness of the stack of plate is equal to the thickness of the single plate and if the distance between the plates of the stack $l_2 > L_{Vac}^{Form}$ when there is no interference between GCTR from various plates of the stack. Indeed, as it has been shown in [7] for GCTR from one layer for the same case when $\lambda \leq 5\cdot 10^{-11}$ cm and $\gamma \approx 4\cdot 10^4$ (see figure 5 of [7]) the dependence of the GCTR yield upon the radiator thickness is linear around $L_{rad} \approx 1$ cm, while for $L_{rad} << 1$ cm this dependence is quadratic. As a result the GCTR yield from M independent plates with thickness l_1 is equal to that from a single plate with thickness Ml_1 only if l_1 is not much less than 1 cm, and the GCTR yield from M independent plates will be less less than that from a single plate with thickness Ml_1 if l_1 is much less than 1 cm. Such a property of GCTR produced in a stack of plates is connected with the fact that in the case of GCTR the absorption length of the produced MeV photons equal to a few cm is much greater than the absorption length of XCR and XTR photons with energies ~ (1–10) keV. This means that in contrast to XTR and XCR, in the case of GCTR it is not reasonable to take stack of plates instead of one plate with the same total thickness. This expectation will be confirmed below by numerical calculations.

3. GCTR produced at an interface between two media

Now let us derive the expression for spectral-angular distribution of the number of photons of GCTR produced at an interface between medium and vacuum substituting the values of χ' and χ'' into the formulae (1.59) of the work [12] for XTR in small θ and γ>>1 approximation. One obtains

$$\frac{d^2N_{GCTR}}{d(\hbar\omega)d\theta} = \frac{2\alpha\theta^3}{\pi\hbar\omega}\frac{(\chi'^2+\chi''^2)}{(\theta^2+\gamma^{-2})^2[(\theta^2+\gamma^{-2}-\chi')^2+\chi''^2]}. \tag{4}$$

Let us note that the formula (4) can be derived from the formula for GCTR produced in a single plate [6, 7] taking $L_{abs} >> L_{rad}$, i.e. when the radiator is much thinner than the absorption length. Let us also note that as it has been shown above the formula (4) has no practical value.

4. Theory of GCTR produced in stack of plates

Following the methods used in [3, 6, 7] we shall consider GCTR produced in a stack of plates with the help of the formulae of the theory of XTR [11–13], substituting the values of χ', χ'' for gamma ray region measured and calculated in [5]. Using the formula (3.15) of the work [12] for the spectral angular distribution of XTR, one obtains the following formulae for GCTR produced in a radiator consisting of M plates with thicknesses l_1 placed in vacuum at distances equal to l_2

$$\left.\frac{d^2N_{GCTR}}{d\omega d\theta}\right|_{MPl} = \left.\frac{d^2N_{GCTR}}{d\omega d\theta}\right|_{Int} F_{Pl}F_{MPl}, \tag{5}$$

where $\left.\frac{d^2N_{GCTR}}{d\omega d\theta}\right|_{Int}$ is GCTR produced at one interface and given by (4) and the factors F_{Pl} and F_{MPl} which are due to interference in one and M plates are given by

$$F_{Pl} = (1 - Q^{1/2})^2 + 4Q^{1/2}\sin^2 Y\,, \tag{6}$$

$$F_{Mpl} = \frac{(1 - Q^{M/2})^2 + 4Q^{M/2}\sin^2 MX}{(1 - Q^{1/2})^2 + 4Q^{1/2}\sin^2 X}\,, \tag{7}$$

where

$$Q = \exp(-\frac{\hbar\omega}{\hbar c}\chi''l_1)\,,\ Y = \frac{\hbar\omega l_1}{4\hbar c}(\theta^2 + \gamma^{-2} - \chi')\,, \tag{8}$$

$$X = \frac{\hbar\omega}{4\hbar V}[l_1(\gamma^{-2} + \theta^2 - \chi') + l_2(\gamma^{-2} + \theta^2)]\,. \tag{9}$$

The purpose of this work is to calculate the 1) spectral, 2) angular distributions and 3) the total number of GCTR photons numerically integrating (5) over θ, ħω and both θ and $\hbar\omega$, respectively .

5. Numerical results and discussion

As in [6, 7] the below results are obtained using the data on χ', χ'' for Si given in [5]. The electron energy is taken equal to E = 20 GeV. To compare with the results of [6,7] the calculations are carried out for three Si radiators with parameters: the total thickness of the three stratified radiators is taken equal to 1 cm. Taking the number of the Si layers in vacuum or air equal to M =10, 100 and 1000, means that l_1=10^{-1}, 10^{-2} and 10^{-3} cm. Taking a l_2=1 mm for which the distances between the plates are greater than the GCTR formation length in vacuum, and the yields of GCTR are the sums from each plate in the stack, one obtains the radiators thicknesses equal to 1, 10 and 100 cm, respectively.

Figure 1 a), b), c) and d) show the spectral distributions of GCTR photon number produced per electron in a single plate (M =1) with L_{rad}=1 cm [7], and in above described three radiators with l_1=0.1 cm (M =10), l_1=0.01 cm (M =100), c) and l_1=0.001 cm (M =1000), respectively.

As it was expected the spectrum (figure 1b) from the first stack of plates, consisting of 10 plates with l_1=0.1 cm, is very similar to that calculated for one thicker radiator with L_{rad}= 1 cm [7], because the taken l_2=1 mm is in the average larger than the formation length in vacuum with maximum $L_{Vac}^{Form} = \lambda\gamma^2$, and GCTR produced in various plates of the stack is summed. The intensities from the other radiators (figure1c and d) are less, than that from one thicker radiator with thickness equal to the sum of the thicknesses of the plates of the second and third stack of plates. This fact is explained with the radiator thickness dependence of the GCTR given in figure 5 of [7] from which it follows that for their thicknesses of the plates (l_1= 0.01 and 0.001 cm) the GCTR yield is less than 0.01 and 0.001 times.

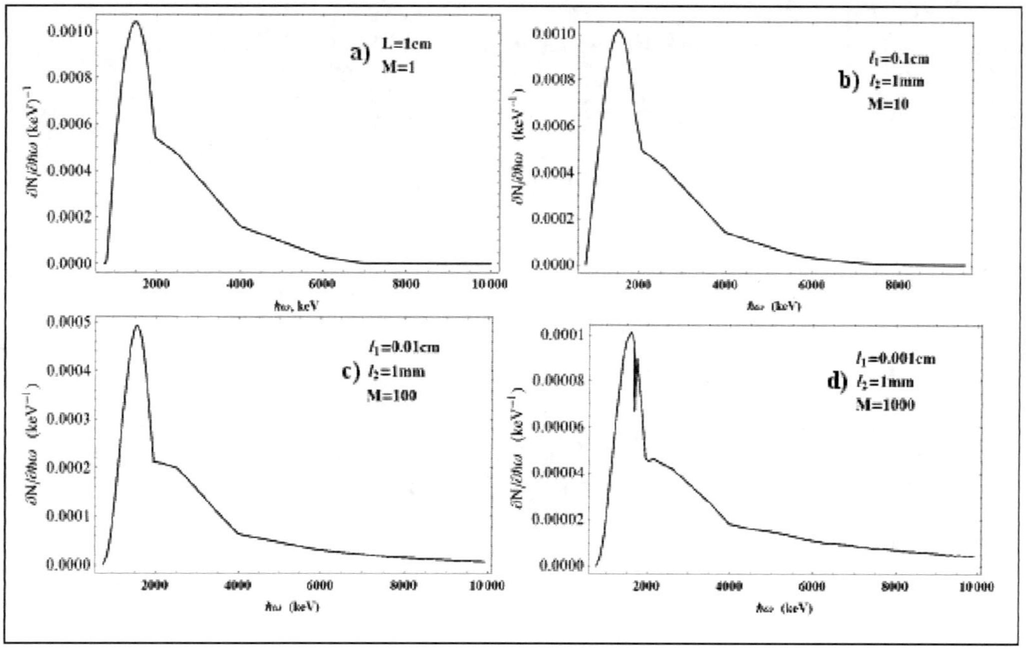

Figure 1. The spectral distributions of GCTR produced in a single plate a) and the 3 radiators described in the text b), c) and d).

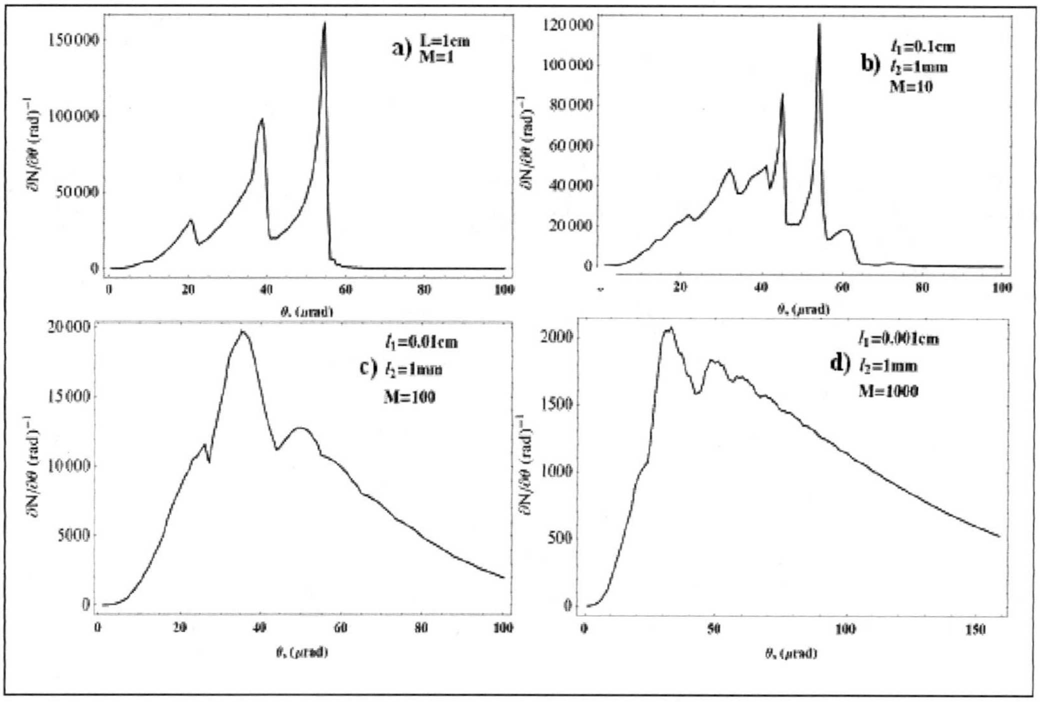

Figure 2. The angular distributions of GCTR produced in a single plate d) and the 3 radiators a), b) and c).

Figure 2 shows the angular distributions of GCTR produced per electron in these same radiators. As it is seen these angular distributions differ from each other and from the angular distribution of GCTR produced in 1 cm thick radiator (figure 3 of [7]) which is explained by the facts that the thicknesses of the plates in the cases a), b), c) and d) are different.

Finally, the total number of GCTR photons are equal to N_{GCTR}=1.6, 1.6, 0.8 and 0.2, for the radiator a), b), c) and d), respectively.

6. Conclusion

Thus, as it has been discussed above in contrast to XTR [12] and XCR [3], in the case of GCTR it is not reasonable to take stack of plates instead of one plate with the same total thickness. This conclusion is explained by the fact that in the case of GCTR the absorption length of the produced MeV photons equal to a few cm is much greater than the absorption length of XCR and XTR photons with energies ~ (1-10) keV. Of course, one can make the stack consisting of centimetre plates. However, in this case due to the energy losses and production of new electron positron pairs the energy of electrons can be less than the GCTR production threshold and accurate Monte Carlo simulations are necessary.

References

[1] Henke B L, Gullikson E M and Davis J C 1993 *At. Data and Nucl. Data Tables* **54** 181; http://www.cxro.lbl.gov/optical_constants.
[2] Zrelov V P 1968 *Izluchenie Vavilova-Cherenkova i Ego Primenenie v Fizike Visokikh Energii* Moscow, Atomizdat (in Russian)
[3] Bazilev V A, Zhevago N K 1987 *Izluchenie Bistrikh Chastits v Veshchestve i vo Vneshnikh Polyakh* Moscow, Nauka (in Russian)
[4] Ispirian K A *X-Ray Cherenkov Radiation*, in Proc. of NATO ARW Advanced photon Sources and Applications (Nor Hamberd, Armenia, 29 August-3 Sep, 2004); NATO Sc. Ser. II, Math. Phys Chem. 2005 **199** 217
[5] Habs D, Gunther M M, Jentschel M and Urban 2012 *Phys. Rev. Lett* **108** 184802
[6] Aginian MA, Ispirian K A and Ispiryan M 2013ArXiv:1302.5208; *Europhysics Lett.* **104** 24002
[7] Aginian MA, Ispirian K A, Ispiryan M and G. Sargsyan To be publ.in 2014 in *Izvestia NAS of Armenia*
[8] Fainberg Ya B and Khijnyak N A 1957 *Zh. Eksp. Teor. Fiz.* **32** 883
[9] Pratesi R *et al Nuovo Cimento* 1962 **25** 756
[10] Toraldo J *et al* 1963 *Alta Frequensia* **32** 1
[11] Ter-Mikaelian M L *The Influence of the Medium on High Energy Processes at High Energies* Publishing House of Academy of Science of Armenia, Yerevan, 1969; *High Energy Electromagnetic Processes in Condensed Media* Wiley Interscience, New York, 1972
[12] Garibian G M and Yan Shi 1983 *Rentgenovskoe Perekhodnoe Izluchenie*, Publishing House of Academy of Sciences of Armenia (Yerevan, Armenia
[13] Artru X, Yodh G B, Mennessier G 1975 *Phys. Rev. D* **12** 1650

Journal of Physics: Conference Series **517** (2014) 012015 doi:10.1088/1742-6596/517/1/012015

Coherent effects in the transition radiation of electron bunches on acoustic superlattices

V V Parazian and A A Saharian

Institute of Applied Problems in Physics, 25 Nersessian Street, 0014 Yerevan, Armenia

E-mail: vparazian@gmail.com, saharian@ysu.am

Abstract. We investigate the spectral-angular distribution of the forward transition radiation from an electron bunch interacting with an acoustic superlattice generated in a finite thickness plate. In the quasiclassical approximation, a formula for the radiation intensity is provided for a general distribution function of electrons in the bunch. We investigate coherence effects in the case of modulated electron distribution. The features of the radiation intensity are described in dependence of the ratio of the acoustic wavelength to the period of the modulation of the bunch. We demonstrate the appearance of the peaks in the radiation intensity for special values of this ratio.

1. Introduction

Transition radiation is produced when a relativistic particle traverses an inhomogeneous medium. Such radiation has a number of remarkable properties and at present it has found many important applications. In particular, transition radiation can be fruitfully exploited for energy detection or mass identification of high energy particles. The detectors based on the transition radiation have been used and are currently being used in a wide range of accelerator based experiments and in astroparticle and cosmic ray experiments.

The intensity of the transition radiation can be increased considerably by using the interference effects in periodic structures (for a review see [1]-[5]). The corresponding emission, called resonant transition radiation, results from constructive interferences between the waves emitted by each element of the periodic structure. In experiments, incident electrons cross stacks of thin foils and the transition radiation is formed at the interfaces between the material and vacuum. For the generation of high frequency transition radiation periodical structures with a small period are needed and the fabrication of this kind of radiator is a difficult task. In addition, irregularities in the spacing between the foils can rapidly destroy the constructive interference. In Ref. [6] it was suggested to generate the periodic radiator for the X-ray transition radiation by using acoustic waves. The parameters of this type of radiator are easily controlled by tuning the wavelength and the amplitude of the acoustic wave. The radiation from a charged particle for a semi-infinite laminated medium has been recently considered in [7]. The optical transition radiation in an ultrasonic superlattice excited in a finite thickness plate is considered in [8] under normal incidence of a relativistic electron and in [9] the results are generalized for the oblique incidence. It is shown that the acoustic waves generate new resonance peaks in the spectral and angular distributions. The heights and the location of these peaks can be controlled by choosing the parameters of the acoustic wave. In the papers given above the transition radiation from a single particle is considered. From the point of view of the experimental observation it is an important step to investigate the radiation from particle beams. Here we consider the forward transition radiation generated by a monoenergetic electron bunch in a dielectric plate in presence of ultrasonic waves.

2. Electromagnetic field and the radiation intensity

Consider the transition radiation emerging in the forward direction when a monoenergetic bunch of N particles moves at constant velocity $\mathbf{v} = v\mathbf{n}_z$ along the z-axis, enters normally into a plate whose surfaces coincide with planes $z = -l$ and $z = 0$. We assume that longitudinal ultrasonic vibrations are excited in the plate along the normal to its surface (along the axis z), that form a superlattice. The dielectric permittivity inside the plate we shall take in the form

$$\varepsilon(z) = \varepsilon_0 + \Delta\varepsilon \cos(k_s z + \omega_s t + \varphi), \tag{1}$$

for $-l \leqslant z \leqslant 0$. In (1), ω_s, k_s are the cyclic frequency and the wave number of the ultrasound, φ is the initial phase. Under the condition $\nu_s l/v \ll 1$, with $\nu_s = \omega_s/(2\pi)$, during the transit time of the electron the dielectric constant in the superlattice is not notably changed. For relativistic electrons and for the plate thickness $l \lesssim 1$ cm this leads to the constraint $\nu_s \ll 10^{11}$ Hz. In the discussion below we shall assume that the plate is immersed in a homogeneous medium with dielectric permittivity ε_1.

Here we are interested in the radiation with frequencies ω satisfying the condition $\omega \gg k_s c$. The presence of small parameter $k_s c/\omega$ allows one to use the quasi-classical approximation for the evaluation of the radiation field. The current density will be taken in the form

$$\mathbf{j} = ev \sum\nolimits_{j=1}^{N} \delta(x - X_j)\, \delta(y - Y_j)\, \delta(z - Z_j(t))\, \mathbf{n}_z. \tag{2}$$

In the Lorentz gauge, the vector potential of the electromagnetic field for j-th particle can be written as $\mathbf{A}_j = A^{(j)}\mathbf{n}_z$. This condition determines the radiation polarization. The magnetic field intensity is perpendicular to the plane containing $\mathbf{n}_z$ and the photon wave vector.

For the j-th particle in the bunch one has $Z_j(t) = -l + v(t - t_0^{(j)})$. Assuming that $|\Delta\varepsilon| \ll 1$, the Fourier transform of the vector-potential in the region $z > 0$ is presented as

$$\begin{aligned} A(\mathbf{k}_\perp, \omega, z) &= \frac{iee^{ik_3^{(1)}z}}{2\pi^2 c k_3^{(1)}} \sum_{j=1}^{N} e^{-i\omega Z_j/v - ik_x X_j - ik_y Y_j} \left[\frac{e^{i\phi^{(j)} - il(\omega/v - k_3^{(1)})} - 1}{2i(\omega/v - k_3^{(1)})} + e^{ia_1 \sin\varphi^{(j)}} \sqrt{\frac{k_3^{(1)}\varepsilon_1}{k_3^{(0)}\varepsilon_0}} \right. \\ &\times \left. \sum_{m=-\infty}^{+\infty} J_m(a_1) e^{-im\varphi^{(j)} - il(\omega/v - k_{3m}^{(0)})/2} \frac{\sin(l(\omega/v - k_{3m}^{(0)})/2)}{\omega/v - k_{3m}^{(0)}} \right], \end{aligned} \tag{3}$$

where and in what follows $\mathbf{k}_\perp = (k_1, k_2)$, $Z_j = Z_j(0)$, $J_m(x)$ is the Bessel function of the first kind, $k_{3m}^{(0)} = k_3^{(0)} + mk_s$, and

$$\begin{aligned} k_3^{(i)} &= \sqrt{\omega^2\varepsilon_i/c^2 - k_\perp^2},\ i = 0, 1,\ \varphi^{(j)} = \varphi - (Z_j + l)\omega_s/v, \\ \phi^{(j)} &= (k_3^{(0)} - k_3^{(1)})l + 2a_1 \sin(k_s l/2)\cos(\varphi^{(j)} - k_s l/2),\ a_1 = \omega^2\Delta\varepsilon/(2c^2 k_s k_3^{(0)}). \end{aligned} \tag{4}$$

The spectral-angular density of the radiation intensity during the electron transit time in the forward direction ($z \gg l$), averaged over the phase φ, is given by the formula

$$I_N(\omega, \theta, \phi) = 2\pi\varepsilon_1^{3/2} \frac{\omega^4}{c^3} \sin^3\theta \cos^2\theta \int_0^{2\pi} d\varphi\, |A(\mathbf{k}_\perp, \omega, z)|^2, \tag{5}$$

where θ is the angle between the z-axis and the wave vector of the radiation,

$$k_x = (\omega/c)\sqrt{\varepsilon_1}\sin\theta\cos\phi,\ k_y = (\omega/c)\sqrt{\varepsilon_1}\sin\theta\sin\phi, \tag{6}$$

with ϕ being the azimuthal angle for the radiation direction. By taking into account the expression (3), we get

$$I_N(\omega,\theta,\phi) = \frac{e^2\beta_1^2\sin^3\theta}{2\pi^2 c\sqrt{\varepsilon_1}}\sum_{j,j'=1}^{N} e^{-2iq_{jj'}\omega/\omega_s - ik_xX_{jj'} - ik_yY_{jj'}}\left\{U\sum_{m,m'=-\infty}^{+\infty} B_m(\theta,\omega)\right.$$
$$\times\left[UB_{m'}(\theta,\omega)J_{m-m'}(2a_1\sin q_{jj'})e^{i(m'-m)(k_sl-\pi)/2}e^{i(m'+m)q_{jj'}}\right.$$
$$\left.+\frac{J_{m-m'}(a_1)}{1-\beta_1\cos\theta}J_{m'}(b_1)e^{2im'q_{jj'}}\sin[(k_3^{(0)}-\omega/v)l/2+(m'-m)k_sl/2+m'\pi/2]\right]$$
$$\left.-U\sum_{m=-\infty}^{+\infty}\frac{W_mB_m^2(\theta,\omega)}{1-\beta_1\cos\theta}+\frac{1+J_0(2b_1\sin q_{jj'})-2\cos[(k_3^{(0)}-\omega/v)l]J_0(b_1)}{4(1-\beta_1\cos\theta)^2}\right\}, \tag{7}$$

where $\beta_i = v\sqrt{\varepsilon_i}/c$, $U_{jj'} = U_j - U_{j'}$, $U = X, Y$, $q_{jj'} = (Z_j - Z_{j'})\omega_s/2v$, $b_1 = 2a_1\sin(k_sl/2)$, and

$$B_m(\theta,\omega) = J_m(a_1)\frac{\sin(l\omega W_m/2v)}{W_m},\ W_m = W + mv\frac{k_s}{\omega},\ W = 1-\beta_1\sqrt{\frac{\varepsilon_0}{\varepsilon_1}-\sin^2\theta}. \tag{8}$$

Now one has $k_3^{(0)} = (\omega/c)\sqrt{\varepsilon_0-\varepsilon_1\sin^2\theta}$,

$$a_1 = \frac{\omega\Delta\varepsilon}{2ck_s\sqrt{\varepsilon_0-\varepsilon_1\sin^2\theta}},\ U = \left(\frac{\varepsilon_1\cos\theta/\varepsilon_0}{\sqrt{\varepsilon_0/\varepsilon_1-\sin^2\theta}}\right)^{1/2}. \tag{9}$$

For $N = 1$ the expression (7) reduces to the one derived in [8].

3. Averaging over the coordinates of particles

For the radiation intensity averaged over the coordinates of particles in the bunch we get

$$I_N(\omega,\theta,\phi) = NI_1(\omega,\theta) + N(N-1)I_c(\omega,\theta,\phi), \tag{10}$$

where the contribution of the coherent effects is given by the second term on the right-hand side with

$$I_c(\omega,\theta,\phi) = \frac{1}{2}\int d^3R_j\int d^3R_{j'}I_{jj'}(\omega,\theta,\phi)f(R_j)f(R_{j'}),\ j'\neq j. \tag{11}$$

(Note that the formula (10) is valid for various types of radiation processes, including the transition radiation, the diffraction radiation and the synchrotron radiation (see [10]).) We assume that the distribution function is normalized in accordance with $\int d^3R_jf(R_j) = 1$. In (11), $I_{jj'}(\omega,\theta)$ is given by the expression (7) if we remove the summation sign and $R_j = (X_j, Y_j, Z_j)$. Assuming that $f(R_j) = f_x(X_j)f_y(Y_j)f_z(Z_j)$, with the functions $f_u(U_j)$ separately normalized to 1, and introducing the notation $F_u(\tau) = \left|\int_{-\infty}^{+\infty}ds f_u(s)e^{i\tau s}\right|^2$, with $u = x, y, z$, we find

$$I_c(\omega,\theta,\phi) = \frac{e^2\beta_1^2\sin^3\theta}{2\pi^2c\sqrt{\varepsilon_1}}F_x(k_x)F_y(k_y)\left\{2U\sum_{m,m'=-\infty}^{+\infty}B_m(\theta,\omega)\left[UB_{m'}(\theta,\omega)F_{mm'}(\omega,\theta)\right.\right.$$
$$\left.\times\cos\left(lk_s\frac{m-m'}{2}\right)-\frac{J_{m'}(a_1)J_{m'+m}(b_1)}{1-\beta_1\cos\theta}\sin\left(\frac{l\omega}{2v}W_{m'}+\pi\frac{m'-m}{2}\right)F_{m+m'}(\omega)\right]$$
$$\left.-U\sum_{m=-\infty}^{+\infty}\frac{W_mB_m^2(\theta,\omega)}{1-\beta_1\cos\theta}F_z(\omega/v)+\frac{[1-2\cos(l\omega W/v)J_0(b_1)]F_z(\omega/v)+2F(\omega,\theta)}{4(1-\beta_1\cos\theta)^2}\right\}. \tag{12}$$

In (12) the following notations are introduced:

$$\begin{aligned}
F_{mm'}(\omega,\theta) &= \int_0^{\infty} d\tau G(\tau) J_{m'-m}\left(2a_1 \sin\left(\frac{\tau\omega_s}{2v}\right)\right) \cos\left(\left(\omega_s\frac{m'+m}{2}-\omega\right)\frac{\tau}{v}-\pi\frac{m-m'}{2}\right), \\
G(\tau) &= \int_{-\infty}^{+\infty} du\, f_z(u-\tau) f_z(u),\ F_m(\omega) = \int_0^{\infty} d\tau G(\tau)\cos((2m\omega_s-\omega)\tau/v), \\
F(\omega,\theta) &= \int_0^{\infty} d\tau G(\tau)\cos(\omega\tau/v) J_0(2b_1\sin(\tau\omega_s/(2v))). \qquad (13)
\end{aligned}$$

Note that we can write $F_z(u)$ in terms of the function $G(\tau)$: $F_z(u) = 2\int_0^{\infty} d\tau G(\tau)\cos(u\tau)$.

Note that the dominant contribution in the integrals for the form-factors in (13) comes from the region of the integration $\tau \lesssim L_z$, where L_z is the longitudinal size of the bunch. Now one has $\tau\omega_s/v \lesssim 2\pi\,(L_z/\lambda_s)\,(v_s/v)$, where λ_s and v_s are the wavelength and the velocity for the acoustic wave. If $L_z \lesssim \lambda_s$ one has $\tau\omega_s/v \ll 1$ for relativistic electrons. In this case the form-factor $F_{mm'}(\omega,\theta)$ is small for $m \neq m'$ and the main contribution in the corresponding part in (12) comes from the term $m = m'$ with $F_{mm}(\omega,\theta) \approx F_{m/2}(\omega)$. We also have $F(\omega,\theta) \approx F_z(\omega/v)/2$. If in addition, m is not too large, so $m\omega_s \ll \omega$, we see that $F_m(\omega) \approx F_z(\omega/v)/2$. In this case the summation over m' in the second term in the square brackets of (12) is done explicitly by using the addition theorem for the Bessel function and one finds

$$I_c(\omega,\theta,\phi) \approx F_x(k_x)F_y(k_y)F_z(\omega/v)I_1(\omega,\theta,\phi). \qquad (14)$$

Hence, under the conditions $L_z \lesssim \lambda_s$ and $m\omega_s \ll \omega$ the influence of the bunch structure is separated in the form of the form-factor $F_x(k_x)F_y(k_y)F_z(\omega/v)$.

The general formula (12) is further simplified for large values of the plate thickness. Introducing the quantity $I_c^{\infty}(\omega,\theta) = \lim_{l\to\infty} I_c(\omega,\theta)/l$, from (12) we get

$$I_c^{\infty}(\omega,\theta,\phi) = F_x(k_x)F_y(k_y)\frac{e^2\beta_1\omega}{2\pi c^2}\frac{\varepsilon_1\sin^3\theta\cos\theta}{\varepsilon_0\sqrt{\varepsilon_0/\varepsilon_1-\sin^2\theta}}\sum_{m=-\infty}^{+\infty} C_m(\omega)J_m^2(a_1)\delta(W_m), \qquad (15)$$

with $\delta(x)$ being the Dirac delta function and $C_m(\omega) = F_{mm}(\omega,\theta)$. This expression for the radiation intensity is valid under the condition $l \gg 2\pi v/\nu$, where $\nu = \omega/(2\pi)$. Note that one has $\nu \gg \nu_s$ and this condition does not contradict the condition $l \ll v/\nu_s$ (see the paragraph after formula (1)) we have assumed before. In (15), the term with $m = 0$, which we denote by $I_{c,0}^{\infty}(\omega,\theta,\phi)$, describes the radiation which propagates along the direction $\theta_0 = \arcsin(\sqrt{\varepsilon_0/\varepsilon_1}\sin\theta_{\mathrm{Ch}})$, $\theta_{\mathrm{Ch}} = \arccos(1/\beta_0)$ is the Cherenkov angle in the medium with permittivity ε_0. This term corresponds to the Cherenkov radiation in the plate which propagates in the region $z > 0$ after the refraction at $z = 0$. Integrating over θ we find

$$I_{c,0}^{\infty}(\omega,\phi) = \int_0^{\pi} d\theta\, I_{c,0}^{\infty}(\omega,\theta,\phi) = \frac{e^2\omega}{\pi c^2}F_x(k_x)F_y(k_y)C_0(\omega)\left(1-1/\beta_0^2\right)J_0^2(a_1), \qquad (16)$$

where now $a_1 = \omega v\Delta\varepsilon/(2c^2k_s)$ and in the expressions (6) one has $\theta = \theta_0$.

Now we consider the terms with $m \neq 0$. For a given m, the frequency for the radiation propagating along the direction θ is given by the relation $\omega = mvk_s/W$, where $m > 0$ for $W > 0$ and $m < 0$ for $W < 0$. After the integration over ω, for $I_{c,m\neq 0}^{\infty}(\theta,\phi) = \int d\omega\, I_c^{\infty}(\omega,\theta,\phi)$ we find

$$I_{c,m\neq 0}^{\infty}(\theta,\phi) = \frac{e^2\beta_1^3k_s^2\sin^3\theta\cos\theta}{2\pi\varepsilon_0\sqrt{\varepsilon_0/\varepsilon_1-\sin^2\theta}\,|W|^3}\sum_{mW>0} F_x(k_x)F_y(k_y)C_m(mvk_s/W)m^2J_m^2(a_1), \qquad (17)$$

where in the expressions (6) the substitution $\omega = mvk_s/W$ should be made.

4. Example of a microbunched beam

As an example of a microbunched beam we consider a beam with Gaussian distribution functions in x and y and with modulated Gaussian distribution in the longitudinal direction:

$$f_u(\tau) = \frac{e^{-\tau^2/(2\sigma_u^2)}}{\sqrt{2\pi}\sigma_u}, \ u = x, y, \ f_z(z) = \frac{e^{-z^2/(2\sigma_z^2)}}{\sqrt{2\pi}\sigma_z} \sum_{n=-\infty}^{\infty} d_n e^{ink_r z}, \tag{18}$$

where k_r is beam modulation wave number and $d_{-n} = d_n^*$, $d_0 = 1$. For these functions one has

$$F_u(k_u) = e^{-\sigma_u^2 k_u^2}, \ u = x, y, \ F_z(u) = |\sum_{n=-\infty}^{\infty} d_n e^{-\sigma_z^2(nk_r+u)^2/2}|^2. \tag{19}$$

For the function $G(\tau)$ one gets the expression

$$G(\tau) = \frac{e^{-\tau^2/(4\sigma_z^2)}}{2\sqrt{\pi}\sigma_z} h(k_r\sigma_z, \tau/(2\sigma_z)), \ h(x,u) = \sum_{n,n'=-\infty}^{\infty} d_n d_{n'} e^{-(n+n')^2 x^2/4} \cos[(n-n')xu]. \tag{20}$$

With the use of (20), the expression for the function $F_m(\omega)$ takes the form:

$$F_m(\omega) = \frac{1}{2} e^{-(2m\omega_s-\omega)^2 \frac{\sigma_z^2}{v^2}} \sum_{n,n'=-\infty}^{\infty} d_n d_{n'} e^{-(n^2+n'^2)k_r^2 \frac{\sigma_z^2}{2}} \cos\left[(n-n')(2m\omega_s-\omega) k_r \frac{\sigma_z^2}{v}\right]. \tag{21}$$

In the simple case $d_n = 0$, $n = \pm 2, \pm 3, \ldots$, defining $d_1 = d_b e^{i\beta}$, the distribution function is given by

$$f_z(z) = \frac{e^{-z^2/(2\sigma_z^2)}}{\sqrt{2\pi}\sigma_z} [1 + 2d_b \cos(k_r z + \beta)]. \tag{22}$$

For the function $h(x,u)$ we find

$$h(x,u) = 1 + 2d_b^2 \cos(2xu) + 4d_b e^{-x^2/4} \cos\beta \cos(xu) + 2d_b^2 e^{-x^2} \cos(2\beta). \tag{23}$$

For the numerical example we consider large values of the plate thickness. For the angular density of the number of quanta radiated at a given frequency $\omega_m = mvk_s/W$, per unit trajectory of the particle, one has

$$\mathcal{N}_{N,m}(\theta,\phi) = N\mathcal{N}_{1,m}(\theta,\phi) + N(N-1)\mathcal{N}_{c,m}(\theta,\phi). \tag{24}$$

For large values of the plate thickness, for the coherent part $\mathcal{N}_{c,m}^{\infty}(\theta,\phi) = \lim_{l\to\infty} \mathcal{N}_{c,m}(\theta,\phi)/l$, from (17) one gets

$$\mathcal{N}_{c,m}^{\infty}(\theta,\phi) = \frac{e^2}{\hbar c} \frac{\varepsilon_1 \beta_1^2 k_s \sin^3\theta \cos\theta}{2\pi\varepsilon_0 \sqrt{\varepsilon_0 - \varepsilon_1 \sin^2\theta} W^2} F_x(k_x) F_y(k_y) C_m(\omega_m) |m| J_m^2(a_1). \tag{25}$$

In figure 1 we have plotted the quantity (25) for $m = -1$ as a function of $\sigma_z k_s$ and θ for the electron energy 17.5 MeV and for the longitudinal distribution function (22) with $d_b = 0.5$, $\beta = 0$, $\sigma_z = 3 \times 10^{-2}$cm. We assumed that the plate is made from fused quartz and $\varepsilon_1 = 1$. For the velocity of the acoustic wave one has $v_s = 5.6 \times 10^5$cm/s and we have taken $\Delta\varepsilon = 0.02$. For the left plot $\sigma_z k_r = 5$ and for the right one $\sigma_z k_r = 10$. For the transverse part of the distribution function we have taken $F_x(k_x) F_y(k_y) = 1$. With increasing $\sigma_z k_s$ the separation between the peaks and the height of the right peak increase. In addition, the right peak exhibits microstructure. We have numerically checked that the contributions of the higher harmonics $m = -2, -3, \ldots$ are small compared to the one for $m = -1$.

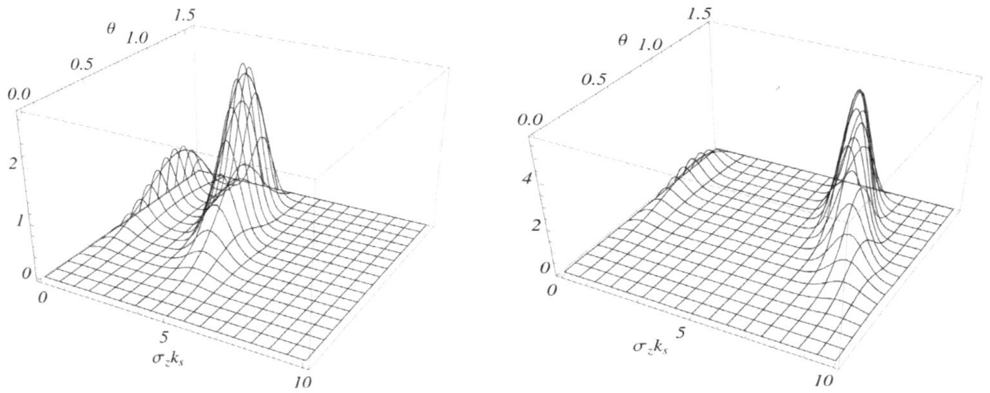

Figure 1. Coherent part in the number of the radiated quanta, $10^8\mathcal{N}_{c,m}^{\infty}$ (see (25)) at a given harmonic corresponding to $m = -1$.

5. Conclusion

In the present paper we have considered the coherence effects in the transition radiation of a microbunched electron beam on an acoustic superlattice excited in a finite thickness dielectric plate. Assuming that the amplitude of the modulation for dielectric permittivity is small, the spectral-angular density of the radiation intensity in the forward direction is given by formula (7). After averaging over the coordinates of particles in the bunch, it is presented as (10), where for the contribution of coherent effects one has the expression (12). Under the conditions $L_z \lesssim \lambda_s$ and $m\omega_s \ll \omega$, the influence of the bunch structure on the radiation intensity is separated in the form of the form-factor (see (14)). For a thick plate, the general formula is further simplified to (15). In this formula, the $m = 0$ term corresponds to the Cherenkov radiation in the plate which propagates in the region $z > 0$ after the refraction at the plate boundary. As an application of the general expression for the transition radiation intensity, in section 4 we have considered a beam with Gaussian distribution functions in transverse directions and with modulated Gaussian distribution in the longitudinal direction. The spectral-angular distribution of the radiation intensity can be controlled by tuning the parameters of two periodic structures: the beam modulation wave number k_r and the acoustic wave number k_s.

References

[1] Ter-Mikaelian M L 1972 *High Energy Electromagnetic Processes in Condensed Media* (New York: Wiley Interscience)
[2] Gharibian G M and Yan S 1983 *Rentgenovskoye Perekhodnoye Izluchenie* (Yerevan: Izdatelstvo AN Arm. SSR, in russian)
[3] Ginzburg V L and Tsytovich V N 1990 *Transition Radiation and Transition Scattering* (Bristol: Adam Hilger)
[4] Rullhusen P, Artru X and Dhez P 1998 *Novel Radiation Sources Using Relativistic Electrons* (Singapore: World Scientific)
[5] Potylitsin A P 2009 *Izluchenie Electronov v Periodicheskikh Structurakh* (Tomsk: Izdatelstvo Nauchno-Tekhnicheskoi Literaturi, in russian)
[6] Grigoryan L Sh, Mkrtchyan A H and Saharian A A 1988 *Nucl. Instr. Meth.* **B145** 197
[7] Grigoryan L Sh, Mkrtchyan A R, Khachatryan H F, Arzumanyan S R and Wagner W 2010 *J. Phys.: Conf. Ser.* **236** 012012
[8] Mkrtchyan A R, Parazian V V and Saharian A A 2010 *Mod. Phys. Lett. B* **24** 2693
[9] Mkrtchyan A R, Parazian V V and Saharian A A 2012 *Int. J. Mod. Phys. B* **26** 1250036
[10] Watanable T et al 2002 *Int. J. Appl. Electrom.* **14** 191

Simulations of Transition Radiation from a Flat Target using CST Particle Studio

K V Lekomtsev[1], A S Aryshev[1], P V Karataev[2], M V Shevelev[1], A A Tishchenko[3] and J Urakawa[1]

[1]High Energy Accelerator Research Organization (KEK), 1-1 Oho, Tsukuba, Ibaraki, 305-0801, Japan
[2]John Adams Institute at Royal Holloway University of London, Egham, Surrey, TW20 0EX, UK
[3]National Research Nuclear University "MEPhI", Kashirskoe shosse 31, Moscow, 115409, Russia

E-mail: konstlek@post.kek.jp

Abstract. Numerical simulations of Backward Transition Radiation originating from a flat ideally conducting target inclined at 45 degrees with respect to a charge propagation direction are presented in this paper. The simulation results are compared with the theory for the wave and pre-wave zones. The main characteristics of the simulated radiation distributions are discussed in detail.

1. Introduction

In accelerator physics research various target geometries such as screens and gratings can be used as sources of radiation [1 - 4] and for beam diagnostics [5, 6]. Most of the target geometries have well developed analytical models describing characteristics of the radiation processes such as Transition, Diffraction and Smith – Purcell radiation [7]. However, these models are usually idealised and have applicability limitations when compared with experimental data. Advanced electromagnetic simulations provide an opportunity to perform simulations describing a real experiment in a more efficient manner, e.g. to take into account surrounding hardware of an experimental setup and, for example, identify potential sources of backgrounds.

Computer Simulation Technology (CST) Particle Studio (PS) is a specialist tool for a fast and accurate analysis of charged particle dynamics in 3D electromagnetic fields [8, 9]. The time domain particle-in-cell (PIC) solver of CST PS can perform a fully consistent simulation of particles and electromagnetic fields. It is based on the Finite Integration Method which adopts a self-consistent algorithm combining full Maxwell's equations along with the classical and relativistic equations of motion.

The main goal of this study is to establish a certain level of confidence in the simulations by calculating well understood phenomenon such as Transition Radiation (TR) in CST PS and comparing it with the theory. Once a good agreement is achieved the radiation characteristics for more complex target geometries and radiation phenomena may be considered [4, 10]. In this report we present initial simulations of Backward Transition Radiation (BTR) originating from a flat target positioned at 45 degrees with respect to a charge propagation direction. The calculations are performed for the millimetre wavelength range and a relativistic beam of electrons consisting of a single bunch.

2. PIC solver

In this section we will introduce the key concepts of the field calculation in the CST PS. The PIC solver calculates the development of the fields generated by a charged particle at discrete time samples. The fields are estimated at discrete points whereas the particles are tracked in continuous phase space. The field calculation is performed by the "Leap Frog" updating scheme. This method remains stable if the integration time step does not overcome the maximum usable time step which is related to the minimum mesh step used in the discretization of the considered geometry. The PIC solver is a self-consistent simulation method for particle tracking, therefore it is necessary to interpolate fields to provide feed-back on the particle motion and calculate particle currents for the fields computation [8].

2.1 Finite Integration Theory

The PIC solver is based on MAFIA4, which is a general purpose electromagnetic simulation software package. It is based on the Finite Integration Theory [11]. This theory is a discretisation scheme which defines Maxwell's equations in integral form on a dual non-co-ordinate grid doublet $\{G, G'\}$. The non-co-ordinate discretisation scheme was introduced as an extension of the standard co-ordinate scheme in order to improve the capability of the grid to approximate curved boundaries. A discretisation of R^3 space into the doublet $\{G, G'\}$ results in the so-called Maxwell's Grid Equations [12]:

$$\begin{matrix} -\frac{\partial}{\partial t}\mathbf{b} = \mathbf{Ce} & \mathbf{j} + \frac{\partial}{\partial t}\mathbf{d} = \mathbf{C'h} \\ \mathbf{Sb} = 0 & \mathbf{S'd} = \mathbf{q} \end{matrix} ; \quad (1)$$

where **e** and **h** are electric and magnetic grid voltage vectors, **d**, **j** and **b** are the corresponding facet flux vectors. Discrete permeability, permittivity and conductivity are defined as:

$$\mathbf{b} = \mathbf{D}_\mu \mathbf{h}; \ \ \mathbf{d} = \mathbf{D}_\varepsilon \mathbf{e}; \ \ \mathbf{j} = \mathbf{D}_k \mathbf{e}. \quad (2)$$

The material matrices $\mathbf{D}_\mu$, $\mathbf{D}_\varepsilon$ and $\mathbf{D}_k$ are sparse for a generalised case and diagonal for the dual orthogonal grid-doublet.

3. Problem definition.

A description of the considered geometry is presented in figure 1. In the top view one may see that a bunch of electrons enters the calculation domain along axis z, it travels all the way to the target, hits the target surface and as a consequence BTR is generated in the direction of specular reflection. All planes confining the calculation domain act as free spaces, so the radiation leaves the domain without any additional reflections. The fields are calculated in the spherical wave approximation, i.e. the distance from the radiation source to the observation plane is larger compared to the considered wavelength. It is important to point out the difference between the pre-wave zone ($\gamma^2\lambda > \mathrm{d} >> \gamma\lambda$) and the wave zone ($\mathrm{d} >> \gamma^2\lambda$) of the radiation formation, where d is the distance between the target centre and the observation plane. In case of the wave zone the radiation source is point-like and the radiation wave may be considered as a plane wave. In case of the pre-wave zone the radiation source cannot be represented as point-like any more, because one has to consider the interference between the spherical waves emitted from individual points on the target surface. In the CST context for relativistic and ultra-relativistic beam energies the spherical wave approximation is more applicable for the field calculations in the pre-wave zone, rather than the wave-zone. The BTR distributions are derived from the field intensities calculated at the observation plane for individual mesh cells.

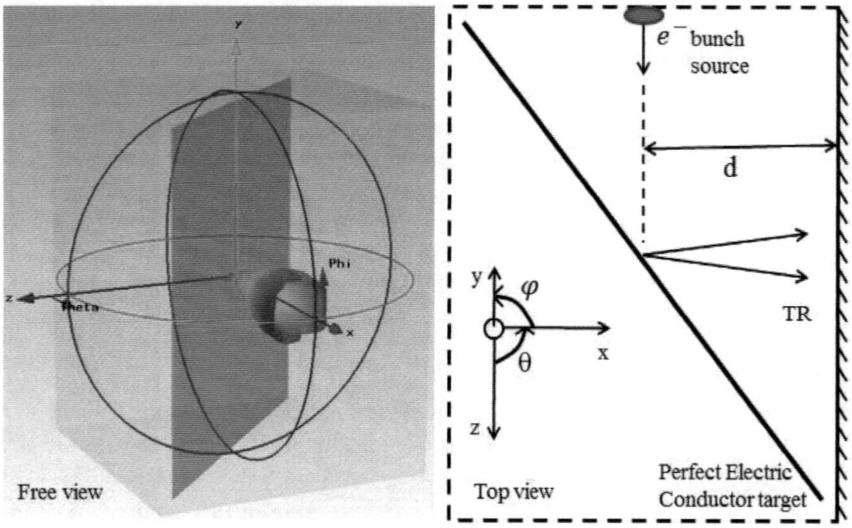

Figure 1. Schematic view of the considered geometry.

The parameters used in the simulation are shown in table 1. The target is made of perfect electric conductor and positioned in vacuum at 45 degrees with respect to the beam propagation direction. Electron bunch is point-like in the transverse plane (size of the source is much smaller than the considered wavelength and the other characteristic dimensions). Longitudinally the charge is distributed over Gaussian with σ shown in table 1. The characteristic dimensions of the target (height and width) are larger than $2.5\times\gamma\lambda$ for the radiation considered in the pre-wave zone and $5\times\gamma\lambda$ for the radiation in the wave zone. This should limit the edge effects in the simulated radiation distributions [13, 14], especially for the wave zone due to the fact that the Fourier components of the transverse particle field are spatially limited within a circle of the radius $\gamma\lambda$ [15].

Table 1. Model parameters.

Parameter	Value, units
Beam energy (γ)	5(wave zone) / 10 (pre-wave zone)
Wavelength (λ)	10 – 12 mm
Target dimensions	300 x 300 x 1 mm
Target material	Perfect Electric Conductor
Calculation domain material	Vacuum
Target inclination angle	45 deg.
Dimensions of the integration domain	(1125/625/500/375) x 310 x 250 mm
Target centre – observ. plane dist. (d)	1000/500/375/250 mm
σ of longitudinal charge distr. (σ_{bunch})	0.03 mm
Bunch charge (Q_{bunch})	0.5 nC

4. Simulation results

In this chapter test simulations of the BTR in the wave and pre-wave zones of radiation formation will be compared with the existing theories. Firstly we shall consider the pre-wave zone. For comparison with the simulation results we shall use the theory developed in [16] using the Kirchhoff's method for a flat target. In this paper analytical distribution of BTR from a finite ideally conducting target inclined at arbitrary angle to a charge trajectory was obtained for the distances from the target $\gamma^2\lambda > d >> \gamma\lambda$.

The CST radiation distributions are calculated at the observation plane and then may be plotted as 2D or 3D graphs. In order to identify the key properties of the BTR in the pre-wave zone three different volumes (calculation domains) were chosen. Each volume corresponded to a certain distance, d, from the target centre to the observation plane.

The radiation distributions plotted for three different distances from the target to the observation plane and the wavelength of 12 mm are shown in figure 2. According to the theory the radiation distribution in the Oxz plane has to demonstrate an asymmetry between peaks [7]. From figure 2 one may conclude that there is a clear asymmetry, well reproduced by the simulations for the distances 125 and 250 mm, also the larger the distance d the smaller the distance between peaks. The latter is in agreement with the theory which states that for the TR distribution in the pre-wave zone the distance between peaks is larger compared to the wave zone [7].

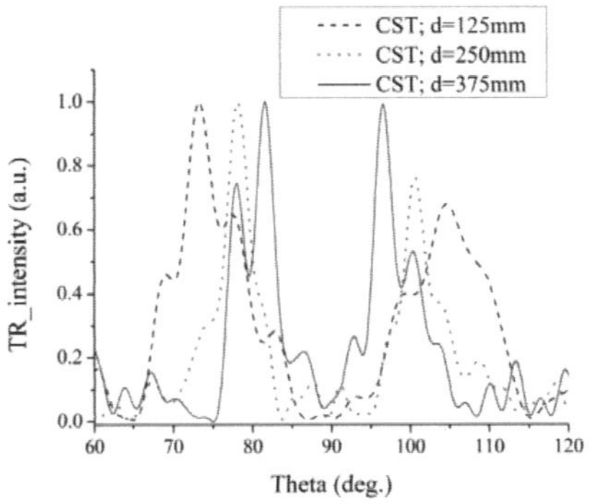

Figure 2. BTR distributions in Oxz plane for three distances from the target to the observation plane, other parameters: $\gamma = 10$ and $\lambda = 12$ mm.

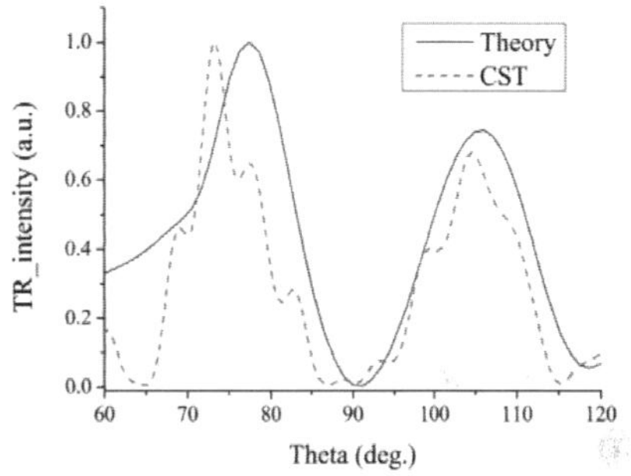

Figure 3. BTR distributions in Oxz plane for $d = 125$ mm, $\gamma = 10$ and $\lambda = 12$ mm.

Figures 3, 4 and 5 show a direct comparison between the simulated BTR distributions and the theoretical ones. For the distances d = 125 mm and d = 250 mm (figure 3 and 4) the asymmetry is in agreement with the theory, however the positions of the left hand side peaks are slightly shifted away from the centre of the distribution. The simulated curves are confined within the theoretical envelopes, but their overall shape is rather approximate. For the distance d = 375 mm the simulation demonstrates large intensity variation, which causes the disagreement in terms of peaks asymmetry as well as overall integrated intensity.

There might be several factors which cause this type of disagreement between the theory and the simulations. The most obvious one may appear due to incorrect meshing. It is important to point out that the CST automatic meshing tool should provide a reasonable result already at the first meshing iteration, especially for such simple geometries. The target edge effects may also contribute in the approximate shape of the simulated BTR distributions. More detailed studies will be required to investigate these issues.

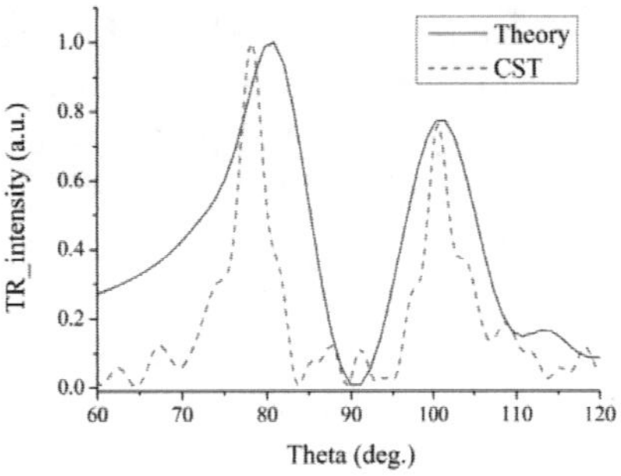

Figure 4. BTR distributions in Oxz plane for $d = 250\,\text{mm}$, $\gamma = 10$ and $\lambda = 12\,\text{mm}$.

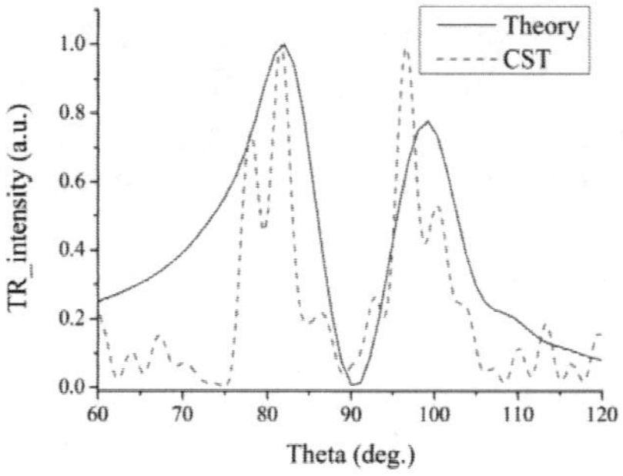

Figure 5. BTR distributions in Oxz plane for $d = 375\,\text{mm}$, $\gamma = 10$ and $\lambda = 12\,\text{mm}$.

Q^2 dependence, where Q is the bunch charge, shall be tested since in this paper we consider coherent radiation which is observed when the radiation wavelength much larger than the bunch length and as a result radiation intensity is proportional to the number of electrons in the bunch squared [7]. Figure 6 shows the TR intensity distribution corresponding to four different bunch charges and normalised by the maximum value of the radiation distribution corresponding to the bunch charge of 2 nC. The shape remains the same for all radiation distributions; there is only change in the maximum intensity. By averaging the distributions in figure 6 over the range of angles from 45 to 135 deg. and normalising obtained values by the value corresponding to the average TR intensity for the bunch charge of 0.5 nC; one may obtain the intensity dependence as a function of a bunch charge. The Q^2 dependence is shown in figure 7, it is confirmed for four values of a bunch charge.

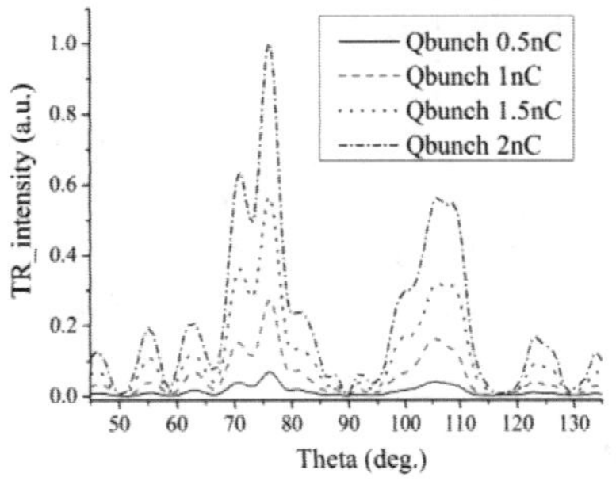

Figure 6. TR intensity distribution in the pre-wave zone ($d = 250\text{mm}$) for four values of bunch charge, other parameters: $\lambda = 15\,\text{mm}$, $\sigma_{\text{bunch}} = 0.03\,\text{mm}$ and $\gamma = 10$.

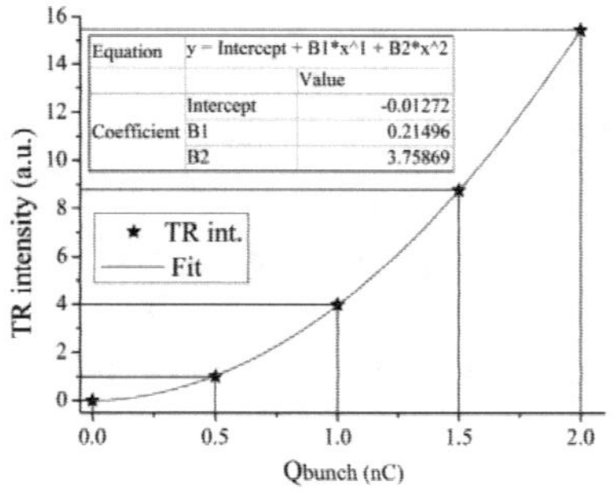

Figure 7. TR intensity integrated over the Theta angular range from 45 to 135 deg. and normalised by the average TR intensity corresponding to the $Q_{\text{bunch}} = 0.5\,\text{nC}$.

Now we shall move on to the study for the wave zone and consider larger vacuum cube corresponding to the distance d = 1 m from the target to the observation plane. The theory developed by Ter-Mikaelian within macroscopic approach in the wave zone for an inclined ideally conducting plate of a limited thickness was used for comparison with the simulations [15, p.223, eq. (25.27)-(25.28)]. Three chosen wavelengths in combination with the electron beam energy of $\gamma = 5$ satisfy the condition for the wave zone $d \gg \gamma^2 \lambda$. Figure 8 demonstrates the comparison of the theory with the simulation for the wavelength of 12 mm. The distribution within the angular region of 60 to 90 degrees demonstrates an oscillating shape confined within the theoretical envelope. By slightly changing the wavelength to 11 mm (figure 9) we observe the change in the simulated radiation distribution. The dash-dot curve reproduces the asymmetry of the distribution, but the shape of the distribution is very approximate in relation to the theoretical curve with a large intensity variation. At the wavelength of 10 mm (figure 10) the simulated radiation distribution demonstrates less oscillating shape, but the left peak is much narrower compared to the theoretical curve.

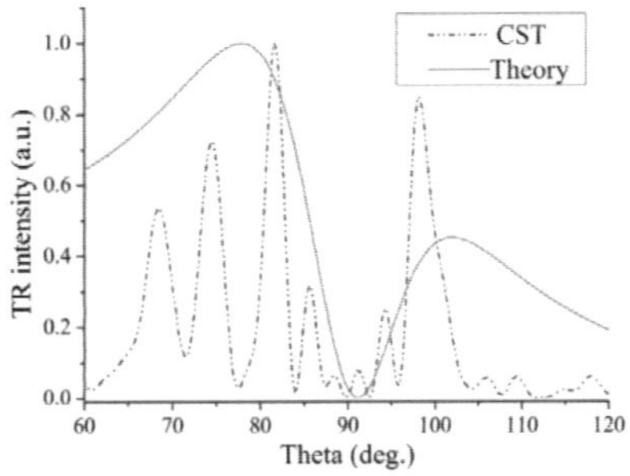

Figure 8. BTR distributions in Oxz plane for $d = 1\,\mathrm{m}$, $\gamma = 5$ and $\lambda = 12\,\mathrm{mm}$ in the wave zone.

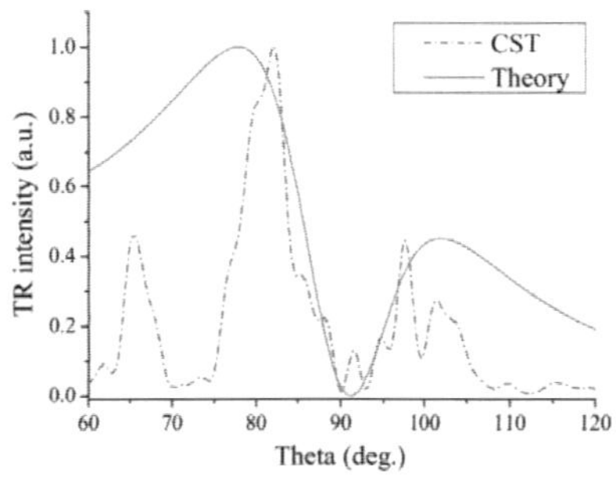

Figure 9. BTR distributions in Oxz plane for $d = 1\,\mathrm{m}$, $\gamma = 5$ and $\lambda = 11\,\mathrm{mm}$ in the wave zone.

Figure 11 shows shift in the distributions depending on a chosen wavelength. The peaks are displaced while the wavelength is changed; however this should not take place for the radiation distribution in the wave zone. On the contrary this behaviour is characteristic of the pre-wave zone. Further studies might be required, but for the moment we may conclude that CST PS does not perform extrapolation of the radiation distribution into the wave zone, despite of the condition $d >> \gamma^2\lambda$.

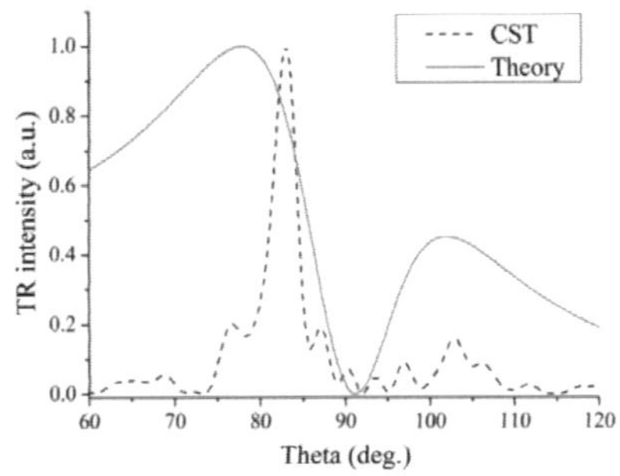

Figure 10. BTR distributions in Oxz plane for $d = 1\,\mathrm{m}$, $\gamma = 5$ and $\lambda = 10\,\mathrm{mm}$ in the wave zone.

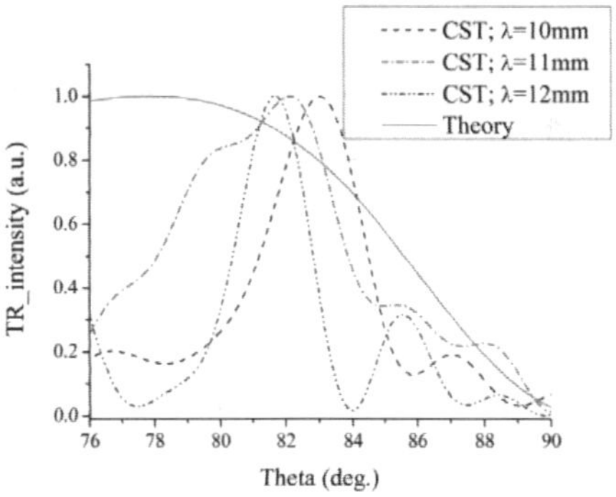

Figure 11. Shift of the simulated radiation distributions in figures 8 – 10.

5. Discussion and conclusions

The BTR from ideally conducting target was simulated and compared with the theory for the wave and the pre-wave zones. A reasonable qualitative agreement is achieved for the radiation distributions in the pre-wave zone. The asymmetry between peaks is in agreement with the theory; however calculation uncertainties or possibly target edge effects cause oscillating behaviour in the obtained distributions. Evaluation of the target edge effects in the wave zone will require additional studies.

The major issues that have to be addressed in the future work are the following. For the simulation results, especially in the wave zone, the distribution shapes are rather approximate and not in full agreement with the theory. The peaks asymmetry in the wave zone is not consistent for all frequencies.

Providing that the identified issues are resolved, this work may be considered to be the first step towards the usage of CST PS for simulations of other phenomena such as Diffraction radiation and Smith - Purcell radiation. This will be very important for the THz program at LUCX facility at KEK [17], where SPR gratings and other more advanced structures for THz radiation generation will be studied experimentally [3, 4].

Acknowledgments

Our work is under the program for "Fundamental Technology Development for High Brightness X-ray Source and the Imaging by Compact Accelerator". This research was supported by Photon and Quantum Basic Research Coordinated Development Program of Japanese Ministry of Education, Culture, Sports, Science and Technology (MEXT) and by JSPS KAKENHI Grant number 23226020 and 24654076. The work of one of the authors (A.T., comparison of the results with existing analytical models) was supported by the Russian Ministry of Education and Science, Agreement 02.A03.21.0005. We would also like to acknowledge RFBR Grant No. 14-02-31642-mol_a.

References

[1] Smirnov A V 2008 *A high performance, fir radiator based on laser driven e-gun* (In: Photonics Research Developments, ed V.P. Nilsson, Nova Science Publishers)
[2] Bashmakov Y A, Bondarenko T V, Komarov D A, Polozov S M, Rashchikov V I, Shchedrin I S, Smirnov A V, Voronkov A V and Tishchenko A A 2012 *Problems of Atomic Science and Technology* **79** (№3) 92
[3] Konoplev I, Aryshev A, Urakawa J, Lekomtsev K, Shevelev M and Seryi A 2013 Linac based broadband source of THz coherent Smith-Purcell radiation *Proc. 38th International Conference IRMMW - THz* (Mainz, Germany)
[4] Ponomarenko A A, Ryazanov M I, Strikhanov M N and Tishchenko A A 2013 *NIM B* **309** 223
[5] Muto T, Araki S, Hamamatsu R, Hayano H, Hirose T, Karataev P, Naumenko G, Potylitsyn A and Urakawa J 2003 *Phys. Rev. Lett.* **90** 104801
[6] Doucas G, Blackmore V, Ottewell, Perry C, Huggard P, Castro-Camus E, Johnston M, Lloyd Hughes J, Kimmitt M, Redlich B and Van der Meer A 2006 *Phys. Rev. STAB* **9** 092801
[7] Potylitsyn A P, Ryazanov M I, Strikhanov M N and Tishchenko A A 2011 *Diffraction radiation from relativistic particles* (Springer)
[8] http://www.cst.com/Content/Products/PS/Overview.aspx
[9] Clemens M, Drobny S, Kruger S, Pinder P, Podebrad O, Schillinger B, Trapp B, Weiland T, Wilke M, Bartsch M, Becker U and Zhang M 1999 The electromagnetic simulation package MAFIA 4 *Proc. ICCEA'99* pp 565 - 568
[10] Lekomtsev K V, Strikhanov M N and Tishchenko A A 2010 *J. Phys.: Conf. Ser.* **236** 012023
[11] Weiland T 1977 *Electronics and Communication (AEU)* **31** 116
[12] Weiland T 1996 Time domain electromagnetic field computation with finite difference methods *International Journal of numerical modelling: electronic networks, devices and fields* **9** 295
[13] Shul'ga N F and Dobrovol'skii S N 2000 *JETP* **90** 579
[14] Tishchenko A A, Strikhanov M N and Potylitsyn A P 2005 *NIM B* **227** 63
[15] Ter-Mikaelian M L 1972 *High-energy electromagnetic processes in condensed media* (Wiley-Interscience)
[16] Karlovets D V and Potylitsyn A P 2008 *NIM B* **266** 3738
[17] Aryshev A, Araki S, Fukuda M, Lekomtsev K, Shevelev M, Urakawa J, Potylitsyn A and Sakaue K 2013 Development of Advanced THz Generation Schemes at KEK LUCX Facility *Proc. Particle Accelerator Society of Japan (Nagoya, Japan)*

Journal of Physics: Conference Series **517** (2014) 012017 doi:10.1088/1742-6596/517/1/012017

X-ray imaging based on small-angle X-ray scattering using spatial coherence of parametric X-ray radiation

Y Hayakawa[1], K Hayakawa[1], M Inagaki[1], T Kaneda[2], K Nakao[1], K Nogami[1], T Sakae[2], T Sakai[1], I Sato[3], Y Takahashi[4] and T Tanaka[1]

[1] Laboratory for Electron Beam Research and Application (LEBRA), Institute of Quantum Science, Nihon University, Narashinodai 7-24-1, Funabashi 274-8501, Japan

[2] School of Dentistry at Matsudo, Nihon University, Sakaecho-Nishi 2-870-1, Matsudo 271-8587, Japan

[3] Advanced Research Institute for the Science and Humanities, Nihon University, 12-5, Gobancho, Chiyoda-ku, Tokyo 102-8251, Japan

[4] Institute of Materials Structure Science, High Energy Accelerator Research Organization (KEK-PF), 1-1 Oho, Tsukuba, Ibaraki 305-0801, Japan

E-mail: yahayak@lebra.nihon-u.ac.jp

Abstract. X-ray imaging based on small-angle X-ray scattering (SAXS) was carried out using the parametric X-ray radiation (PXR) source at the Laboratory for Electron Beam Research and Application (LEBRA) of Nihon University. The experimental setup employed in this novel imaging approach is the same as that employed in diffraction-enhanced imaging (DEI), a kind of X-ray phase-contrast imaging method. In SAXS-based imaging, the image contrast is correlated with the broadening of the rocking curve peak due to the scattering from micron- or sub-micron-sized grains in the sample material. An experiment using the 25.5-keV PXR beam demonstrated that SAXS-based imaging with PXR provides a substantially strong contrast for granular materials despite the extremely low density of the material.

1. Introduction

A tunable and monochromatic X-ray source based on parametric X-ray radiation (PXR), a phenomenon caused by the interaction between a relativistic electron and a crystal medium, was developed at the Laboratory for Electron Beam Research and Application (LEBRA), Nihon University [1, 2]. The PXR beam was obtained using a 100-MeV electron beam from the linac of LEBRA, and the LEBRA-PXR source has a double-crystal system for a wide tunability [3]. A PXR beam from 4 to 34 keV is available using Si(111) or Si(220) planes as the radiator. The current specifications of the LEBRA-PXR source are listed in Table 1. Because the PXR irradiation field is large (100 mm in diameter), X-ray imaging experiments have been the most popular application of the PXR source [4, 5]. In particular, the results of diffraction-enhanced imaging (DEI), which is a sort of X-ray phase-contrast imaging, have demonstrated the high performance of PXR with respect to spatial coherence [6].

Recently, a new imaging technique based on small-angle X-ray scattering (SAXS) has been of special interest for next-generation X-ray imaging. This technique is derived from phase-contrast imaging, and the same setup can be used for measurement. For instance, SAXS-based imaging using a Talbot interferometer has successfully been demonstrated [7]. In principle, information on the SAXS effect due to an object material can be obtained using a DEI measurement system. Therefore, the DEI experiment was carried out to demonstrate the applicability of PXR to SAXS-based imaging.

Table 1. Typical parameters of the linac and the PXR source of LEBRA.

Electron energy	100 MeV (typ.)
Accelerating frequency	2856 MHz (S-band)
Macropulse beam current	120 – 135 mA (typ.)
Macropulse duration	4 – 5 μs (typ.)
Macropulse repetition rate	2–3.5 pps
Average beam current	1 – 5 μA
Beam size on the target	0.5 – 2 mm in dia. (FWHM)
PXR energy range	
Si(111) target:	4.0 – 20.5 keV
Si(220) target:	6.5 – 34 keV
Irradiation field at the X-ray exit	100 mm in diameter
Total X-ray photon rate	$\sim 10^7$ /s @17.5 keV

2. Diffraction-enhanced imaging using PXR

The radiation cone of PXR depends on the energy of the incident electron, as is the case with other radiation phenomena. In the case of the LEBRA-PXR source, the actual aperture angle of the PXR cone-beam extracted from the X-ray exit is determined by the size of the exit window 7.3 m distant from the radiator crystal irradiated with the electron beam, and is estimated to be 13.7 mrad. On the other hand, a PXR beam has spatial chirp, which means a slight energy shift according to the change in the X-ray emitting direction, along the horizontal plane. Combining the cone like divergence and spatial chirp of PXR allows the diffraction of the whole PXR beam when an analyzer crystal for DEI (third crystal) is set up in the $(+, -, +)$ arrangement as shown in Fig. 1. The rocking curves observed at the third crystal have almost the same shapes as in the case of plane-wave-like X-ray beams. The narrow diffraction widths for the PXR beam mean that the DEI measurement system using the LEBRA-PXR source has an angular resolution of 10^{-6} rad despite the cone-beam aperture of the mrad order [8].

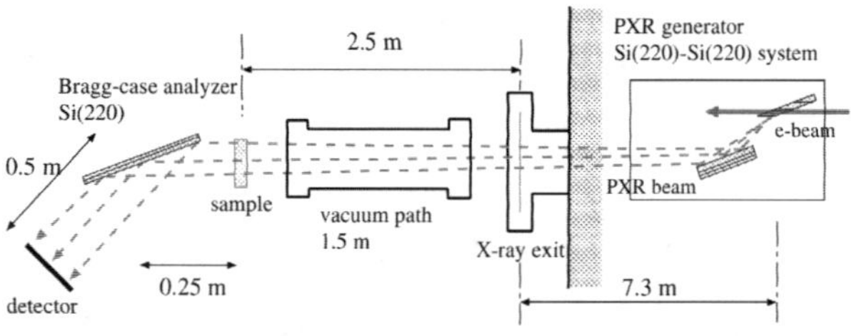

Figure 1. Schematic top view of the setup of the DEI experiment using the LEBRA-PXR source. The same crystal plane as the PXR double-crystal system is used for the analyzer crystal.

There are three considerable interactions between X-rays and materials as the rays pass through the sample: absorption corresponding to X-ray amplitude attenuation; refraction resulting from X-ray phase-shift; and, when the sample material is granular, small-angle scattering due to micro structures of the sample materials. These effects can be evaluated using the result of DEI measurement, which corresponds to the rocking-curve measurement for each pixel of the image detector. Here, we refer to an angular distribution of the X-rays diffracted by the analyzer including the influence of samples as a rocking curve. The X-ray absorption effect is estimated from the reduction of the rocking-curve peak area. The angle of X-ray refraction is measured as the angular shift of the peak center of the rocking curve. Finally, the peak broadening of the rocking curve reflects the effect of SAXS [9]. The angular resolution for the refraction and the scattering depends on the diffraction width of the analyzer crystal.

3. Experimental results of DEI and SAXS-based imaging

To demonstrate SAXS-based imaging using PXR, a DEI experiment for samples of light materials was carried out in the setup shown in Fig. 1. Since the rocking curves for higher-energy X-rays are narrower, the higher energy of X-rays is advantageous in detecting slight refraction and/or scattering of X-rays in DEI experiments. Thus, 25.5-keV PXR from the Si(220) double-crystal system was used, though the photon rate is estimated to be one order of magnitude less than that of the 17-keV PXR from the Si(111) double-crystal system. The X-ray photon density at the sample was roughly estimated to be on the order of 10^2 mm^{-2} s^{-1}.

Figure 2. Image of the samples for the DEI experiment using a 25.5-keV PXR beam. From left to right, an acrylic rod, a styrene-foam rod, and a polystyrene rod with respective diameters of 3 mm, 6 mm, and 3 mm and respective densities are 1.17, 0.16, and 0.986 g/cm^3.

Acrylic-resin, styrene-foam, and polystyrene cylindrical rods were selected as samples and are shown in Fig. 2. The X-ray absorption in such materials is extremely small, especially in styrene-foam. The DEI analyzer crystal was a 5-mm thick Si(220) plate having a dimension of 160 mm × 35 mm and was mounted on a gonio stage having an angular step of 0.4625 μrad. Since the Bragg angle is 7.25° for 25.5 keV, the imaging area was limited to 20 mm × 35 mm.

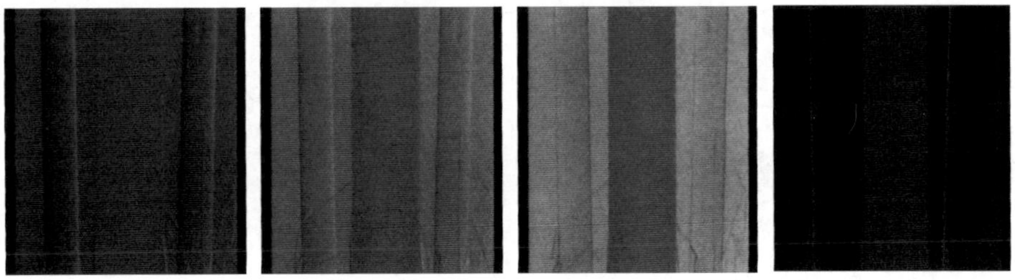

Figure 3. Results of the DEI experiment for the samples as shown in Fig. 2. The DEI contrast varies according to the analyzer angle θ.

In the DEI experiment, 13 DEI images were obtained as a function of the analyzer angle θ, where the peak of the rocking-curve for the direct PXR beam is defined as $\theta = 0$. Figure 3 shows several examples of DEI images taken by an X-ray CCD with a resolution of 24 μm × 24 μm pixels. Because the quantum efficiency of the detector is 10 % at most for 25.5-keV X-rays, 15-min exposure was necessary for each image measurement, and the total measurement time was longer than 3 hours. The contrast of the DEI images strongly depends on the analyzer angle. The rocking curves at the direct beam region and at the styrene-foam region in the DEI images are plotted as shown in Fig. 4. The rocking curve of the styrene-foam region is an order of micro-radian broader than that of the direct beam region. The peak broadening effect seems to be caused by SAXS due to the micro-structure of styrene-foam.

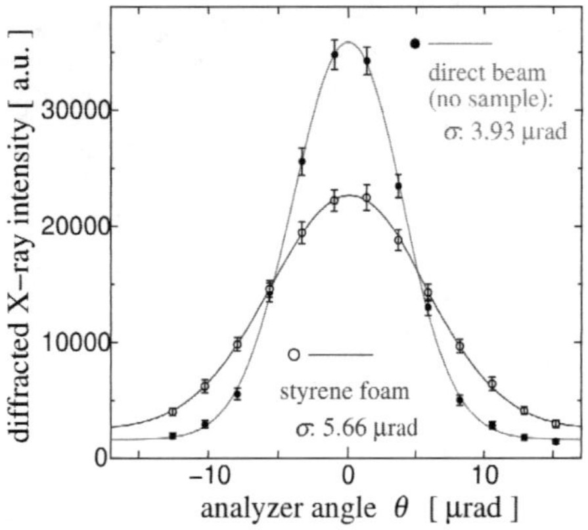

Figure 4. Comparison of the rocking curves at the direct beam region and that at the styrene-foam region.

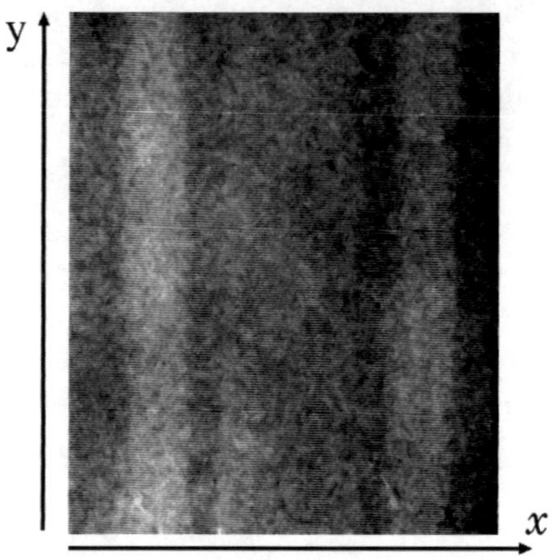

Figure 5. The absorption-coefficient map calculated from the sum of the DEI images

3.1. Absorption-contrast image

The absorption-contrast image is obtained by an integral of the DEI image $I(x, y, \theta)$ with respect to the analyzer angle θ,

$$I_{abs}(x, y) = \int I(x, y, \theta)\, \mathrm{d}\theta = \sum_i I(x, y, \theta_i). \tag{1}$$

The logarithmic function $\ln(I_{abs}/I_0)$ means the absorption coefficient, where $I_0(x, y)$ denotes the intensity map of the direct X-ray beam. Figure 5 shows the absorption-coefficient map calculated from the DEI images. In general, a complex refraction index $n(x, y)$ at the position (x, y) in a two-dimensional coordinate system is expressed as $n(x, y) = 1 - \delta(x, y) + i\beta(x, y)$, where δ and β denote the phase-shift coefficient and the extinction coefficient, respectively. Except in the anomalous dispersion region, both $\delta(x, y)$ and $\beta(x, y)$ are almost proportional to the material density $\rho(x, y)$. Since β is proportional to the absorption coefficient, the absorption-contrast image contains information about the density map $\rho(x, y)$. The contrast, however, is very weak for such light materials, especially styrene-foam.

3.2. Phase-contrast image

The phase-gradient map obtained by the calculation

$$I_{grad}(x, y) = \int \theta\, I(x, y, \theta)\, \mathrm{d}\theta \Big/ \int I(x, y, \theta)\, \mathrm{d}\theta = \sum_i \theta_i\, I(x, y, \theta_i) \Big/ \sum_i I(x, y, \theta_i), \tag{2}$$

corresponds to the partial differential of the phase-shift $\delta(x, y)$ with respect to the x direction, and is also referred to as a refraction-contrast or differential phase-contrast image [10]. The result of the process for the phase-gradient map is shown in Fig. 6. The phase-shift itself is obtained by the integral of the phase-gradient map along x,

$$\delta(x, y) = \int \partial\delta(x, y)/\partial x\, \mathrm{d}x \propto \int I_{grad}(x, y)\, \mathrm{d}x, \tag{3}$$

and Fig. 7 shows the result of the image calculation. The phase map essentially has the same information as the absorption-contrast image, but the sensitivity is drastically improved.

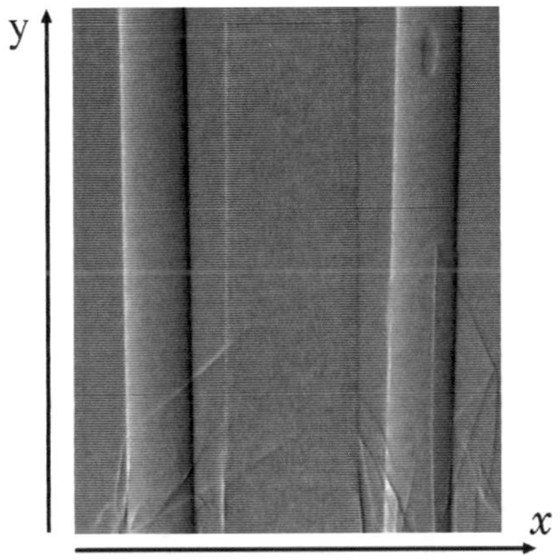

Figure 6. The phase-gradient map obtained as the image calculation by eq. (2).

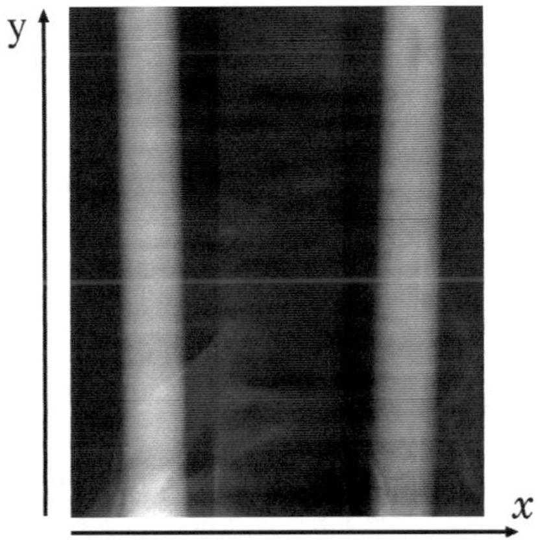

Figure 7. The phase map obtained as the integral of the phase-gradient map shown in Fig. 6.

3.3. SAXS-based imaging

When the X-ray attenuation in the sample is negligible, the peak-height reduction of the rocking curve is almost equivalent with the peak broadening due to the SAXS effect. Thus, there should be several methods to quantify the effect as the image contrast. For simplicity, we defined the contrast due to the SAXS effect as the difference between the image of the analyzer angle $\theta = 0$ and the image of $\theta = 2\sigma$, namely, $I_{saxs}(x, y) = I(x, y, 0) - I(x, y, 2\sigma)$, where σ denotes the standard deviation of the rocking curve for the direct PXR beam and is estimated to be 3.93 μrad in this case. This definition can provide a comparatively strong contrast without complicated calculation. Figure 8 shows the result of this process. Here, the average of $I(x, y, -7.93\ \mu\text{rad})$ and $I(x, y, +8.27\ \mu\text{rad})$ was substituted for $I(x, y, 2\sigma)$, since the angle θ was a discrete parameter. The SAXS-based contrast is extremely sensitive to the styrene-foam region independently of the material density and the sample macro-shape. The contrast is thought to be attributable to the micrometer-ordered micro structures in the sample material, and is essentially equivalent to the visibility contrast in the case of the Talbot-interferometer method [11].

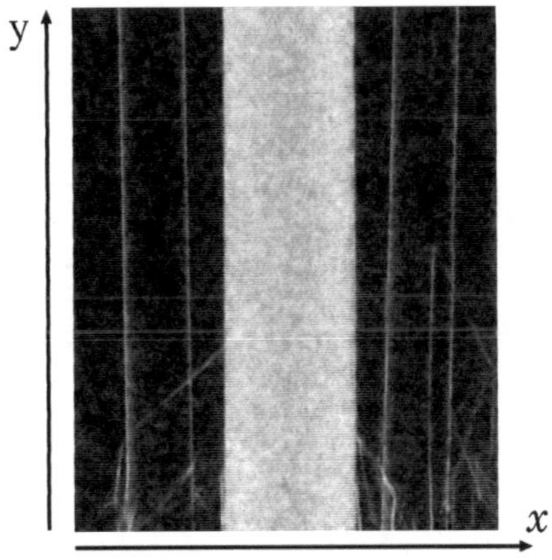

Figure 8. The result of the calculation based on the definition of the SAXS-contrast.

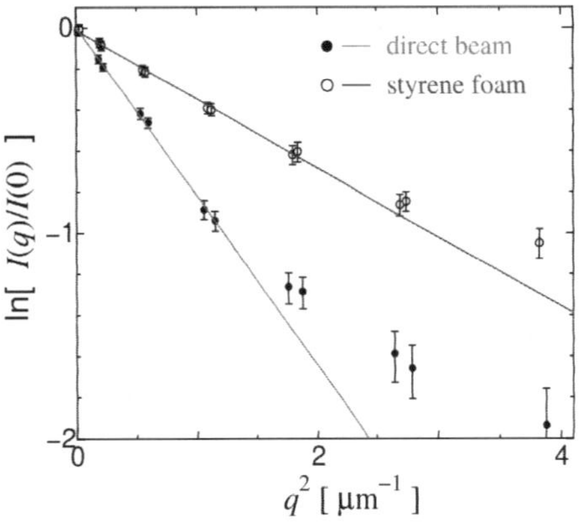

Figure 9. Guinier plots as another expression of curves in Fig. 4.

Here, the distribution of small-angle scattering contains information about the particle-domain size. Figure 9 shows another expression of the rocking curves in Fig. 4, using the variable conversion $q = (4\pi/\lambda)\sin(\theta/2)$, where λ is the X-ray wavelength. Such a logarithmic graph as a function of q^2 is referred to as a Guinier plot in the field of typical SAXS analysis. At the region of small q^2, the SAXS distribution can be approximated by $\exp(-R_g^2 q^2/3)$, where R_g means the inertial radius of particles in the sample material. Thus, the average of R_g is estimated from the slope of the Guinier plot [12]. When DEI data are used, however, the intrinsic diffraction width of the analyzer-crystal itself restricts the detection of the effect due to large particles. The slope of the Guinier plot for the direct beam (red line) corresponds to R_g=1.57 μm, which is the upper limit of the detectable structure size under this condition. From the slope for the styrene-foam region (blue line), R_g is estimated to be 1.00 μm. Since this value contains the effect of the analyzer diffraction width, the actual inertial radius of the styrene-foam structure may take on a value between 1.00 and 1.57 μm. Nevertheless, we can conclude that the SAXS effect in DEI is sensitive to structures much smaller than the 24 μm pixel size of the image sensor. For a more exact estimation, the sample thickness has to be optimized and a lot of measurement points are necessary.

4. Summary

Combining the cone-like divergence and the spatial chirp of PXR allows DEI using a PXR beam, which is sensitive to the refraction and the small-angle scattering of X-rays passing through the sample materials with an angular resolution of 10^{-6} rad order. The DEI experiment using PXR successfully demonstrated that SAXS-based imaging is sensitive to micro-structures of sample materials smaller than the pixel size of the image sensor. The results proved that the PXR beam has a sufficient spatial coherence to detect scattering angles in the micro-radian range.

In addition to the sensitivity to granular materials with low densities, SAXS-based imaging could be used for the quantitative analysis of material structures of the sub-micron order. Thus, the novel imaging technique would be a useful method for material science and fundamental medical research.

Acknowledgments

This work was partly supported by MEXT. KAKENHI (24560069, 24651105 and 25286087) and by Grants-in-Aid for Scientific Research, Nihon University (S12-19, Takashi Kaneda).

References

[1] Hayakawa Y, Sato I, Hayakawa K and Tanaka T 2005 *Nucl. Instrum. and Meth.* B **227** 32
[2] Hayakawa Y, Sato I, Hayakawa K, Tanaka T, Mori A, Kuwada T, Sakai T, Nogami K, Nakao K and Sakae T 2006 *Nucl. Instrum. and Meth.* B **252** 102
[3] Tanaka T, Hayakawa K, Hayakawa Y, Kuwada T, Sakai T, Nakao K, Takahashi Y, Nogami K, Inagaki M and Sato I 2010 *AIP Conf. Proc. 1234: SRI 2009, 10TH INTERNATIONAL CONFERENCE ON RADIATION INSTRUMENTATION* vol 1234 (Melbourne, Australia) pp 587–590
[4] Hayakawa Y, Hayakawa K, Inagaki M, Kuwada T, Nakao K, Nogami K, Sakae T, Sakai T, Sato I, Takahashi Y and Tanaka T 2008 *Nucl. Instrum. and Meth.* B **266** 3758
[5] Hayakawa Y, Hayakawa K, Inagaki M, Kaneda T, Nakao K, Nogami K, Sakae T, Sakai T, Sato I, Takahashi Y and Tanaka T 2013 *Nucl. Instrum. and Meth.* B **309** 230
[6] Takahashi Y, Hayakawa Y, Kuwada T, Tanaka T, Sakae T, Nakao K, Nogami K, Imagaki M, Hayakawa K and Sato I 2012 *X-Ray Spectrom.* **41** 210–215
[7] Pfeiffer F, Bech M, Bunk O, Kraft P, Eikenberry E F, Bronnimann C, Grunzweig C and David C 2008 *Nature Materials* **7** 134
[8] Hayakawa Y, Takahashi Y, Kuwada T, Sakae T, Tanaka T, Nakao K, Nogami K, Inagaki M, Hayakawa K and Sato I 2013 *Journal of Instrumentation* **8** C08001
[9] Levine L E and Long G G 2004 *Journal of Applied Crystallography* **37** 757
[10] Koyama I, Momose A, Wu J, Lwin T T and Takeda T 2005 *Jpn. J. of Appl. Phys.* **44** 8219
[11] Yashiro W, Harasse S, Kawabata K, Kuwabara H, Yamazaki T and Momose A 2011 *Phys. Rev. B* **84** 094106
[12] Glatter O and Kratky O (eds) 1982 *Small angle X-ray scattering* (London: Academic Press)

Journal of Physics: Conference Series **517** (2014) 012018 doi:10.1088/1742-6596/517/1/012018

Spectral distribution in the reflection of parametric X-rays

Yu A Chesnokov[1], A V Shchagin[2,3], N F Shul'ga[2], A S Kubankin[3], A P Potylitsyn[4], A S Gogolev[4], S R Uglov[4], Yu M Cherepennikov[4], P Karataev[5]

[1] Institute of High Energy Physics, Moscow Region, Protvino, Russia;
[2] National Science Center "Kharkov Institute of Physics and Technology", Kharkov, Ukraine;
[3] Belgorod State University, Belgorod, Russia;
[4] Tomsk Polytechnic University, Tomsk, Russia;
[5] John Adams Institute at Royal Holloway College, University of London, Egham, UK;

E-mail: kubankin@bsu.edu.ru

Abstract. The parametric X-ray radiation (PXR) spectra are measured on condition when the angular size of PXR cone is smaller than the angular resolution of the experiment. The PXR is generated under interaction of 50GeV proton beam with silicon crystal in Bragg geometry. The comparison of experimental data with results of developed theoretical model is presented and discussed.

1. Introduction

Parametric X-ray radiation (PXR) appears due to the Bragg diffraction of a fast charged particle Coulomb field on a system of atomic planes in a crystal [1-5]. The PXR of relativistic charged particles in crystals has been studied theoretically and experimentally (see, for example, reviews of some experimental research in Refs. [1, 2]). In particular, the PXR excited by relativistic protons in crystals has been observed in Refs. [3 – 5].

The PXR spectrum in symmetrical diffraction geometry consists of peaks corresponding to different orders of the diffraction. The shape and spectral width of peaks are determined by angular sizes of detector, and transverse cross-section of incident particle beam, and the incident beam divergence. These parameters determine the angular resolution in PXR experiments. The shape of the PXR spectral peaks can be substantially modified when the angular resolution is comparable with the PXR cone angle.

The presented research is devoted to the study of PXR spectral distribution under conditions when the experimental angular resolution is comparable with the angular size of the PXR reflection. The experiment was performed at the proton accelerator U70 of the Institute of High Energy Physics, Protvino, Russia. The methods for PXR measuring are developed for conditions of intense background radiation of fast charged particles and X-rays. The measurement results and calculations are in good accordance to each other. The PXR model is developed on the base of kinematical theory of PXR [6]. The relativistic system of units $\hbar = c = 1$ is used in the theoretical part of the work.

It is should be noted that the presented results are the preparation stage for the comprehensive research of effects of focusing of the PXR emitted from a bent crystal [7].

2. Calculations

The kinematical model of PXR [6] is used below in calculations of the PXR properties, in spite that is does not describe the asymmetry of the yield in the PXR reflection [8]. According to calculations the position of the PXR peak is about at energy $\omega \approx 10.9$keV for the observation angle of the beam 20° that corresponds to the present experimental setup geometry which is described in details below. The mechanisms of the transition radiation and bremsstrahlung are realized also in process of protons interaction with the target. The yield of the transition radiation is negligible in condition $\omega >> \gamma\omega_p \approx$ 1.8 keV, where $\gamma\omega_p$ determines the threshold of intense transition radiation manifestation (ω_p is the plasma frequency of the target, γ is the Lorentz factor of an emitting particle). This circumstance means that the contribution of the transition radiation is negligible. The contribution of the bremsstrahlung is also negligibly because of its small intensity for heavy particles.

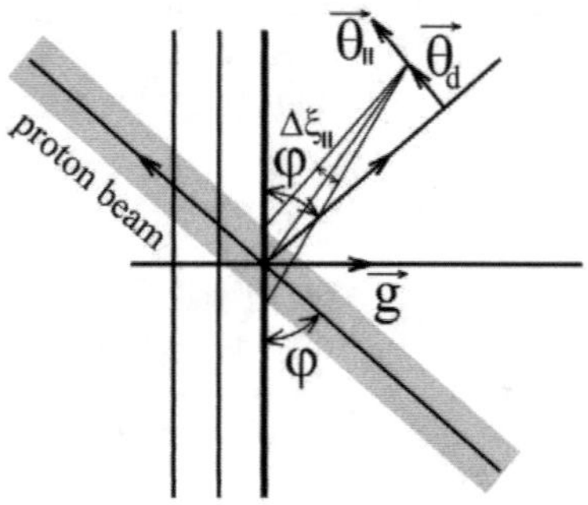

Figure 1. The radiation process geometry.

The radiation process geometry for the finite transverse size of the particle beam can be considered as presented in figure 1, where φ is the incidence angle of the particle beam on the crystallographic plane determined by reciprocal lattice vector **g**, the angles θ_d and $\theta_{\|}$ determine the detector position and observation angle in the incidence plane relative to the direction of mirror reflection of the particle beam, $\Delta\xi_{\|}$ is the angular size of the beam cross section in the incidence plane. The variables $\theta_{\perp}$ and $\Delta\xi_{\perp}$ mean the same that $\theta_{\|}$ and $\Delta\xi_{\|}$ but at transverse direction (normally to the incidence plane). The spectral-angular distribution of PXR for a single proton can be presented in a convenient form for the future calculations

$$\omega\frac{d^3N}{d\omega\, d\theta_{\|}\, d\theta_{\perp}} = \frac{e^2\omega_g^4}{\pi g^2}\frac{1}{\omega\chi_0''(\omega)}\frac{\tan(\varphi)}{\omega_B}\mathrm{F}\left(\theta_{\|},\theta_{\perp},\omega,\varphi\right)\delta\left[\theta_{\|}+\left(1-\frac{\omega}{\omega_B}\right)\tan(\varphi)\right], \quad (1)$$

$$\mathrm{F}\left(\theta_{\|},\theta_{\perp},\omega,\varphi\right) = \frac{\theta_{\perp}^2+\theta_{\|}^2\cos^2(2\varphi)}{(\gamma^{-2}+\omega_0^2/\omega^2+\theta^2)^2},$$

where $\varepsilon(\omega)=1-\omega_p^2/\omega^2+i\chi''(\omega)$ is the average dielectric permeability of the crystal, $\omega_B = g/2\sin(\varphi)$

is the Bragg frequency, $\omega_g^2 = \omega_p^2 (F(g)/Z)(S(\vec{g})/N_0)\exp\left(-g^2 u_T^2\right)$, $F(g)$ is the formfactor of an atom, Z is the number of electrons in the atom, $S(\mathbf{g})$ is the structure factor of an elementary cell containing N_0 atoms, u_T is the mean-square amplitude of thermal vibrations of atoms, other variables are determined in figure 1.

The PXR is realized in conditions of large transverse size of proton beam and distribution (1) must be averaged by the detector observation angles for each proton in the beam. The averaging can be performed taking into account that angular size of PXR cone is substantially larger than the angular size of the detector aperture determined by the ratio d/L, where d is the aperture size, L is the distance from the detector to the point of PXR process. The maximum of the PXR angular distribution is realized at angles $\pm\sqrt{\gamma^{-2}+\omega_0^2/\omega^2} = \pm 19$ mrad in the given geometry, whereas $d/L \approx 9.5$ mrad. This feature allows to substitute the integration of the PXR spectral-angular distribution over angular detector size $\Delta\theta_\perp$ by multiplication of the distribution on $\Delta\theta_\perp$.

The averaged spectral-angular distribution of PXR can be presented in the form

$$\left\langle \omega \frac{\mathrm{d}^3 N}{\mathrm{d}\omega \mathrm{d}\theta_\parallel \mathrm{d}\theta_\perp} \right\rangle = \frac{e^2\omega_g^4}{\pi g^2} \frac{1}{\omega\chi_0''(\omega)} \frac{1}{\omega_B \cot(\varphi)} \frac{\Delta\theta_\perp}{\Delta\xi_\perp \Delta\xi_\parallel} \times$$
$$\int_{-\frac{\Delta\xi_\parallel}{2}}^{\frac{\Delta\xi_\parallel}{2}} \sigma\left(\frac{\Delta\theta_\parallel}{2} + \xi_\parallel - \theta_d + \theta^*\right) \sigma\left(\frac{\Delta\theta_\parallel}{2} - \xi_\parallel + \theta_d - \theta^*\right) \int_{-\frac{\Delta\xi_\perp}{2}}^{\frac{\Delta\xi_\perp}{2}} \mathrm{F}\left(\theta^*, \xi_\perp, \omega, \varphi\right) d\xi_\parallel \, d\xi_\perp \,. \qquad (2)$$

where $\theta^* = \left(1-\omega/\omega_B\right)\tan(\varphi)$, $\Delta\theta_\perp$ and $\Delta\theta_\parallel$ are the transverse and longitudinal angular sizes of the detector aperture, $\sigma(x)$ is the Heaviside step function. The expression (2) is valid for the rectangular beam cross-section. For other shapes of the cross-section the averaging must be performed with a function of cross-section current density.

The shape of PXR spectral peak can be modified substantially on condition when $\Delta\xi_\parallel$ size of order $\sqrt{\gamma^{-2}+\omega_0^2/\omega^2}$. The figure 2 presents dependences of the shape on the angular beam size $\Delta\xi_\parallel$. The curve 2 corresponds to case when the angular sizes of the beam cross section in the incidence plane equal to PXR cone angular size ($\Delta\xi_\parallel = 2\sqrt{\gamma^{-2}+\omega_0^2/\omega^2}$). It is shown the modification of the shape of PXR peak to two-hump shape. The size $\Delta\xi_\perp$ determines the magnitude of the effect and it is decreased if $\Delta\xi_\perp$ increases, that is presented in figure 3.

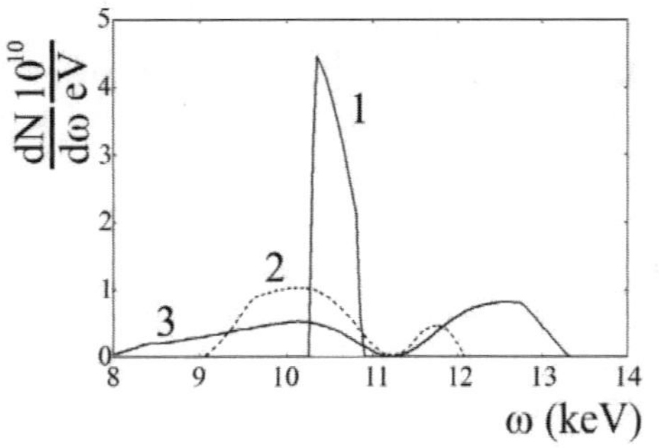

Figure 2. The dependence of PXR on radiating particle beam size $\Delta\xi_{\|}$. The curves are calculated at parameters $\Delta\theta_{\perp}$=8.4 mrad, $\Delta\theta_{\|}$ = 8.4 mrad, φ=170 mrad, $\Delta\xi_{\perp}$=5 mrad, γ=54, θ_d=-10 mrad. 1 – $\Delta\xi_{\|}$=1 mrad, 2 – $\Delta\xi_{\|}$ =38 mrad, 3 – $\Delta\xi_{\|}$=76 mrad.

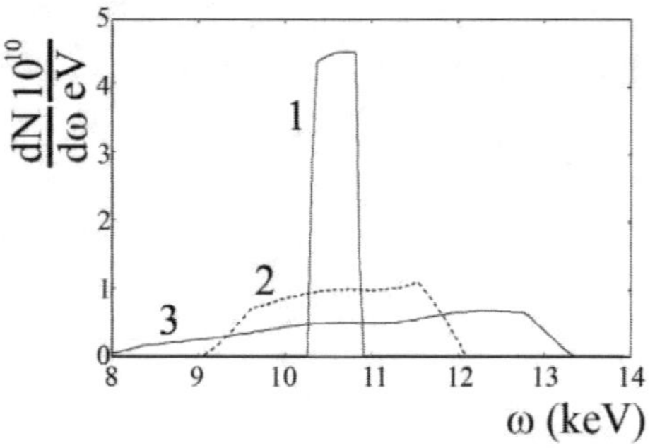

Figure. 3. The influence of angular transverse size of the beam cross-section on PXR. The parameters are the same that for the figure 2 except $\Delta\xi_{\perp}$ = 92 mrad.

The proton beam dimensions (FWHM) in the setup geometry were about 25 mm in the vertical direction and 10 mm in the horizontal direction that corresponds to angular sizes of beam observation $\Delta\xi_{\perp}$=100 mrad and $\Delta\xi_{\|}$=40 mrad in geometry of the calculations for (2). Both angular sizes exceed typical angular size of the PXR manifestation for 50 GeV protons.

3. Experimental

The experiment has been performed at extracted 50 GeV proton beam in accelerator U70. The setup is presented schematically in figure 4. The beam was extracted from U70 orbit by a Si crystal bent on 90 mrad. The intensity of the extracted beam was about 10^7 protons per pulse. The beam time structure

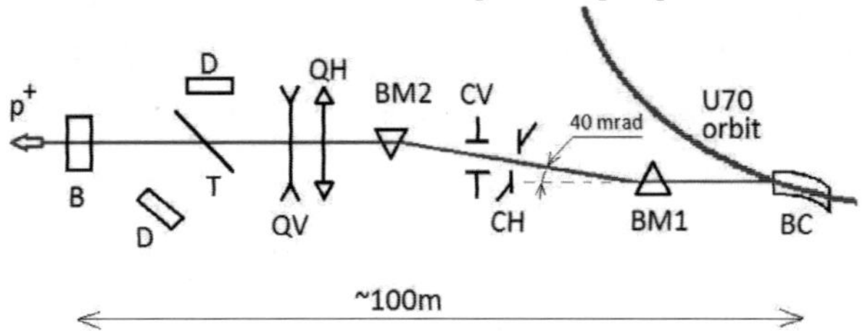

Figure 4. The experimental setup.

was one pulse duration 1s every 10s. The extraction is well described in [9]. The extracted beam is parallel shifted by the 6-m long bending magnets BM1 and BM2 for the beam clearing of the background generated during extraction. The copper collimators CH and CV of length 75 cm form a rectangular shape of the proton beam. The 2-m long quadrupole lenses (QH and QV) provide the beam divergence of about 100 μrad in both horizontal and vertical planes. The PXR was produced under interaction of the formed beam with the 300 μm thick Si (111) target T, mounted on a goniometer with step of $2.18\cdot10^{-5}$ rad. The goniometer, the target and the X-ray detector were assembled in air. The proton beam propagated from

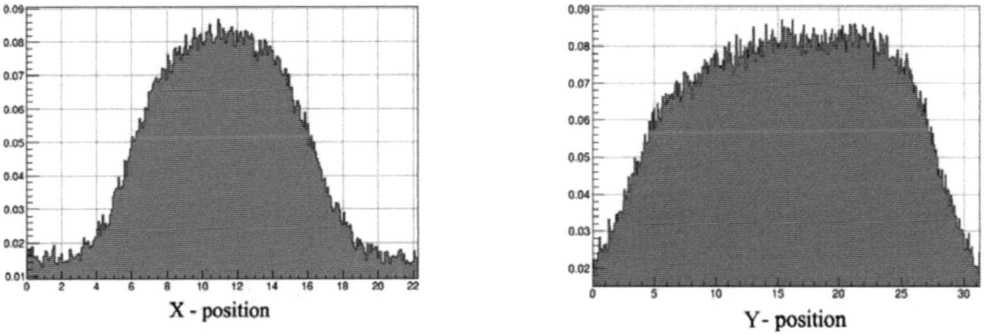

Figure 5. The proton beam profile in the vicinity of the target.

the vacuum tube to air through a Mylar foil. The distance from the Mylar foil to the target was about 1 m. The proton beam profile measured in the vicinity of the target is shown in figure 5. The measurements were performed using the Dosimetry Film EBT2 and a scanner with a resolution of 20 μm.

The number of protons in the beam is counted by a fast scintillator counter B. The energydispersive PIN detector XR-100CR and a digital pulse processor PX4 (both from Amptek) were used for PXR and background measurements. The detector has 500 μm thick Si crystal with the area of 6 mm^2 equipped with a 0.3 mil Be window. An inner collimator of detector crystal has the aperture of 4.4 mm^2. The collimator is opaque for soft X-rays but transparent for hard X-rays and relativistic particles. Due to intense background radiation we applied the shortest available in the PX4 peaking time 0.8 μs. The energy resolution of the spectrometer is about 260 eV at X-ray energy 5.9 keV. The applying of the small gain in spectrometer allows observing the intense background occurring as a consequence of the energy losses of charged particles in the detector crystal.

At the first stage, the radiation of the Cu characteristic lines at energies 8.04keV and 8.93keV was measured to test the spectroscopy equipment characteristics and to measure the background level relative to Cu characteristic radiation. A 20 μm thick Cu foil was used as the target. The surface of the Cu foil was oriented at angle 45° relative to the proton beam. The detector was installed at angle 90° relative to the proton beam at the 60cm distance from the target. The measurement showed a continuous background appearing in soft and hard X-ray regions of PX4 multichannel analyzer. The soft background part corresponds to radiation processes at interaction of primary protons and secondary charged particles with a medium. The spectrum of this part is in the range up to 10keV. The background mechanism at the hard region corresponds to ionization losses of charged particles in the detector crystal. The intensity of this background is depended on the distance between the detector and the proton beam and in general does not depend on the target material.

The Si (111) target 300 μm thick was used for PXR observation. The detector D was installed at observation angle 19.9° at distance 25 cm away from the target. The measurement accuracy of the incidence angle and the distance was in the range ±1.2° and ±15mm that is explained by the large size of the particle beam (these estimations for the transverse FWHM of the beam in incidence plane 10 mm). The angle and position of the detector installation were chosen for PXR observation on condition of weak background manifestation that allows to measure clear signal in PXR spectral range. The Si target was mounted on the goniometer and preliminarily aligned for observation of the PXR in Bragg geometry. The experimental setup for the PXR studies is shown in figure 6.

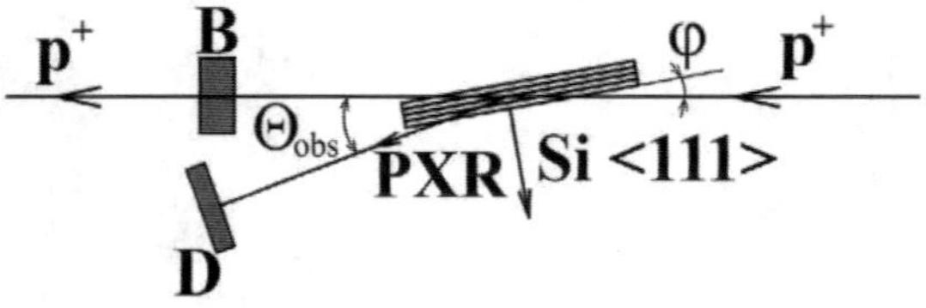

Figure 6. The setup for the PXR measurements.

4. Results and discussion

The PXR spectrum was measured for two orientations of the crystal. The first measurement was done for φ=9.95° and Θ_{obs}=19.9°. The comparison of the measurement with calculation is shown in figure 7. The dotted graph includes the possible correction for the observation and incidence angles in frames of the measurement accuracy mentioned above (φ=11.2° and Θ_{obs}=22.4°).

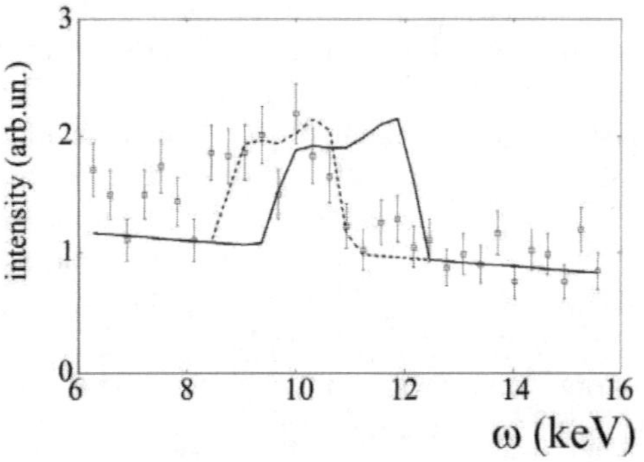

Figure 7. The comparison of the PXR measurement with calculation for the angles Θ_{obs}=19.9°, φ=9.95° and Θ_{obs}=22.4°, φ=11.2° (dotted graph).

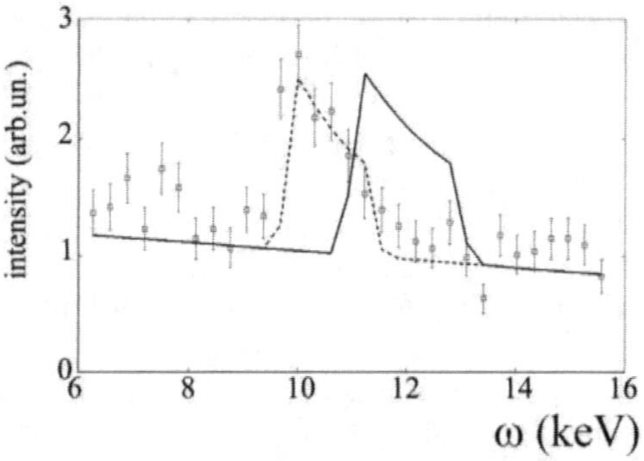

Figure 8. The comparison of the PXR measurement with calculation for the angles Θ_{obs}=19.9°, φ=11.45° and Θ_{obs}=22.4°, φ=12.7° (dotted graph).

The second measurement was done at the same observation angle Θ_{obs}, but for φ=11.45°. The result is presented in figure 8. The dotted graph is in condition of Θ_{obs} correction like to the dotted graph in figure 7 (Θ_{obs}=22.4°, φ=12.7°).

In present paper we demonstrated both theoretically and experimentally that the shape of the PXR spectral distribution can posses a complicate form if the experimental angular resolution exceeds the angular size of the PXR reflection. In particular, the distribution can have two maxima and the energy in the maxima can be different from the PXR energy calculated for average observation angle. This situation is completely different from the case when the experimental angular resolution is well below of the angular size of the PXR reflection and the PXR spectral distribution is quasi-monochromatic.

Presented here preliminary calculated and experimental data on the PXR spectral distributions are in satisfactory agreement one to another at deviation of the observation angle for 2.5°. But we have to note poor statistics and high level of radiation background in the experiment as well as that the asymmetry of the yield in the PXR reflection was not taken into account in calculations. In more detail research one has to obtain the experimental data with reduced level of background radiation, and with better statistics, and with more exactly determined experimental geometry and compare results of measurements to calculations performed with proper account of the asymmetry of the yield in the PXR reflection [8].

Acknowledgements
The joint research is supported by the directorates of the IHEP and KIPT. The activity of participants from Protvino and Kharkov was supported by joint Russian-Ukrainian Grants by the RFBR Grant № 13-02-90434 (Russia) and by the SFFR Grant № F53.2/107 (Ukraine). The participants from Tomsk acknowledge the partial support by the Ministry of education and science of the Russian Federation grant № 14.B37.21.0912 and the RFBR Grant № 13-02-31090.

References
[1] Shchagin A V, Maruyama X K 1997 *Parametric X-rays Accelerator-Based Atomic Physics Techniques and Applications (AIP Press, New York)* ed S.M. Shafroth and J.C. Austin *279*
[2] Potylitsyn A P 2011 *Electromagnetic Radiation of Electrons in Periodic Structures* (Springer-Verlag, **243**)
[3] Afanasenko V P et al. 1992 *Phys. Lett.* A **170** *315*
[4] Adishchev Yu N, Artemov A S, Afanasiev S V et al. 2005 *JETP Lett.* **81** *241*
[5] Scandale W et al. 2011 *Phys. Lett.* B **701** *180*
[6] Feranchuk I, Ivashin A 1985 *J. Phys.* (Paris) **46** *1981*
[7] Shchagin AV 2004 *JETP Lett.* **80** *469*
[8] Shchagin A V, Takabayashi Y 2013 *Nucl. Instr. and Meth. in Phys. Res.* B **309** *198*
[9] Arkhipenko A A et al. 2008 *JETP Lett.* **88** *229*

Quantum Features of PXRC Angular Distributions From Relativistic Channeled Electrons in a Crystal

K B Korotchenko, Yu L Pivovarov

Department of Theoretical and Experimental Physics, National Research Tomsk Polytechnic University, 30, Lenin Avenue, Tomsk, 634050, Russia

E-mail: korotchenko@tpu.ru

Abstract. We predict quantum features in angular distributions of parametric X-radiation from channeled relativistic electrons (PXRC). The effect is connected with the number of quantum states of channeled electrons, form-factors of the transverse quantum channeling states and initial populations of these quantum states. The main motivation of this work is theoretical prediction for the future experiment at the SAGA-LS facility.

1. Introduction

Recent experiments [1, 2, 3] on studies of angular distributions of parametric X-radiation from 255 MeV channeled electrons (PXRC) in a Si crystal showed some differences between of that distribution and standard angular distribution of parametric X-radiation (PXR) from non-channeled electron. According [1], this difference is connected with manifestation of two quantum effects: a) form-factors of the transverse quantum states of channeled electrons; b) initial populations of these quantum states. The band structure of transverse quantum states in a periodic planar channeling potential sufficiently influences the form-factors and initial populations of these states. As a result, the angular distribution of PXRC changes with appearance of every new quantum channeling state following increase of the electron beam energy.

2. Basic equations

According [1], the formula for angular distribution of PXRC intensity from planar channeled electrons captured into a transverse quantum state with a number n reads:

$$I^n_{\rm PXRC} = \frac{d^3 N_{nn}}{d\theta_x d\theta_y dz} = I_{\rm PXR}\,|F_{nn}(q)|^2, \tag{1}$$

here $q = (\boldsymbol{\kappa} - \mathbf{g})_y = \omega_B\theta_y/c$ and $I_{\rm PXR}$ is the standard angular distribution of PXR intensity near the Bragg direction calculated in a two-wave approximation:

$$I_{\rm PXR} = \frac{\alpha\omega_B}{16\pi c\sin^2\theta_B}\left[\frac{\theta_x^2}{1+W_\pi^2} + \frac{\theta_y^2}{1+W_\sigma^2}\right], \tag{2}$$

θ_x, θ_y – the angular coordinates counted from the Bragg direction $\boldsymbol{\kappa}_B$ and θ_B – the Bragg angle defined by [4] $\sin\theta_B = |\mathbf{v}_\parallel \mathbf{g}|/(\mathrm{v}_\parallel g)$ – see, in figure 1, $\hbar\omega_B = 2\pi\hbar c/d$ – energy of PXRC photon (we consider the first allowed order of the Bragg reflection), d – spacing between the channeling planes, $\boldsymbol{\kappa}$ – the wave vector of PXRC photon and $\mathbf{g}$ – reciprocal lattice vector (see, in figure 1).

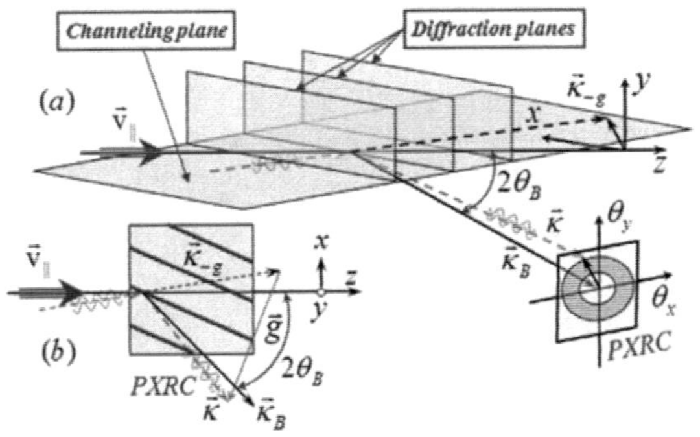

Figure 1. The angular coordinates θ_x and θ_y, describing angular distribution of PXRC near the Bragg direction $\boldsymbol{\kappa}_B$, and mutual arrangement of the vector of longitudinal velocity $\mathbf{v}_{\parallel}$, wave vector $\boldsymbol{\kappa}$ of PXRC photon and $\mathbf{g}$ – reciprocal lattice vector: (a) 3D view; (b) the schematics of mutual arrangement of the vectors.

The important quantity entering Eq.(1) is the form-factor of the transverse quantum channeling state with the number n:

$$F_{nn}(q) = \int\limits_{-d/2}^{d/2} \varphi_n^*(y) exp(-iqy)\varphi_n(y)dy, \tag{3}$$

Here, $\varphi_n(y)$ is the wave function of channeled electron in a quantum state with the number n. Other quantities entering Eq.(2) are [5, 6]:

$$\begin{aligned} W_\tau &= \frac{1}{2|\chi_g|\Gamma_\tau}\left(R - \frac{|\chi_g|^2\Gamma_\tau^2}{R}\right), \\ R &= \theta_x^2 + \theta_y^2 + \theta_{kin}^2,\ \theta_{kin}^2 = \gamma^{-2} + |\chi_0|, \\ \Gamma_\pi &= \cos 2\theta_B,\ \Gamma_\sigma = 1,\ \tau = (\pi, \sigma), \end{aligned} \tag{4}$$

here $\gamma = E/mc^2$ is the relativistic factor of electron, χ_0, χ_g – the Fourier components of the local electric susceptibility of a crystal.

The angular distribution of PXRC from electrons planar channeled in a crystal is calculated using the following formula [1, 2, 3]:

$$\begin{aligned} I_{\text{PXRC}} &= \sum_n^N I_{\text{PXRC}}^n P_n(\theta_0) \\ &= I_{\text{PXR}} \sum_n^N P_n(\theta_0)|F_{nn}(q)|^2, \end{aligned} \tag{5}$$

where N is number of bound channeling states and

$$P_n(\theta_0) = \frac{1}{d}\left|\int_{-d/2}^{d/2} e^{\mathrm{i}p\theta_0 y/\hbar}\varphi_n(y)dy\right|^2, \tag{6}$$

is the the initial population [7] of the n-th quantum state depending on the angle of incidence θ_0 of electron beam with respect to the channeling planes (p is initial momentum of electrons). Looking at Eqs.(3, 5, 6) one may conclude that the angular distribution of PXRC depends on the number of quantum channeling states, on the wave functions of these states and on the angle of incidence θ_0.

Let us further consider the simplest case of PXRC, when the angle of incidence $\theta_0 = 0$. In this case, only sub-barrier (bound channeling states) are populated. As is known [7], the number of those channeling states increases proportionally $\sqrt{\gamma}$. When the beam energy increases, the new quantum channeling states appear and contribute to the sum in Eq.(5), every with its own form-factor F_{nn} and population $P_n(\theta_0)$. Therefore, an appearance of every new quantum channeling state will lead to a change in the PXRC intensity.

3. Beyond the separate plane approximation

The "true" wave functions for (220) planar channeling in Si crystal used in further numerical calculations have been obtained using the method described in [6, 8].

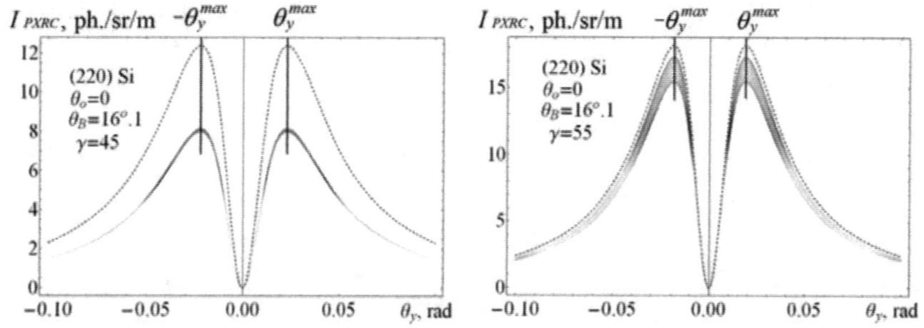

Figure 2. PXRC and PXR (dashed line) angular distribution calculated with the "true" periodic planar (220) Si channeling potential: a) for $\gamma = 45$ and the number of bound channeling states $N = 4$; b) for $\gamma = 55$ and $N = 5$. The increase of the number of bound states N manifests itself as $\sim 200\%$ increase in I_{PXRC} intensity. The maximums of I_{PXRC} are located at $\theta_y = \pm\theta_y^{max}$. In fact, the PXRC intensity graphs are not single lines but rather broad bands. We illustrated this drawing 10 lines (inside the band) corresponding to 10 positions inside transverse energy band of planar channeled electrons. In figure 2a the width of the band is small and these 10 lines are invisible, while in figure 2b the lines are visible and form the broad band.

In figure 2 we presented PXRC angular distributions calculated with the periodic planar (220) Si channeling potential for two values of relativistic factor in a vicinity of the "quantum jump": $\gamma = 45$ corresponds to the number of bound channeling states $N = 4$ and $\gamma = 55$ corresponds to $N = 5$.

To each line on the graph of PXRC in figure 2 corresponds a certain part of energy band of the channeled electron transverse motion. For calculation the each band was divided into 10 equal parts (by the transverse momentum) as it described in [6, 8]. As the result, the PXRC intensity graphs are not single lines but rather broad bands. We illustrated this drawing 10 lines (inside the band) corresponding to 10 positions inside transverse energy band of planar channeled electrons.

The difference in magnitudes of I_{PXRC} in figure 2 is explained by the influence of the band structure of energies and corresponding "true" wave functions on the transverse form-factors and initial populations.

At the same time the PXRC intensity increases by almost a factor of two, following increase of γ from 45 up to 55. The reason is that into Eq.(1), besides form-factors $|F_{nn}|^2$, enter initial populations $P_n(\theta_0)$.

In figure 3 we present PXRC angular distributions calculated with the periodic planar (220) Si channeling potential for two values of relativistic factor: $\gamma = 460$ corresponds to the number of bound channeling states $N = 14$ and $\gamma = 480$ corresponds to $N = 15$.

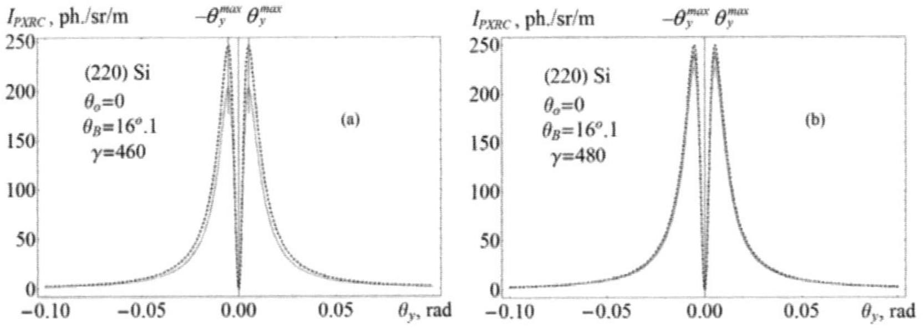

Figure 3. PXRC and PXR (dashed line) angular distribution calculated with the "true" periodic planar (220) Si channeling potential: a) for $\gamma = 460$ the number of bound channeling states $N = 14$; b) for $\gamma = 480 \Longrightarrow N = 15$. Here, the increase of the number of bound states N manifests itself as $\sim 20\%$ increase in I_{PXRC} intensity. The maximums of I_{PXRC} are located at $\theta_y = \pm\theta_y^{max}$. As in figure 2, the PXRC intensity graphs are not single lines but rather broad bands.

The electron relativistic factor in the figure 2a and 3 was taken equal to $\gamma = 45, 55$ and $\gamma = 460, 480$, corresponding to the program of planned experiments on PXRC studies with $20 - 255$ MeV electrons at the SAGA-LS accelerator center [1, 2, 3].

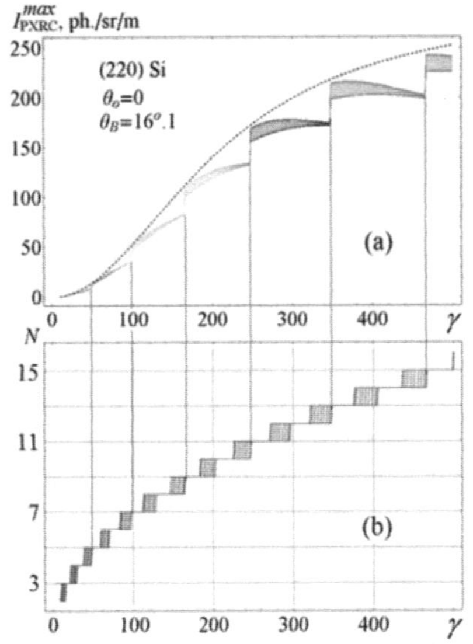

Figure 4. a) The maximums of PXRC and PXR (dashed line) as the function of γ in the (220) Si (along the plane $\theta_x = 0$). b) The number N of bound channeled states as the function of γ. As in figure 2 and figure 3, the graphs are not single lines but rather broad bands. We illustrated this drawing 10 lines (inside the band) corresponding to 10 positions inside transverse energy band of planar channeled electrons.

The maximums of angular distributions $I^{max}_{PXRC}(\gamma) = I_{PXRC}(\gamma)|_{\theta_y=\theta_y^{max}}$ (in the plane $\theta_x = 0$) as the functions of beam energy are presented in figure 4. The correlation of the number of quantum bound states $N = N(\gamma)$ and values of the maximums of $I^{max}_{PXRC}(\gamma)$ manifests itself at definite γ values. Here, for comparison, we plotted also the maximums of PXR angular distribution (see, in details the Section 4).

4. Comparison of PXRC and PXR

To characterize the difference in angular distributions of PXR and PXRC, we introduce the quantity

$$\begin{aligned} \delta &= \delta(\theta_y^{max};\gamma) = \frac{I^{max}_{PXR} - I^{max}_{PXRC}}{I^{max}_{PXR}} \equiv \\ &\equiv 1 - \sum_n P_n(\theta_0)\left|F_{nn}(\theta_y^{max})\right|^2, \end{aligned} \quad (7)$$

which depends on the number of quantum states N (a function of relativistic factor γ), initial population of every n-th state (a function of angle of incidence θ_0 and wave function $\varphi_n(y)$). Both $\varphi_n(y)$ and form-factor $F_{nn}(q)$ depend on the relativistic factor.

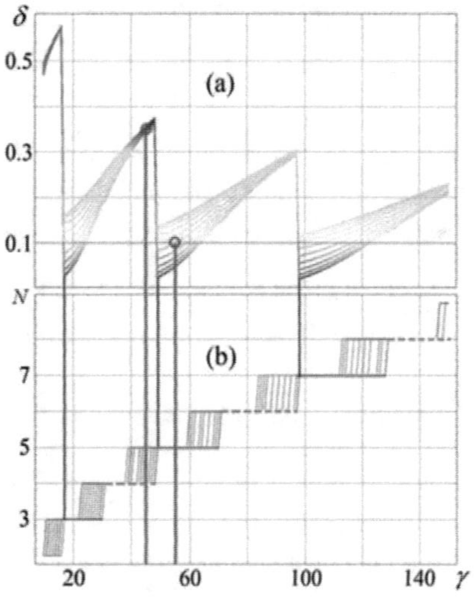

Figure 5. Dependences on relativistic factor γ of: a) maximum values of the $\delta = \delta(\theta_y^{max};\gamma)$ (formula (7)); b) number N of bound channeling states for (220) planar channeling in a "real" 1D periodic potential Si crystal. Here, the wave functions are obtained numerically following procedure described in [6, 8] and are used to calculate the initial populations $P_n(\theta_0)$ and transverse form-factors $|F_{nn}(\theta_y;\gamma)|^2$. The vertical lines correspond to the values of the relativistic factor $\gamma = 45$ and $\gamma = 55$ (see, figure 2) and the circles at the ends of these lines mark the values of δ, corresponding $I^{max}_{PXRC}(\gamma) = I_{PXRC}(\gamma)|_{\theta_y=\theta_y^{max}}$ in figure 2. The origin of 10 lines is explained in caption to figure 2.

As is well known in quantum mechanics, at least one bound state always exists in a 1D potential well of finite depth. We would like to stress, that in the problem of planar electron channeling there is an additional parameter - relativistic factor of the beam γ, which allows the change of the number N of quantum channeling states.

The evolution of the quantity δ (formula (7)) with the beam energy increasing is presented in figure 5, where we plotted both the number of quantum states $N = N(\gamma)$ and the values

$\delta = \delta(\theta_y^{max}; \gamma)$ taken at the maximums of angular distributions I_{PXR}^{max} and I_{PXRC}^{max} (along the plane $\theta_x = 0$). As one can see, the correlation of the number of quantum bound states $N = N(\gamma)$ and $\delta = \delta(\theta_y^{max}; \gamma)$ manifests itself at the definite γ values.

In figure 3 – figure 5, to each line on the graph corresponds a certain part of energy band of the channeled electron transverse motion (as in figure 2).

In figure 5, the band structure of transverse quantum channeling states (sequence of periodicity of planar channeling potential) decreases the value of transverse form-factor squared $|F_{nn}|^2$ and, in addition, sufficiently changes the values of initial populations $P_n(\theta_0)$ of quantum states. Both two factors increase the difference in angular distributions of PXRC and PXR, i.e. increase the δ value.

5. Conclusions

To summarize, as it was first suggested in [1], the difference in angular distributions of the standard Parametric X-Radiation (PXR) from plane-wave electrons (no channeling) and Parametric X-Radiation from channeled electrons (PXRC) is connected with manifestation of two quantum effects: form-factors of transverse quantum channeling states and their initial populations. In the present work we showed that this difference has non-trivial dependence on the electron beam energy: it undergoes some quantum jumps with appearance of the new quantum channeling states following increase of the electron beam energy. The magnitude of the jump depends on form-factors and initial populations of these quantum states.

The experimental studies of predicted quantum features in PXRC angular distributions are planned with 20 .. 255 MeV electrons at the accelerator center SAGA-LS [1, 2, 3].

6. Aknowledgements

The authors are grateful to Prof. V. G. Baryshevsky and Dr. Y. Takabayashi for stimulating discussions.

References

[1] Korotchenko K B, Pivovarov Yu L and Takabayashi Y 2012 JETP Lett. **95** 8 433
[2] Korotchenko K B, Pivovarov Yu L and Takabayashi Y 2013 Nucl. Instrum. Methods Phys. Res. **B**309 25
[3] Takabayashi Y, Korotchenko K B, Pivovarov Yu L and Tukhfatullin T A 2013 Nucl. Instrum. Methods Phys. Res. **B**315 105
[4] Baryshevsky V G, Feranchuk I D and Ulyanenkov A P 2005 *Parametric X-ray Radiation in Crystals* (Berlin: Springer)
[5] Yabuki R, Nitta H, Ikeda T and Ohtsuki Y H 2001 Phys. Rev. B **63** 174112
[6] Bogdanov O V, Korotchenko K B and Pivovarov Yu L 2007 JETP Lett. **85** 11 555
[7] Bazylev V A, Zhevago N K 1987 *Radiation of Fast Particles in Matter and External Fields* [in Russian] (Moscow: Nauka)
[8] Bogdanov O V, Fiks E I, Korotchenko K B and Pivovarov Yu L 2010 Journal of Physics: Conference Series 1 236

Influence of real photons diffraction contribution on parametric X-ray observed characteristics

S A Laktionova, O O Pligina, M A Sidnin and I E Vnukov
Belgorod State University, 14 Studencheskaya str., 308007 Belgorod, Russia
E-mail: vnukov@bsu.edu.ru

Abstract.
It is proposed and implemented a simple method of calculating of the output of the diffracted real photons of bremsstrahlung and transition radiation from perfect crystals of arbitrary thickness up to a primary extinction lengths. It is shown that for small thickness of crystals and observation angles concerning the center of the diffraction of reflex, contribution of diffraction of real photon is comparable with the output of parametric X-radiation. Influence of experimental equipment characteristic on observed characteristics of radiation is discussed.

1. Introduction
Parametric X-ray radiation (PXR) generated by passing fast charged particles through crystals had been actively studied till the end of the last century (see, for example, [1] and references therein). The interest to this type of radiation was mainly due to looking for tunable compact sources of intensive X-ray radiation for medical applications as an alternative to storage rings. Nowadays it is considered to be established (see, for example, [2]), that the PXR intensity in perfect crystals is not sufficient for medical applications. However, studies aimed at increasing the radiation intensity through the use of multi-crystal target [3] and multiple passing of particles through a thin target in circular accelerators [4] are being discussed till today. There have been recent suggestions to use PXR in thin crystals for diagnosing of electron beams parameters [5], and a new result for a thin crystal [6], which is not described by PXR theory [1].

In the first approximation PXR can be considered as coherent scattering of its own electromagnetic field of a particle on the electron shells of periodically arranged atoms of a target [1]. Diffracted in crystal X-ray and γ-rays, born directly inside the target or on its surface can be spread in the same direction as PXR. In the first case we mean the diffracted bremsstrahlung (DBS), and in the second one we mean the diffracted transition radiation (DTR). Apparently, first contribution of diffraction of real photons in the radiation yield at the Bragg angles of perfect crystals was confidently observed in experiments [8] for the (111) reflection of a single crystal silicon and [9] for the (110) of a diamond crystal. In the last paper the method of calculating the contribution of real photon diffraction in crystals with a thickness much greater than the length of the primary extinction for the Laue geometry is proposed. This limitation methods [9] and availability of experimental data about PXR which are not described theoretically excluding diffraction of real photons, provide an opportunity to consider the study of the influence of real photons diffraction on the observable characteristics up-to-date and of importance.

2. Theoretical consideration

In the general case in the experiment all the mechanisms of generation of radiation at the Bragg angles are implemented at the same time, so we introduce the basic formulae and approaches for each of them that we used during the calculation, following mainly the work of [9]. As it was shown in several experimental studies (see, for example, [10] and references therein), the kinematic PXR theory describes the results of measurements for energy electrons from a few MeV to several GeV with an accuracy better than 10-15 %. Thats why in order to calculate the PXR yield we used the formula obtained in the kinematic approximation by [11]:

$$\frac{d^2N}{dZd\Omega} = \frac{\sum_\alpha \alpha\omega^3|\chi_{\mathbf{g}}|^2}{2\pi\varepsilon_0^{3/2}\beta(1-\sqrt{\varepsilon_0}\boldsymbol{\beta}\mathbf{n})}\left[\frac{(\omega\boldsymbol{\beta}-\mathbf{g})\mathbf{e}_{\mathbf{k}_\alpha}}{(\mathbf{k}_\perp+\mathbf{g}_\perp)^2+\frac{\omega^2}{\beta^2}\{\gamma^{-2}+\beta^2(1-\epsilon_0)\}}\right]^2. \quad (1)$$

Here and further we use the system of units $\hbar = m_e = c = 1$. $\varepsilon_o = 1-\omega_p^2/\omega^2$, where ω_p is plasma frequency of the medium. $\boldsymbol{\beta} = \beta\mathbf{n}_0$ is the electron velocity vector, $\mathbf{n}_0$, $\mathbf{n}$ is a unit vector in the direction of the incident electron and the emitted photon (with energy ω and momentum $\mathbf{k}$), $\mathbf{g}$ is a vector of the reciprocal lattice, $\mathbf{e}_{\mathbf{k}_\alpha}$ are polarization vectors, $\perp$ is an index denoting the projections of a plane perpendicular to $\mathbf{n}_0$. The other symbols we use are common. By $|\chi_{\mathbf{g}}|^2$ we denote the magnitude of:

$$|\chi_{\mathbf{g}}|^2 = |S(\mathbf{g})|^2\exp(-2W)\left[-\frac{\omega_p^2}{\omega^2}\frac{f(\mathbf{g})}{z}\right]^2. \quad (2)$$

In (2) $|S(\mathbf{g})|^2$ is a structure factor, $\exp(-2W)$ is Debye-Waller factor, $f(\mathbf{g})$ is Fourier component of the spatial distribution of the electrons in the atom of the crystal.

The divergence of the electron beam, the electrons multiple scattering in the crystal and other experimental factors were taken into account according to the procedure described in [12].

For medium-energy electron the radiation in the X-ray range of photon energy ($\omega \leq 100$ keV) except PXR, is generated through the mechanisms of the transition radiation (TR) and the bremsstrahlung (BS). The TR spectral and angular distribution at the vacuum-medium interface can be represented in the form [13]:

$$\frac{d^2I_{TR}}{d\omega d\theta_\gamma} = \frac{2e^2\theta^3}{\pi}\left|\frac{1}{\theta_\gamma^2+\gamma^{-2}} - \frac{1}{\theta_\gamma^2+\gamma^{-2}+\omega_p^2/\omega^2}\right|^2 \quad (3)$$

where θ_γ is a photon emission angle according to the direction of the electron.

It is known (see, for instance [7]), that the bremsstrahlung in a dense medium in the energy range $\omega \leq \gamma\omega_p$ is suppressed due to the effect of medium polarization, and the degree of inhibition depends on both the energies of the photon and electron, and the angle of a photon emission. Therefore we used the expression for the spectral-angular distribution of the soft component of the bremsstrahlung ($\omega \ll E_e$), obtained as a result in [14] and taken into account this effect.

$$\frac{d^2I_{BS}}{d\omega d\Omega} = \frac{\gamma^2}{\pi L}\frac{(1+\frac{\gamma^2\omega_p^2}{\omega^2})^2+\gamma^4\theta_\gamma^4}{(1+\frac{\gamma^2\omega_p^2}{\omega^2}+\gamma^2\theta_\gamma^2)^4}, \quad (4)$$

where L is the radiation length.

The angle θ_γ in expressions (3), (4) is measured from the direction of the electron motion. Whereas in the experiment it is measured the distribution of the resulting radiation yield $Y(\omega,\theta)$ depending on the observation angle θ with the aperture ϑ_c.

The spectral and angular distribution of bremsstrahlung at the depth t when multiple scattering of electrons is taken into account $d^2I^*_{BS}(\omega,\theta,\varphi,t)/d\omega d\Omega$ is determined by convoluting the expression (4) with the angular distribution of the electrons $P(t,\theta_e,\varphi_e)$ at this depth:

$$\frac{d^2 I^*_{BS}(\omega,\theta,\varphi,t)}{d\omega d\Omega} = \int P(t,\theta_e,\varphi_e)d\Omega_e \int \frac{d^2 I_{BS}(\omega,\theta_\gamma,\varphi_\gamma)}{d\omega d\Omega_\gamma} f(\mathbf{n},\mathbf{n}_e,\mathbf{n}_\gamma)d\Omega_\gamma. \quad (5)$$

Here $\mathbf{n}_e(\theta_e,\varphi_e)$ and $\mathbf{n}_\gamma(\theta_\gamma,\varphi_\gamma)$ are vectors describing the direction of electrons and photons in the laboratory frame of reference and in a system connected with the direction of motion of the electron, respectively. $\mathbf{n}(\theta,\varphi) = \mathbf{n}_e + \mathbf{n}_\gamma$ is a vector of the photon direction in the laboratory frame, and $f(\mathbf{n},\mathbf{n}_e,\mathbf{n}_\gamma)$ is a function describing the connection between these vectors, see [15] for details.

In the experiments it is measured the dependence of a radiation yield into a fixed collimator from the angle of the crystal orientation Θ or an angular distribution, that is the dependence of the yield from the angles θ accordingly the center of reflex. This dependence can be written as:

$$Y_{DTR}(\theta) = \int d\omega \int \frac{d^2 I^*_{TR}}{d\omega d\Omega} R(\omega,\mathbf{n},\mathbf{g},\Theta_D) S^*(\omega,\mathbf{n},T)d\Omega. \quad (6)$$

$$Y_{DBS}(\theta) = \int_0^T dt \int d\omega \int \frac{d^2 I^*_{BS}}{d\omega d\Omega} R(\omega,\mathbf{n},\mathbf{g},\Theta_D) S(\omega,\mathbf{n},t)d\Omega, \quad (7)$$

where $d^2 I^*_{TR}/d\omega d\Omega$ is TR spectral-angular distribution, taking into account the divergence of the primary electron beam, $R(\omega,\mathbf{n},\mathbf{g},\Theta_D)$ reflecting the ability for the direction vectors $\mathbf{n}$ and $\mathbf{g}$, determined by the crystal orientation angles Θ and the location of the detector Θ_D, $S^*(\omega,\mathbf{n},T)$ and $S(\omega,\mathbf{n},t)$ are functions taking into account the photons absorption in the crystal and the geometry of the experiment, T is thickness of the crystal. The integration into (6), (7) is taking place over all angles and photon energies with the hit of the photons in the collimator.

3. Description of the computational model

To determine the output of the diffracted radiation we need information about $R(\omega,\mathbf{n},\mathbf{g},\Theta_D)$. The work [9] is based on the fact that if the thickness of perfect crystals T, significantly exceeding the primary extinction length l_{ex}, then in a narrow angular cone $\Delta\Theta$ (in so-called Darwins table) near the Bragg direction in absence of absorption reflectivity is equal to 100%, while outside of this range it is negligible [16]. Therefore, as a result of photons multiple reflections in a crystal of big thickness ($T \gg l_{ex}$) at any point of it a half of the radiation that satisfies the Bragg condition is spreading along the initial direction, and the other half in the Bragg direction.

According to this approach for a fixed photon direction $\vec{n}$ from the beam with the spectral-angular distribution of $d^2 I^*_{BS}/d\omega d\Omega$ or $d^2 I^*_{TR}/d\omega d\Omega$ satisfying Bragg's condition for photons with energies of ω only photons in the energy range $\Delta\omega = \omega\cos(\Theta_B)/\sin(\Theta_B)\Delta\Theta$ will be reflected. In accordance with [16] for unpolarized radiation and lack of absorption $\Delta\Theta = 2\cdot\gamma\Delta\theta_0$, where $\Delta\theta_0 = 2\cdot\delta/\sin 2\Theta_B$ is an amendment to the Bragg angle Θ_B because of the refraction of waves in a crystal, $\delta = (\omega_p/\omega)^2/2$ is a difference between the refractive index from 1, and $\gamma = f(\mathbf{g})(1+\cos(2\Theta_B))/2f(0)$.

As an estimate of the characteristic parameter of the model that is the length of the primary extinction we can use the expression [16]:

$$l_{ex} = d/(2\bar{\xi}\sin\Theta_B), \quad (8)$$

where d is a distance between planes, and $\exp(-2\bar{\xi})$ is an impairment of the intensity of the primary wave as it passes through a plane with the reciprocal lattice vector $\vec{g}$:

$$2\bar{\xi} = \frac{\pi d^2 N S(\mathbf{g}) f(\mathbf{g})}{n} \frac{e^2}{mc^2}, \quad (9)$$

where N is a concentration of the scattering centers, n is the order of reflection.

To calculate the reflectivity $R(\omega, \mathbf{n}, \mathbf{g}, \Theta_D)$ we use three of the coordinate system proposed in [17], in which Z-axis of two of them is turned around a vertical axis concerning Z axis of the laboratory system, right on the angle Θ_D (system of a "detector") and $-(\pi/2 + \Theta_B)$ (system of "a crystal"). The boundary directions of hit of radiation in the collimator $\theta_x, \theta_y, \theta_z$ are given in the system of "a detector". By means of transition into a system of "a crystal" we define the direction of movement of the photons after reflection. As in the system of "a crystal" the reflection of photons is equivalent to changing the sign of θ_z, after such transformation we learn the direction of motion of photons prior to diffraction.

The transition from the system of "a detector" into the laboratory system gives us an opportunity to determine the direction of the photons which can reach the detector after the process of diffraction and find the diffracted emission yield. The photon energy and intensity of the reflected radiation is determined in accordance with Bragg's law and the expressions of (3)-(7). The integration of equations (5)-(7) to obtain the final angular distribution or orientation dependence of the radiation yield for each order of reflection is performed numerically.

As it is noted in [18], in crystals with weak absorption the establishment of a stationary mode of transmission of photons through the crystal required for the legitimacy of the use of techniques [9], occurs for a photon path in a crystal of about 10 l_{ex}. A typical value of the primary extinction length is about 10 microns, while we have observed quite differences from the predictions of the PXR theory with a crystal thickness of 50 μm [5] and 20 μm [6].

For the part of the crystal with a thickness much smaller than l_{ex}, the probability of reflection of photons with an energy of ω and direction $\mathbf{n}$, for which the Bragg condition is satisfied, is proportional to the number of planes crossed by them [16]. Therefore, the dependence of the number of photons, which have not undergone the reflection, on the way length in the crystal t can be written as $N_\gamma(t) = N_\gamma(0)\exp(-t/l_{ex})$ [16], where $N_\gamma(0)$ is the number of photons at a starting point. The possibility of such notation allows us to use a well-known in experimental physics method of statistical simulation of photon transmission through a matter and for the conditions of photon diffraction in perfect crystals, as it was done in [19] for mosaic ones.

Taking into account the multiple Bragg reflection, photon absorption and scattering on atoms the dependence of the number of photons on passable way can be rewritten in the following way:

$$N_\gamma(\omega, \mathbf{n}, t) = N_\gamma(0)\exp(-\mu_{tot}(\omega, \mathbf{g}, \mathbf{n})t), \qquad (10)$$

where $\mu_{tot}(\omega, \mathbf{g}, \mathbf{n}) = \mu(\omega) + \mu_{dif}(\omega, \mathbf{g}, \mathbf{n})$ is a total coefficient of linear absorption of radiation with energy ω, for the direction of the reflecting plane of the crystal $\mathbf{g}$ and the direction of the photon velocity $\mathbf{n}$. We consider $\mu(\omega)$ as a linear absorption coefficient of photons due to the all processes on separate atoms and $\mu_{dif}(\omega, \mathbf{g}, \mathbf{n}) = 1/l_{ex}(\omega, \mathbf{g}, \mathbf{n})$ is due to diffraction.

The process of photons propagation through a crystal was being simulated in the following way. There were defined the values $\mu(\omega)$ and $\mu_{dif}(\omega, \mathbf{g}, \mathbf{n})$ for the photon with the energy of ω and the direction of motion $\mathbf{n}$ satisfying the Bragg condition. Then it was played out the photon path to the interaction point $t = -\ln\xi/\mu_{tot}$, where ξ is a random number between zero and one. Then the coordinates of the interaction point were being determined. If the point didn't belong to the crystal, the process with the photon was over and the new process started again.

If the interaction occurred inside the crystal, then we found out what kind of process took place. We were interested if it was diffraction or any other process on a separate atom. In the last case, the process with the photon was over and the new process started again. In the case of diffraction, the simulation was being repeated until the photon left the crystal or absorbed in it. This method is not limited concerning thickness and geometry of a using crystal.

4. Comparison of the calculated results with the experimental data

As it is noted above, the main and true objective of the article is, firstly, the analysis of results of new experimental works, in which the contribution of diffraction of real photons into an output of resulting radiation has become of great importance and secondly, the development of the calculation procedure, giving us an opportunity to consider the contribution of diffraction of real photons in crystals of any thickness. In order to define limits of applicability of a technique [9] it was carried out a number of calculations of an output of the diffracted radiation from thin crystals with the use of both the technique [9], and the one described above, which takes into account more correctly repeated Bragg's reflection of photons inside a perfect crystal.

Vertical angular distribution PXR of is represented in figure 1 (○), distributions of the DTR and DBS and their sum for experimental conditions [6] and the first order reflection calculated by a technique [9], so these are the curves 1-3. The ratio of the DTR angular distributions, calculated by both techniques, is also represented in the figure.

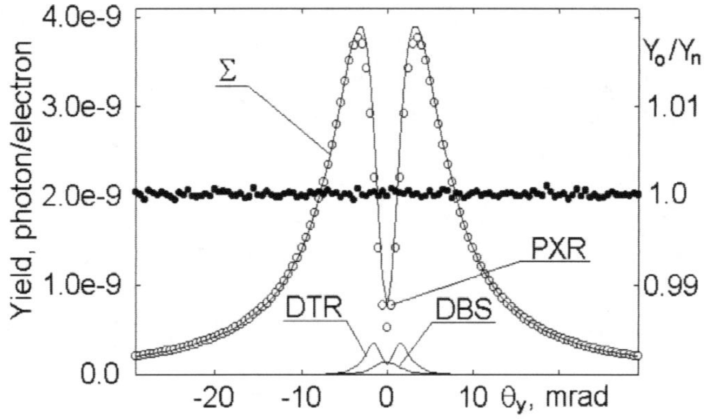

Figure 1. Vertical distribution of the X-ray yield for [6] condition and the first reflection order. • - ratio calculated DTR yields for both techniques

As can be seen from the figure, PXR has bigger intensity than DBS and DTR, and its angular distribution is broader. In the center of PXR angular distribution there is a dip, whereas the output of DTR and DBS is concentrated near the Bragg's direction. As a result, the diffracted photons yield gives the main contribution into the radiation yield in the reflex center.

The ratio of DTR yields, calculated by means of both techniques, is not different from 1, and the straggling concerning it, is caused by statistics of simulation. According to (8) for ω=11.65 keV and reflection (220) the length of primary extinction in silicon is $l_{ex} \sim 5\ \mu$. That is a simpler technique [9] can be used up to thickness of crystals of $T \sim$ 4-5 l_{ex}. The distinction of angular distributions, calculated by both techniques, begins with the a crystal thickness of $T \sim 2\ l_{ex}$.

In the paper mentioned above [5], [6] measurements of radiation angular distributions were carried out by means of the ProxiVision HR25 X-Ray camera [5] and the position-sensitive detector on a basis, of so-called, imaging plate (IP)[6]. In other words, as a matter of fact it wasn't measured an angular distribution of a photon beam, but a response of a detector on the energy of radiation which arrived on it. The first device registers the light output in this or that point of a thin scintillating plate at hit of a X-ray beam on it; as for the second one, it measures the absorbed dose in this or that point of the IP.

The results of calculation of horizontal angular distributions of resultant radiation are represented in figure 2, these are PXR+DTR+DBR for three orders of reflection and conditions of experiments [5], [6], see figures 2a and 2b. So for the reasons stated above, we summed up not the angular distributions of photons, but the angular distributions of intensity of radiation.

As can be seen from figures, the contribution of real photons diffraction significantly increased the yield of radiation in a minimum of PXR angular distribution (see fig.1). Nevertheless the output of radiation in a minimum of calculated dependence $\sim$ 40% from a maximum for figure 2a and $\sim$ 55% for figure 2b, is less than in the cited works, $\sim$ 50% and $\sim$ 75%, respectively. We believe that more consistent accounting of the experimental equipment characteristics and experiment layout will lead to an accord of experimental and calculated results.

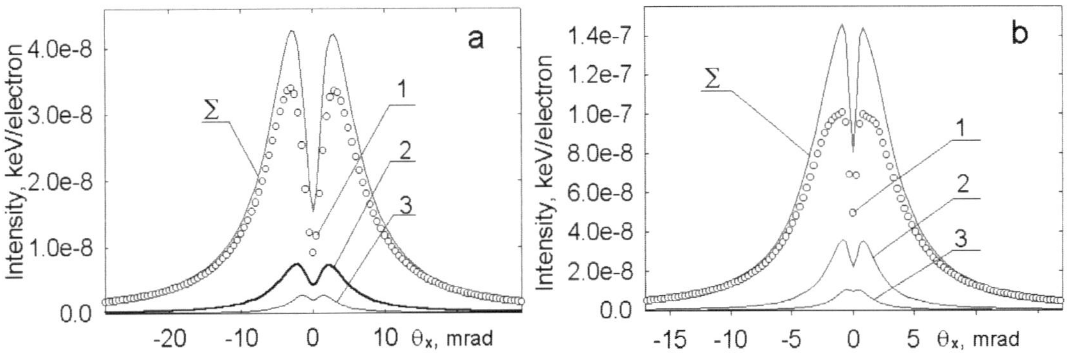

Figure 2. Horizontal angular distribution of X-ray intensity for three reflection orders and the experimental condition [6] (fig. 2a) and [5] (fig. 2b).

5. Summary and Conclusions

The results of the present research could be briefly stated as follows:

1) It is proposed and implemented a simple method of calculating of output of diffracted real photons in perfect crystals of arbitrary thickness up to the primary extinction lengths.

2) The method of computing of the contribution of diffraction of real photons, proposed in the work [9] is efficient up to the thickness of the crystals of an order of several lengths of primary extinction.

3) In order to explain the experimental results [5], [6] you should rationally take into account the characteristics of the used registration equipment.

The authors are grateful to the co-authors of works [9], [19] for their participation in the development and implementation of methods, used in the research. The present work was partially supported by the program of internal grants of Belgorod National Research University.

References

[1] Rullhusen R, Artru X and Dhez P *Novel Radiation Sources Using Relativistic Electrons* 1999 (Singapor: World Scientifi)
[2] Freudenberger J, Hell E, Knupher W 2001 *NIM* A **466** 257
[3] Takashima Y *et al* 1998 *NIM* B **145** 25
[4] Kaplin V V *et al* 2002*Applied Physics Letters* **80** 3427
[5] Cube G *et al Proceedings of IPAC2013, Shanghai, China* 491
[6] Takabayashi Y Shchagin A V 2012 *NIM* B **278** 78
[7] Ter-Mikaelian M L *High-Energy Electromagnetic Processes in Condensed Media* 1972 (New York: Wiley - Interscience)
[8] Brenzinger K-H *et al* 1997 *Z. Phys.* A **358** 107
[9] Baldin A N, Vnukov I E, Kalinin B N, Karataeva E A 2006 *Poverkhnost'* No. 4, 72 (in Russian)
[10] Brenzinger K-H *et al* 1997 *Phys. Rev. Lett.* **79** 2462
[11] Nitta H 1991 *Phys. Lett.* **158** 270
[12] Bogomazova E A *et al* 2003 *NIM* B **2001** 276
[13] Bazylev V A Zhevago N K *Radiation of Relativistic Particles in External Fields and in Matter* 1987 (Moscow: Nauka Pub.)(in Russian)
[14] Kleiner V P, Nasonov N N, Shliakhov N A 1992 *Ukr. Fiz. J.* **57** 48 (in Russian)
[15] Vnukov I E, Kalinin B N and Potylitsin A P 1991 *Sov. Phys. Journal* **34** 481
[16] James R *The optical principles of the diffraction of X- rays* 1958 (London: G. Bell and Sons)
[17] Potylitsin A 1998 *arXiv:cond-mat/9802279 v1*
[18] Pinsker Z *Dynamical Scattering of X-rays in Crystals* 1984 *Berlin: Springer*
[19] Baklanov D *et al* 2010 *Journal of surface investigation. X-ray, synchrotron and neutron techniques* **5** 317

Journal of Physics: Conference Series **517** (2014) 012021 doi:10.1088/1742-6596/517/1/012021

Smith-Purcell radiation from surface waves

A R Mkrtchyan, L Sh Grigoryan and A A Saharian

Institute of Applied Problems in Physics, 25 Nersessian Street, 0014 Yerevan, Armenia

E-mail: malpic@sci.am, levonshg@mail.ru, saharian@ysu.am

Abstract. We consider the radiation from an electron in flight over a surface wave of an arbitrary profile excited in a plane interface. For an electron bunch the conditions are specified under which the overall radiation essentially exceeds the incoherent part. In particular, it is shown that the radiation from the bunch with asymmetric density distribution of electrons in the longitudinal direction is partially coherent for waves with wavelengths much shorter than the characteristic longitudinal size of the bunch. The quantum radiation due to the dynamical Casimir effect induced by surface waves excited on a plane boundary is discussed.

1. Introduction

Surface waves have wide applications in various fields of science and technology. In the present talk, based on [1]-[4], we discuss the radiation from a charged particle flying over surface acoustic wave. The physics of this phenomenon is similar to that for the radiation of particle flying over a diffraction grating (Smith-Purcell radiation, for a review see [5]). The latter is used for the generation of the radiation in the range of millimeter and submillimeter waves.

2. Radiation from a single electron

Let a charge q move with constant velocity $\mathbf{v}$ parallel to a surface wave excited in the plane boundary of a homogeneous medium with permittivity ε (see figure 1). If the axis z is aligned with the particle trajectory, then the equation of the interface has the form

$$x = x_0(z,t) = -d + f(k_0 z \mp \omega_0 t), \tag{1}$$

where k_0 and ω_0 are the wave number and cyclic frequency, d is the distance from the non-excited interface, and $f(u)$ is the function describing the wave profile, $f(u+2\pi) = f(u)$. We assume that the particle moves in the vacuum, $x > x_0(z,t)$, and will consider the radiation in this region. From the symmetry properties of the problem it follows that the emission angle θ (with respect to $\mathbf{v}$) and the frequency ω of the emitted photon are related by (m is an integer)

$$\omega = m(k_0 v \mp \omega_0)/(1 - \beta\cos\theta), \ \beta = v/c. \tag{2}$$

The dependence of the radiation intensity on d is given by the factor $e^{-2\omega d\sigma/v}$, where

$$\sigma = \sqrt{(1-\beta^2)\omega_1^2/\omega^2 + \beta^2\sin^2\theta\sin^2\varphi}, \ \omega_1 = \omega \pm m\omega_0, \tag{3}$$

with φ being the polar angle with respect to the x axis in the plane perpendicular to the particle trajectory. This factor does not depend on the wave profile in (1) and is determined by the dependence on distance d of the field spectral components for a uniformly moving particle.

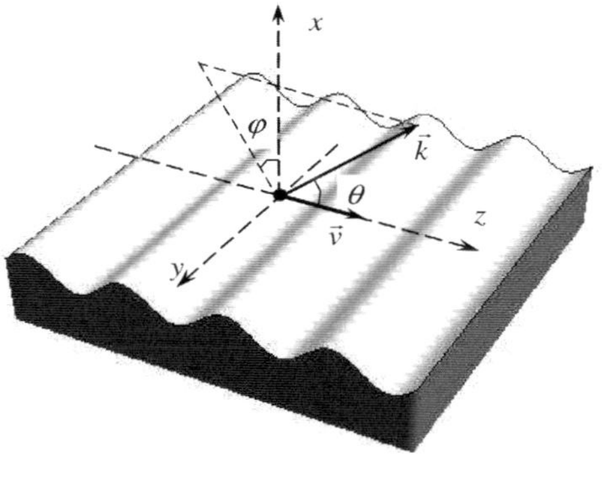

Figure 1. Electron flying over a surface wave excited on a plane boundary.

In order to determine the radiation intensity we assume that the amplitude of the surface wave is small. Under the condition $\beta\sqrt{\varepsilon-1}>1$, the spectral-angular distribution of the radiation intensity (per unit path length) in the region $x>0$ is given by the expression

$$\frac{dW}{d\omega d\Omega}=\frac{2q^2(\varepsilon-1)}{\pi c^2 v^2}\sum_m \omega^3\sin^2\theta\cos^2\varphi|f_m|^2 e^{-2\omega\sigma d/v}$$
$$\times\frac{A_1\sigma^2+A_2\sin^2\theta\sin^2\varphi+A_3\beta^2\sin^4\theta\sin^4\varphi}{[\beta^2+(\varepsilon+1)\sigma^2](\sigma_2+\varepsilon_1\sin\theta\cos\varphi)^2}\delta\left(\cos\theta-\frac{1}{\beta}+\frac{mk_0c}{\omega}\right), \quad (4)$$

where $d\Omega=\sin\theta d\theta d\varphi$, and we have assumed that $v\gg\omega_0/k_0$. In (4), $f_m=\frac{1}{2\pi}\int_{-\pi}^{+\pi}du\,f(u)e^{-imu}$ is the Fourier transform of the profile function,

$$\sigma_1=\sqrt{\beta^2(\varepsilon-\sin^2\theta\sin^2\varphi)-1},\ \sigma_2=\sqrt{\varepsilon-1+\sin^2\theta\sin^2\varphi}, \quad (5)$$

$$\begin{aligned}A_1&=[\sigma_1\sigma_2\cos\theta-\varepsilon(1-\sin^2\theta\cos^2\varphi)]^2+\sin^2\theta[\sigma_1(\sigma_2\cos\varphi+\sin\theta\sin^2\varphi)-\varepsilon\cos\theta\cos\varphi]^2,\\ A_2&=\sigma^2(\sigma_1\cos\theta+\varepsilon\sin\theta\cos\varphi)^2+\beta^2(\sigma_2\sin\theta\cos\varphi+\beta\cos\theta\sin^2\theta\sin^2\varphi+\cos^2\theta)^2,\\ A_3&=\sigma_2^2(1-\beta\cos\theta)^2+[\cos\theta+\beta\sin\theta(\sigma_2\cos\varphi+\sin\theta\sin^2\varphi)]^2. \quad (6)\end{aligned}$$

The radiation intensity for the case of a perfectly reflecting surface is obtained from (4) taking the limit $\varepsilon\to\infty$. The Smith-Purcell radiation from this type of static surfaces has been widely discussed in the literature.

For a sinusoidal surface wave, $f(u)=a\sin u$, one has $f_m=\pm(a/2i)\delta_{m,\pm1}$. In this case the only contribution to the radiation intensity comes from the harmonic $m=1$. The parameters of the radiation may be effectively controlled by tuning the characteristics of the surface wave. For an illustration we plot in figure 2 the spectral distribution of the radiation intensity, $dW/d\omega=\int d\Omega\,dW/d\omega d\Omega$, in the case of a sinusoidal wave excited on the surface of a quartz plate ($\varepsilon=3.75$) as a function of the frequency of the radiated photon for various values of the ratio d/λ_0 (numbers near the curves), where $\lambda_0=2\pi/k_0$ is the wavelength of the surface wave. The full (dashed) curves correspond to the electron energy $E_e=10$ MeV ($E_e=50$ MeV). Note that $\omega/k_0c=\lambda_0/\lambda$ with λ being the wavelength of the radiated photon. The corresponding radiation angle is related to the frequency by the relation $\cos\theta=1/\beta-k_0c/\omega$. In accordance with this relation, for relativistic electrons, large values of $\omega/(k_0c)$ correspond to small angles θ. For $\theta\gg\gamma^{-1}$ the radiation intensity is relatively insensitive to the particle energy, whereas for $\theta\lesssim\gamma^{-1}$ the intensity strongly increases with increasing energy. From figure 2 the suppression of the radiation intensity with increasing d is well seen.

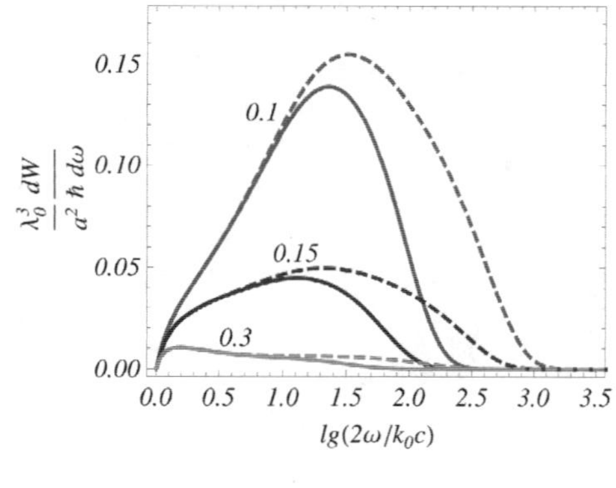

Figure 2. Spectral density of radiation intensity as a function of the frequency of the radiated photon. The numbers near the curves correspond to the values of the ratio d/λ_0.

3. Radiation from an electron bunch

As a source of the radiation, in this section we consider a cold bunch consisting of N electrons and moving with constant velocity v along the z axis (for coherence effects in the Smith-Purcell radiation see also [6]). The density of the current in the bunch can be written in the form $\mathbf{j} = q\mathbf{v}\sum_{j=1}^{N}\delta(\mathbf{r}-\mathbf{R}_j-\mathbf{v}t)$, with $\mathbf{R}_j = (X_j, Y_j, Z_j)$ being the position of the j-th particle at the initial moment $t=0$. The spectral density of the radiation energy flux for a given m, $\mathbf{P}_m^{(N)}(\omega)$, defined by the relation

$$\int_0^\infty d\omega \sum_m \mathbf{P}_m^{(N)}(\omega) = \frac{c}{4\pi}\int_{-\infty}^{+\infty} dt\,[\mathbf{EH}], \tag{7}$$

with $\mathbf{E}$ and $\mathbf{H}$ being the electric and magnetic fields, can be written as $\mathbf{P}_m^{(N)}(\omega) = S_N\mathbf{P}_m^{(1)}(\omega)$. Here, $\mathbf{P}_m^{(1)}(\omega)$ is the corresponding function for the radiation from a single electron and $S_N = |\sum_{j=1}^{N}\exp(-\omega_1\sigma X_j/v - ik_yY_j - i\omega_1 Z_j/v)|^2$. Assuming that the coordinates of the j-th particle are independent random variables and averaging over the position of a particle in the bunch, we obtain

$$\langle\mathbf{P}_m^{(N)}(\omega)\rangle = \langle S_N\rangle\mathbf{P}_m^{(1)}(\omega),\ \langle S_N\rangle = Nh + N(N-1)|h_xh_yh_z|^2, \tag{8}$$

where we have introduced the notations

$$h = g(-2\omega\,\sigma/v),\ h_x = g(\omega\sigma/v),\ h_y = \langle\exp(iK_yY)\rangle,\ h_z = \langle\exp(iK_zz)\rangle, \tag{9}$$

with $g(u) = \langle\exp(uX)\rangle$, $K_y = k_y = (\omega/c)\sin\theta\sin\varphi$, $K_z = \omega_1/v$. In (8), the term proportional to N^2 determines the contribution of coherent effects. Conventionally it is assumed that the coherent radiation is produced at wavelengths longer than the electron bunch length. However, as we shall see below, this conclusion depends on the distribution of electrons in the bunch.

As it is seen from (9), the form factors in y and z directions are determined by the Fourier transforms of the corresponding bunch distributions. For Gaussian distributions:

$$f_y(Y) = (\sqrt{2\pi}b_y)^{-1}\exp\left(-Y^2/(2b_y^2)\right),\ f_z(Z) = (\sqrt{2\pi}b_z)^{-1}\exp\left(-Y^2/(2b_z^2)\right), \tag{10}$$

one finds $|h_y|^2 = e^{-k_y^2b_y^2}$, $|h_z|^2 = e^{-\omega_1^2b_z^2/v^2}$. Assuming that for all particles in the bunch we have $X_j > -d+a$, along the x-direction we take the distribution function

$$f_x(X) = C\exp\left(-X^2/(2b_x^2)\right),\ x > \ b,\ C^{-1} = \sqrt{\pi/2}b_x\,\mathrm{erfc}(-b/(\sqrt{2}b_x)), \tag{11}$$

and $f_x(X) = 0$ for $X < -b$, where $b < d - a$ and erfc(x) is the error function. For function (11) one gets

$$g(u) = e^{u^2 b_x^2/2} \frac{\mathrm{erfc}\left(-b/(\sqrt{2}b_x) - b_x u/\sqrt{2}\right)}{\mathrm{erfc}(-b/(\sqrt{2}b_x))}. \tag{12}$$

In the limit $b_x \ll b$ and $b_x^2 u \ll b$, we obtain a simple result $g(u) \approx e^{u^2 b_x^2/2}$. In this limiting case the form factor is reduced to

$$\langle S_N \rangle = N \exp\left(2\omega^2\sigma^2 b_x^2/v^2\right)\left[1 + (N-1)\exp\left(-\omega^2\sigma^2 b_x^2/v^2 - k_y^2 b_y^2 - \omega_1^2 b_z^2/v^2\right)\right]. \tag{13}$$

As we see, for a Gaussian distribution the relative contribution of coherent effects is exponentially suppressed in the case $b_l > \lambda$, $l = x, y, z$. This result is a consequence of the mathematical fact that for a function $f(x) \in C^\infty(R)$ one has the estimate $F(u) \equiv \int_{-\infty}^{+\infty} f(x)e^{iux}dx = O(u^{-\infty})$, $u \to +\infty$, where $u = 2\pi b_z/\lambda$ for the longitudinal form factor of the relativistic bunch.

It should be noted that due to various beam manipulations the bunch shape can be highly non-Gaussian. In the estimate given above the continuity condition for the function $f(x)$ and for infinite number of its derivatives is essential. It can be seen that when $f(x) \in C^{n-1}(R)$, and the derivative $f^{(n)}(x)$ is discontinuous at the point x_1, we have the asymptotic estimate

$$F(u) = (-iu)^{-n-1}\left[f^{(n)}(x_1+) - f^{(n)}(x_1-)\right], \quad u \to +\infty. \tag{14}$$

Unlike the previous case, now the form factor for the short wavelengths decreases slower, as power-law, $(2\pi b_l/\lambda)^{-n-1}$. In the coherent part of the radiation this form factor is multiplied by a large number of particles per bunch and the coherent effects dominate under the condition $2\pi b_l/\lambda < N^{1/2(n+1)}$. The radiation intensity is enhanced by the factor $N(\lambda/2\pi b_l)^{2(n+1)}$. For instance, in the case of $n = 1$, $N \sim 10^{10}$, the coherent radiation dominates for $100\lambda > b_l$.

As an example, let us consider the case when the distribution function in the z direction has an asymmetric Gaussian form (see also [7])

$$f(z) = \frac{2}{\sqrt{2\pi}(1+p)b_z}\left[\exp\left(-\frac{z^2}{2p^2 b_z^2}\right)\theta(-z) + \exp\left(-\frac{z^2}{2b_z^2}\right)\theta(z)\right], \tag{15}$$

where $\theta(z)$ is the unit step function, $z_0 = (1+p)b_z$ is the characteristic bunch length, parameter p determines the degree of bunch asymmetry. When the transverse size of the bunch is shorter than the radiation wavelength, then the relative contribution of the coherent effects is estimated as $N\left|F(\omega_1/v)\right|^2$, where

$$F(u) = (p+1)^{-1}\left\{e^{-t^2} + pe^{-p^2t^2} - 2i\left[W(t) - pW(pt)\right]/\sqrt{\pi}\right\}, \tag{16}$$

with the notations $W(t) = \int_0^t \exp(l^2 - t^2)dl$, $t = ub_z/\sqrt{2}$. The expression $|F(u)|^2$ is invariant with respect to the replacement $p \to 1/p$, $b_z \to b_z p$ that corresponds to the mirror reversal of the bunch. When the electron distribution is symmetric ($p = 1$), the second summand in the figure braces of (16) vanishes and, as was mentioned earlier, the form factor exponentially decreases for short wavelengths $\lambda < 2\pi b_z/\beta$. For $pub_z \gg 1$ the asymptotic behavior has the form $F(u) \sim i\sqrt{2/\pi}(1-p)/(u^3 b_z^3 p^2)$ and for an asymmetrical bunch the contribution of the coherent effects can be dominant even in the case when the bunch length is greater than the radiation wavelength. Note that for the distribution function (15) one has $n = 3$ and this estimate is in agreement with (14).

Let $f(z, a)$ be a continuous distribution function depending on the parameter a, and $\lim_{a\to 0} f(z, a) = f(z)$. The integral $F(u, a)$ for $f(z, a)$ uniformly converges and hence

$\lim_{a\to 0} F(u,a) = F(u)$. It follows from here that the estimate presented above is valid for continuous functions as well if they are sufficiently close to the corresponding discontinuous function (the corresponding derivative is sufficiently large). The final conclusion can be formulated as follows. If for the distribution function one has

$$\lambda \frac{d^i f}{d(z/z_0)^i} \ll 2\pi, \quad i = 1, ..., n-1; \quad \lambda \frac{d^n f}{d(z/z_0)^n} \gtrsim 2\pi, \tag{17}$$

with z_0 being the bunch length, then the relative contribution of coherent effects into the radiation intensity is proportional to $N(\lambda/2\pi z_0)^{2n}$ and the radiation is partially coherent in the case $\lambda < z_0$ but $\lambda > 2\pi z_0 N^{-1/2n}$. In this case the main contribution into the radiation intensity comes from the parts of the bunch with large derivatives of the distribution function in the sense of the second condition in (17). For example, in the case of asymmetric distribution (15), when $ub_z \gg 1$ and $pub_z \lesssim 1$, with $u = \omega/v$, the main contribution comes from the left Gaussian tail with $z < 0$. For this tail $df/d(z/b_z) \sim u/(pub_z) \gtrsim u$, at $z \sim pb_z$ and therefore $F(u) \sim 1/(ub_z)$. This can be seen directly from the exact relation (16) as well.

4. Quantum radiation from surface waves

The radiation discussed in the previous sections is a classical effect. In quantum field theory the radiation can be present in the absence of charged particles. This radiation arises as a result of the interaction of a surface wave with quantum fluctuations of the electromagnetic vacuum. This phenomenon is an example of the dynamical Casimir effect (for reviews see [8, 9]). The Casimir effect is one of the most interesting manifestations of nontrivial properties of the quantum vacuum and arises due to the imposition of boundary conditions on the field operator.

For simplicity we consider a quantum scalar field $\varphi(x)$ in a spacetime region with the boundary S on which the field operator obeys Dirichlet boundary condition. The field equation and the boundary condition have the form (here and in what follows we use the units $c = 1$, $\hbar = 1$)

$$\left(\nabla_l \nabla^l + m^2\right) \varphi(x) = 0, \; \varphi(x)|_{x\in S} = 0. \tag{18}$$

In the case of a moving boundary, the interaction with quantum fluctuations of the field can lead to the creation of real quanta out of vacuum. Let $\xi^l(x) = n^l \xi(x)$ describes the displacement of the hypersurface S from the static hypersurface S_0: $x^l + \xi^l(x) \in S$ if $x^l \in S_0$. Here n^l is the unit normal to the boundary S_0. Assuming that the displacement is small, in the first approximation, for the number of quanta radiated in the interval of quantum numbers $(\nu, \nu + d\nu)$ one has the formula [10, 11]

$$n(\nu)d\nu = \int d\nu' \left| \int_{S_0} d\Sigma \, n^i n^l \xi(x) [\partial_i \varphi_{0\nu'}(x)][\partial_l \varphi_{0\nu}(x)] \right|^2 d\nu, \tag{19}$$

where $\{\varphi_{0\nu}(x), \varphi_{0\nu}^*(x)\}$ is a complete set of eigenfunctions with the set of quantum numbers ν which are solutions of the boundary-value problem (18) with $S = S_0$. Below we consider applications of the general formula (19) to surface waves excited on a plane boundary, assuming that the excitation vanishes in the asymptotic regions $t \to \pm\infty$.

As the hypersurface S_0 we take the plane $x = 0$. For the density of the number of quanta in the momentum space from (19) we find the expression

$$n(\mathbf{k}) = \frac{1}{16\pi^6} \int d\mathbf{k}' \frac{k_1^2 k_1'^2}{\omega\omega'} |\xi(\omega + \omega', \mathbf{k}_\perp + \mathbf{k}'_\perp)|^2, \; \mathbf{k} = (k_1, \mathbf{k}_\perp), \; \omega = \sqrt{k^2 + m^2}, \tag{20}$$

where $\mathbf{k}_\perp = (k_2, k_3)$, $\xi(\omega, \mathbf{k}_\perp) = \int_{-\infty}^{+\infty} dt dy dz \, \xi(t, y, z) e^{i\mathbf{k}_\perp \mathbf{x}_\perp - i\omega t}$. In the case of a running surface wave excited on the surface of a plane mirror the corresponding displacement function has the form $\xi(t,y,z) = f_s(z - v_0 t)$ and, hence, $\xi(\omega, \mathbf{k}_\perp) = (2\pi)^2 \delta(k_2)\delta(\omega - k_3 v_0) \int_{-\infty}^{+\infty} dz\, f_s(z) e^{ik_3 z}$. One has $k_3 \leqslant \omega$ and for $v_0 < 1$ this expression vanishes. In this case the radiation is absent. This result becomes obvious if we pass to the reference frame moving with the surface wave. In the case of harmonic oscillations, $\xi(t,y,z) = \xi_0 \cos(k_0 z - \omega_0 t)$, with $\omega_0 > k_0$, the quanta are radiated satisfying the condition $\omega_0 - \omega \geqslant \sqrt{k_2^2 + (k_0 - k_3)^2}$. Introducing spherical coordinates in the momentum space, $k_1 = \omega \cos\vartheta$, $k_2 = \omega \sin\vartheta \sin\phi$, $k_3 = \omega \sin\vartheta \cos\phi$, $0 \leqslant \vartheta \leqslant \pi/2$, $0 \leqslant \phi \leqslant 2\pi$, for the spectral-angular density of the number of the radiated quanta we have

$$\frac{dN}{d\omega d\Omega} = \frac{\xi_0^2 \omega^3}{8\pi^3} \cos^2\vartheta (\omega_0^2 - 2\omega_0\omega - k_0^2 + 2k_0\omega \sin\vartheta \cos\phi + \omega^2 \cos^2\vartheta)^{1/2}, \tag{21}$$

where ϑ is the angle between the normal to the mirror and the direction of the radiation. For a given ω, the allowed angular region for the radiation is determined by the non-negativity condition for the expression under the square root in (21). The total radiated energy is obtained from (21) as $E = \int \omega dN$.

The limit $k_0 \to 0$ of formula (21) corresponds to the mirror oscillating as a whole. In this case, the spectral density of the radiation is obtained after the integration over the solid angle:

$$\frac{dN}{d\omega} = \frac{\xi_0^2 \omega_0^4}{32\pi^2} \left[u(1-u)^3 + u^3(1-u) + (1/2)(1-2u)^2 \ln|1-2u| \right], \quad u = \omega/\omega_0. \tag{22}$$

The expression on the right-hand side of (22) takes its maximum value at $u = 1/2$.

Now we turn to the case of a standing surface wave excited on the surface of a plane mirror in the strip $0 \leqslant z \leqslant l$, $-\infty < y < +\infty$:

$$\xi(t,y,z) = \xi_0 \cos(\omega_0 t) \sin(k_0 z), \; k_0 = \pi n/l, \; n = 1, 2, 3, \ldots, \tag{23}$$

for $0 \leqslant z \leqslant l$ and $\xi(t,y,z) = 0$ otherwise. For the number of the radiated quanta per unit time and per unit length along y we find

$$n(\mathbf{k}) = \frac{k_1^2 \xi_0^2}{8\pi^4 \omega} \int_{w_1}^{w_2} du \, \frac{\sqrt{(w - w_1)(w_2 - w)}}{(1 - w^2)^2} \left[1 - (-1)^n \cos(\pi n w)\right], \tag{24}$$

where $w_{2,1} = [k_3 \pm \sqrt{(\omega_0 - \omega)^2 - k_2^2}]/k_0$. The conditions $\omega = \sqrt{k_1^2 + k_2^2 + k_3^2} \leqslant \omega_0$ and $|k_2| \leqslant \omega_0 - \omega$ should be satisfied for the presence of the radiation. In particular, the second inequality imposes a constraint on the angular region for the radiation.

References

[1] Mkrtchyan A R Grigorian L Sh Saharian A A Mkrtchyan A H and Didenko A N 1989 Izvestia AN Arm. SSR. Fizika **24** 62
[2] Mkrtchyan A R Grigorian L Sh Saharian A A and Didenko A N 1991 Zhurnal Tekh. Fiz. **61** 21
[3] Mkrtchyan A R Grigorian L Sh Saharian A A and Didenko A N 1991 Acustica **75**, 184
[4] Saharian A A Mkrtchyan A R Gevorgian L A Grigorian L Sh and Khachatryan B V 2001 Nucl. Instr. and Meth. B **173** 211
[5] Potylitsyn A P 2011 *Electromagnetic Radiation of Electrons in Periodic Structures* (Berlin: Springer)
[6] Mkrtchyan A R Gevorgian L A Grigorian L Sh Khachatryan B V and Saharian A A 1998 Nucl. Instr. and Meth. B **145** 67
[7] Korkhmazian N A Gevorgian L A and Petrosyan M L 1977 Zhurnal Tekh. Fiz. **47** 1583
[8] Bordag M Klimchitskaya G L Mohideen U and Mostepanenko V M 2009 *Advances in the Casimir Effect* (Oxford: Oxford University Press)
[9] Dalvit D Milonni P Roberts D and da Rosa F (Editors) 2011 *Lecture Notes in Physics: Casimir Physics* Vol. 834 (Berlin: Springer)
[10] Grigoryan L Sh and Saharian A A.1997 J. Contemp. Phys. **32** 223
[11] Grigoryan L Sh and Saharian A A.1997 J. Contemp. Phys. **32** 275

Coherent Diffraction and Cherenkov Radiation in Fibers

G Naumenko, A Potylitsyn, V Bleko, V Soboleva
Tomsk Polytechnic University, Tomsk, Russia

E-mail: naumenko@tpu.ru

Abstract. The ability to use radiation of relativistic electrons in optical fibers in beam diagnostics was proposed by X. Artu at symposium RREPS-11. In this work the properties of different types of radiation induced in fibers by electromagnetic field of relativistic electron were considered. In this report we present the results of experimental investigation of this phenomenon in millimeter wavelength region for conditions of complete coherence. The nature and properties of radiation in fibers were analyzed experimentally for different geometry of fiber position with respect to the electron beam.

1. Introduction.

Dielectric fibers are widely used in technic for light transport. In this paper we consider more complicated electromagnetic processes in fibers, namely, generation and propagation of electromagnetic waves in fibers excited by Coulomb field of relativistic electrons. If a relativistic electron crosses a dielectric fiber or moves in vicinity of it, the different types of radiation, such as transition radiation (TR), diffraction radiation (DR) and Cherenkov radiation (ChR) might be generated. Optical fibers are already used as particle detectors, using scintillation or Cherenkov light [1, 2]. Here we present the experimental analysis of the nature of radiation generated by relativistic electron field in fibers in millimetre wavelength region. The geometry, when electrons move in vicinity of fiber, may be applied in a noninvasive diagnostics of relativistic electron beams.

For intuitive understanding of these processes we will use the pseudo-photon description of the interaction of Coulomb field of electron with matter, which was proposed by E. Fermi [3] and developed by E.J. Williams [4]. According to this viewpoint TR and DR is the Coulomb field, refracted at a boundary of fiber, but in case of DR the electros do not cross the fiber. ChR is the result of interference of radiation, induced in fiber by the Coulomb field of electron from each point of electron trajectory. According to the traditional viewpoint ChR is generated if electrons cross a target, but in [5, 6] it was shown that ChR may be generated even when the electrons move in vicinity to the target and don't cross the target material.

The theoretical analysis of radiation of relativistic electrons in fibers was done by X. Artru in [7]. However, experimental results of this phenomenon are absent.

2. Experimental setup

In experiment the radiation intensity, generated and propagated in fibers for different positions of fiber (the angle θ between the fiber and electron beam, and distance h between electron beam and fiber, i.e. impact-parameter) was investigated. Thus two geometries of fiber position are possible:

a) The electron beam and fiber are coplanar. In this case the impact-parameter h is the distance between the electron beam and the end of fiber (see figure 1*a*).

b) The fiber is situated under the electron beam at a distance *h* (figure 1*b*). The end of fiber is at a distance $l > \gamma\lambda$ from the electron beam (γ is the Lorenz-factor of electrons, λ is the wavelength for investigation).

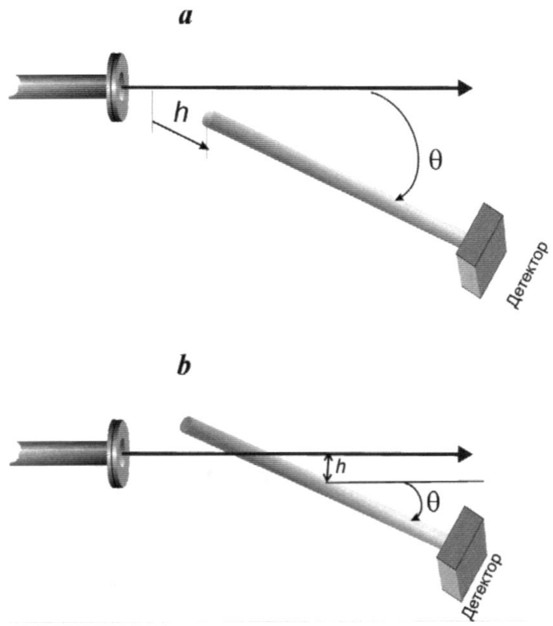

Figure 1. Two geometries of fiber position.

Both these geometries were investigated. In these geometries all types of radiation (ChR, DR and TR) may be generated. ChR satisfies to the condition $\cos\eta = \frac{1}{n\beta}$, where η is the radiation angle with respect to the electron beam direction, β is electron velocity in light velocity units and n is the target refractive index. DR and TR in fibers satisfy the refraction law for transit from vacuum into the fiber. Main condition for evolution of radiation inside the fiber is the requirement $\psi > \psi_{FIR}$, where ψ is the incidence angle of radiation propagation to the normal vector to the fiber internal surface, ψ_{FIR} is the angle of total internal reflection. For ChR, if θ=0, η corresponds to ψ_{FIR}, so for finite electron beam divergence approximately half part of ChR leaves the fiber. DR and TR at the front end of the fiber mainly propagate along the fiber, but DR and TR at the wall of the fiber already leave it.

The experiment was carried out in the extracted electron beam of the Institute of Physics and Technology of Tomsk Polytechnic University microtron with parameters presented in Table 1.

Table 1. Electron beam parameters

Electron energy	6.1 MeV ($\gamma = 12$)	Bunch period	380 psec
Train duration	$\tau \approx 4\,\mu$sec	Bunch population	$N_e = 6\cdot10^8$
Bunches in a train	$n_b \approx 1.6\cdot10^4$	Bunch length	$\sigma \approx$1.9~2.4 mm

For the radiation measurements the room temperature detector DP20M was used, with parameters described in [8]. The detector efficiency in the wavelength region of λ = 3~16 mm is estimated to be constant to *a* ±15 % accuracy. The detector sensitivity is 0.3 V/mW. The high frequency limit ($\lambda_{\min}$ = 9 mm) of the wavelength interval is defined by the bunch form-factor. The coherent radiation intensity for λ > 9 mm is by 8 orders of magnitude larger than the incoherent one. The cylindrical Teflon fibers with diameter of 10 mm, length of 500 mm and refractive index n = 1.41 were used. The range of variation of impact-parameter is defined by effective transversal size of electron field which is $\approx \lambda\gamma$. For experimental conditions $\gamma = 12$, $\lambda = 17$ mm (maximal wavelength for our fibers).

3. Experimental results

3.1. Measurements for the geometry a of the scheme of experiment shown in figure 1a

In this experiment the radiation intensity as a function of fiber angle θ was measured for different values of impact-parameters *h* (figure 2*a*).

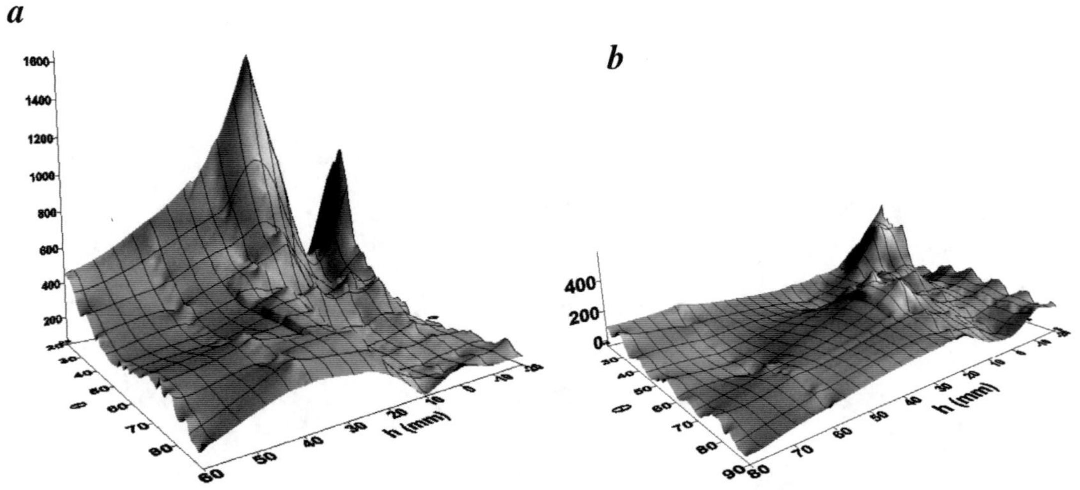

Figure 2. Radiation intensity as a function of fiber angle θ for different values of impact-parameters h. a – the end of fiber is opened; b – the end of fiber is screened by conductive screen

To clarify the nature of the radiation we had measured the same dependence when the edge of fiber was screened by a conductive screen (figure 2*b*).

In figure 2 we can see, that for the open end of fiber in region of angles θ between 20 and 50 degrees the radiation intensity is large and has a deep minimum at *h* = 0 (when the end of fiber crosses the electron beam). On the other hand, in the region $\theta > 50^0$ and 10 mm < *h* < 60 mm the observed radiation intensity does not depend on the screen presence.

3.2. Measurements fort the geometry b of the scheme of experiment shown in figure 1b

In this experiment the fiber is situated under the electron beam so, that the centre of the fiber is strictly under the beam. In this geometry the radiation intensity as a function of fiber angle θ was measured for different values of impact-parameters *h* (see figure 3).

It should be noted, that in this geometry we see the oscillations in the angular dependence. To check possible contribution of radiation reflection from the end of fiber, the similar measurement was performed when the end of fiber was screened by the absorber. These measurements showed the same results, i.e. the reflection contribution is absent.

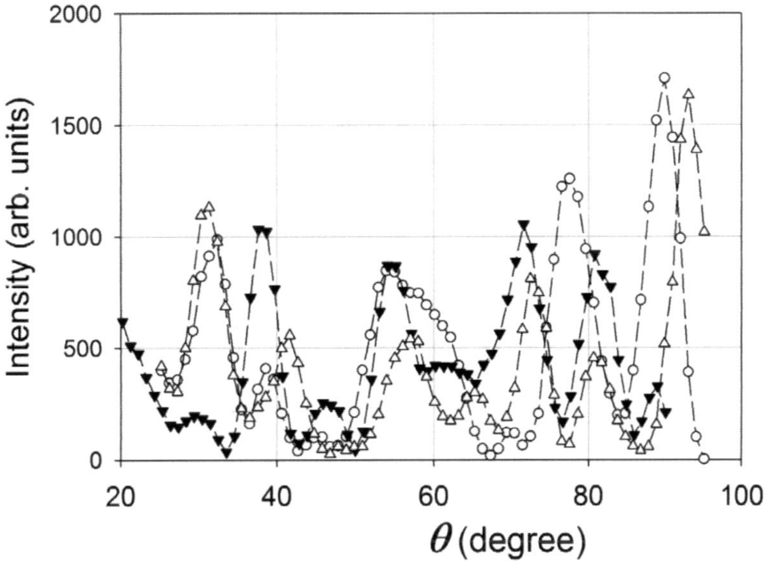

Figure 3. Radiation intensity in geometry *b* as a function of fiber angle θ for different values of impact-parameter *h*: ○ – $h = 0$ mm, △ – $h = 30$ mm, ▼ – $h = 70$ mm.

4. Discussion

In measurements in the figure 1*a* geometry it should be noted, that for a small angle θ pseudo-photons of the electron field are refracted at a small angle to the wall of fiber (less than angle of the total internal reflection, which for Teflon is $\approx 45^0$). This radiation, which may be classified as a DR at the end surface of fiber, propagates along the fiber without losses (see figure 2*a* in region near the point $\theta = 20^0$ and h = 20 mm). However, the angle of propagation of ChR is close to the angle of the total internal reflection, therefore due to the electron beam divergence, some part of ChR leaves the fiber. In this case the DR gives the main contribution to the registered radiation. With the increasing of angle θ (see figures 2*a* and 2*b* in region near the point $\theta = 60^0$ and h = 30 mm) the situation is reversed, and the ChR gives the main contribution to the registered radiation. Let's note that the DR, which is generated on the wall of fiber, may not propagate along the fiber, because in this case the condition of total internal reflection is not satisfied.

In geometry 2*b* in contrast to geometry 2*a*, we see the oscillations in the angular dependence. As was shown above, this isn't the result of reflection from the end of fiber. To understand the reason of such behaviour let's note that the horizontal (in our geometry) component of the electron field at the left and right side of the electron beam is of the opposite direction. This field generates ChR with opposite phases in the left and right halves of the fiber. The interference of this radiation may cause oscillations in angular dependence.

Of course, these explanations are simplified. Really, the propagation of radiation along a fiber is more complicated. It may be single-modal, multi-modal or prohibited, depending on the relation between wavelength and radius of fiber (see [7]), but simplified explanation may be useful for intuitive understanding of these processes. For the future we plan the experimental investigation of the spectral characteristics of radiation in fibres and dependence of this radiation on the fiber curvature, which is important for applications.

Acknowledgment
This work is supported by grant of Russian MES Program "Nauka" number 2456.

References

[1] Coyle P, Kawahara H, Lu A *et al* 1994 *Nucl. Instr. Methods in Phys. Research* A **343** 292
[2] Contin A, De Salvo R, Gorodetzky P *et al* 1994 *Nucl. Instr. Methods in Phys. Research* A **367** 271
[3] Fermi E 1924 *Zh. Phyz.* 29 315
[4] Williams E J 1935 *Kgl. Danske Videnskab. Selskab. Mat.-fys. Medd.* **13** 4
[5] Takahashi T, Shibata Y, Ishi K *et al.* 2000 *Phys. Rev.* E **62** 8606
[6] Shevelev M V, Naumenko G A, Potylitsyn A P, Popov Y A 2012 *Journal of Physics: Conference Series.* **357** 1, Article number 012020
[7] Artru X and Ray C 2013 *Nucl. Instrum. Meth.* B **309**
[8] Kalinin B N, Naumenko G A, Potylitsyn A P *et al* 2006 *JETP Letters* **84** 3

RREPS13 and Meghri13 IOP Publishing
Journal of Physics: Conference Series **517** (2014) 012023 doi:10.1088/1742-6596/517/1/012023

Influence of the Beam Divergence on Diffraction Radiation

V Shpakov[1] **and S B Dabagov**[1,2,3]

[1] RAS P.N. Lebedev Physical Institute, Moscow 119991, Russia
[2] INFN Laboratori Nazionali di Frascati, Frascati 00044, Italy
[3] NRNU MEPhI, Moscow 115409, Russia

E-mail: shpakov@lnf.infn.it

Abstract. Nowadays the achievements in accelerator physics allows us to operate with ever-growing energy of the particles beams. However, the increase of the beam energy requests new issues to be resolved. For instance, it becomes necessary to develop new techniques for beam diagnostics. One of the promising solutions for beam diagnostics is based on diffraction radiation (DR).

In this work similar to the methods of bunch length measurement based on transition radiation, the longitudinal size of a bunch is retrieved from the dependence of the DR intensity on the wavelength of radiation. Such dependence is expressed by so-called form-factor, which in turn depends on the beam parameters. Influence of the beam divergence on the form-factor of a beam has been examined in this work.

1. Introduction

One of the attractive DR features is the fact that it appears without direct contact, interaction, between the particle and the discontinuity, in general case, of the medium. Combined with the low radiation intensity with respect to the particle energy, it makes DR a great candidate for non-invasive beam diagnostics instrument.

DR from one particle passing through the slit system was considered in a number of papers [1–3]. However, in a real experiment we practically always deal with radiation by the bunch of particles [4, 5]. The intensity of radiation emitted by the beam depends on its coherence, on the beam form-factor. Such dependence can be used for the bunch length measurement [6]. The influence of the beam size on its form factor was previously studied both theoretically and experimentally [7, 8]. The present work is dedicated to theoretical investigation of the beam divergence influence on the beam form-factor that can be revealed from the DR spectrum.

2. DR theory

DR from one particle was studied in a number of papers [1, 2] and we do not consider this case in details. In this work we use the approach, which was applied, for example, in [9, 10]. Following those works we will consider one of the components of the DR field. Let us consider the coordinate system associated with the particle trajectory (figure 1). In such coordinate system the unit vector $\mathbf{e_z}$ is directed along the particle velocity $\mathbf{V}$, the unit vector $\mathbf{e_x}$ is parallel to the half-plane edge and $\mathbf{e_y}$ is perpendicular to the edge. In further calculations we consider only one of the components of the DR field, namely E_y. Omitting some known details the E_y component for DR from infinite half-plane can be presented in the form

$$E_{\mathrm{y}}^{DR} = \frac{e}{2\pi^2 V} \frac{e^{-h_1(\sqrt{k_x^2+\alpha^2}+ik_y)}}{(\sqrt{k_x^2+\alpha^2}+ik_y)}, \tag{1}$$

where V is the particle velocity, e is the particle charge, h_1 is the impact parameter half-plane, the coefficient $\alpha = \frac{k}{\gamma\beta}$, where $\mathbf{k} = (k_x, k_y, k_z)$ is the wave vector, $\gamma = (1-\beta^2)^{-1/2}$ is the relativistic parameter, $\beta = V/c$ is the normalized particle velocity, and c is the light velocity.

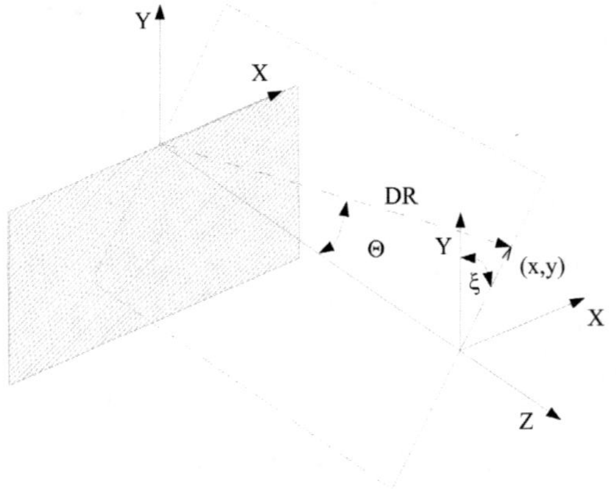

Figure 1. Scheme for DR from a infinite half-plane. The direction of the DR is defined by angles Θ and ξ. Θ is the angle between radiation vector and velocity vector, ξ is the angle between the plane of radiation and the $y0z$-plane.

As above mentioned, in this work we consider DR applied to the longitudinal beam diagnostics. For such diagnostics it is necessary to know the form-factor dependence on the wavelength of radiation. Influence of both longitudinal and transverse sizes of the beam on its form-factor is considered earlier [11, 12]. In present work we consider the beam divergence influence on its form-factor as well as the case when such influence is meaningful.

It is well known that intensity of radiation emitted by the beam (in general case any radiation, synchrotron radiation, transition radiation, etc...) depends on its coherence degree [13] and can be written as

$$I = I_0(N + N(N-1)F), \tag{2}$$

where N is the number of particles in the beam, I_0 is the intensity of radiation from a single particle, and F is the beam form-factor. The coherence degree of the beam expressed by its form-factor, which, in turn, depends on the parameters of the beam and on the wavelength of radiation. Thus, having known the dependence $I(\lambda)$, we can obtain the beam parameters. Such technique is widely used today for longitudinal beam diagnostics of short bunches.

Full form-factor of the beam based on transition radiation [14] can be determined in the following way

$$F_{\text{beam}} = \left| \int_V S(\mathbf{r}) e^{-i\phi(\mathbf{r})} d\mathbf{r} \right|^2 , \tag{3}$$

where $S(\mathbf{r})$ is the normalized particles distribution inside of the beam, and $\phi(\mathbf{r})$ is the phase difference between the particles [2]. The integral is taken over the all volume occupied by the beam. However, as shown by equation (1), for DR the intensity emitted by one electron depends on the impact parameter and, as a consequence, can vary for different particles. For the far-field approximation we can consider DR as emitted by one point (figure 1). Thus, the strength of DR field can be expressed by

$$E(y) \propto E_0 e^{-\dfrac{2\pi y \sqrt{1+\gamma^2\beta^2 \sin^2\Theta \sin^2\xi}}{\gamma\beta\lambda}} , \tag{4}$$

where E_0 is the constant amplitude, y is the impact parameter and λ is the wavelength of radiation. Both angles Θ and ξ define the direction to the observation point (figure 1). Hence, the form-factor of the beam with respect to DR can be presented in the form

$$F^{DR} = \left| \int_V S(\mathbf{r}) e^{-\dfrac{2\pi y \sqrt{1+\gamma^2\beta^2 \sin^2\Theta \sin^2\xi}}{\gamma\beta\lambda}} e^{-i\phi(\mathbf{r})} d\mathbf{r} \right|^2 \tag{5}$$

3. Beam divergence from a form-factor for DR

The influence of the beam divergence on its form-factor were earlier considered for transition radiation [15]. In that work the influence of the beam divergence on the intensity of radiation were separated from that caused by the spatial beam distribution. In our approach below the beam divergence to be taken into account will be reveled from analysis of phase shifts.

Let assume the total particle distribution inside the beam to be defined by two independent distributions in longitudinal and transverse directions

$$S(\mathbf{r}) = S_{\text{long}}(z) S_{\text{tr}}(x, y) \tag{6}$$

In this case the phase difference for various beam particles is defined as follows (figure 1)

$$\phi(\mathbf{r}) = \frac{2\pi}{\lambda}(x \sin\Theta \sin\xi + \frac{z}{\beta}) \tag{7}$$

Taking into consideration equations (6) and (7), equation (5) for the beam form-factor becomes equal to

$$F^{DR} = \left| \int dx dy dz S_{\text{long}}(z) S_{\text{tr}}(x, y) e^{-\dfrac{2\pi y \sqrt{1+\gamma^2\beta^2 \sin^2\Theta \sin^2\xi}}{\gamma\beta\lambda}} e^{-\dfrac{2\pi i}{\lambda}(x \sin\Theta \sin\xi + \dfrac{z}{\beta})} \right|^2 \tag{8}$$

For relativistic beams the longitudinal size is usually much larger than its transverse size for the observer. For this reason, the beam form-factor is often defined as its longitudinal form-factor. Let us consider the beam with Gaussian distribution in longitudinal direction and neglect

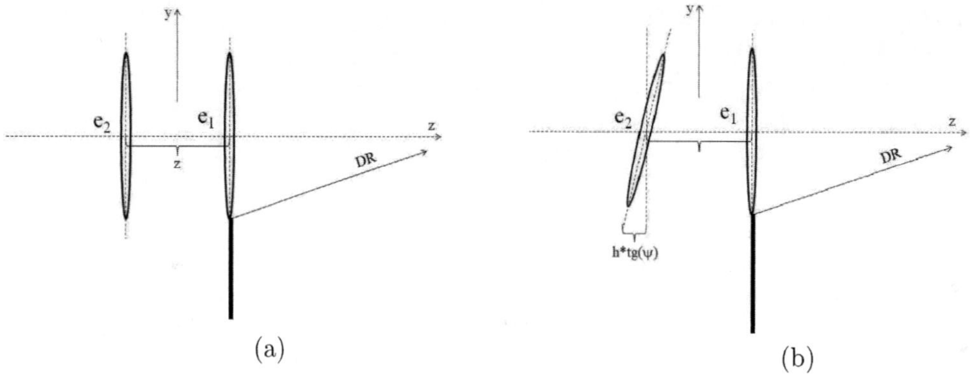

Figure 2. The additional phase difference appearance for particles with different incident angles. Here the impact parameters for both particles are equal to h. In figure b the particle e_2 has an incident angle ψ.

its transverse size. In order to take into account the beam divergence let us consider "z0y" plane as shown in figure 1.

For normally incident particles (figure 2(a)) the phase difference is defined by the distance between particles inside the bunch (see equation (7)). However, if one of the particles passes near the half-plane with non zero incident angle (figure 2(b)) additional phase difference appears. Here we assume that radiation is emitted when the particle field reaches the point of radiation. Thus, new phase difference can be defined as

$$\phi(\mathbf{r},\psi) = \frac{2\pi}{\lambda\beta}(z + h\tan\psi), \tag{9}$$

where h is the impact parameter, which is considered to be constant for all particles inside the beam. Since in a new expression for the phase difference a new variable has appeared, we need to extend the expression for the beam form-factor (8). Let $S_{div}(\psi)$ be the angular distribution of the particles inside the beam. Thus, assuming that spatial and angular distributions are independent, another expression for the beam form-factor can be written

$$F_{\rm div}^{DR} = \left| \int dx dy dz d\psi S_{\rm long}(z) S_{\rm tr}(x,y) S_{\rm div}(\psi) \times \right.$$

$$\left. \times e^{-\frac{2\pi(y-h)\sqrt{1+\gamma^2\beta^2\sin^2\Theta\sin^2\xi}}{\gamma\beta\lambda}} e^{-\frac{2\pi i}{\lambda}(\frac{z}{\beta}+\frac{(y-h)\tan\psi}{\beta})} \right|^2 \tag{10}$$

Let consider the beam angular distribution as Gaussian distribution, and ψ to be small enough, thus $\tan(\psi) \sim \psi$. Then, one can analytically derive the form-factor, in the form

$$F_{\rm div}^{DR} \sim e^{-\frac{2\pi^2}{\lambda^2\beta^2}(\sigma_z^2 + h^2\sigma_{div}^2)}, \tag{11}$$

where $\sigma_{\rm z}$ is the standard particle deviation from the beam center in longitudinal direction, and $\sigma_{\rm div}$ is the standard particle deviation from the normal incidence. We would like to remind that in this approximation the transverse beam size has been neglected. For relativistic beam, $\beta \sim 1$, the divergence could play an essential role in the beam form-factor if the condition $\lambda \sim h\sigma_{\rm div}$ is satisfied. For $h \sim \gamma\lambda$ this condition can be transformed into $\sigma_{\rm div} \sim 1/\sqrt{2}\pi\gamma$. Let estimate the divergence contribution in a beam form-factor for the system of electron beam with $\gamma = 1000$ and divergence $\sigma_{\rm div} = 100$ μrad, passing at the half-plane, assuming that the wavelength of radiation is $\lambda = 800$ nm. The impact parameter thus equals to $h \approx \gamma\lambda = 1$ mm. The divergence contribution in a beam form-factor is expressed by the following exponent $e^{-2\pi^2\gamma^2\sigma_{\rm div}^2} \approx 0.82$. It is not negligible that means, the beam form-factor obtains an additional coefficient (less than unit) due to the beam divergence.

4. Conclusion

In presented work the role of the beam divergence in its form-factor for DR has been analyzed. It is shown how beam divergence could influence on the form-factor as well as when this influence may be significant. However, in the approximation we used the transverse beam size has been neglected. Such assumption is allowed only if the beam size is much smaller than the impact parameter. Moreover, in such approximation, even for relativistic beam its divergence has to be taken into account under the conditions $\sigma_{\rm div} \sim 1/\sqrt{2}\pi\gamma$. The influence of the transverse beam size on the beam form-factor usually can be neglected, that allows us to equate the longitudinal form-factor(which depends only on longitudinal size) with its total form-factor.

In this work the evaluation of the beam form-factor for radiations, such as synchrotron or transition, to the beam form-factor for DR taking into account the beam divergence has been considered. More accurate expression for DR beam form-factor without neglecting the transverse beam size will be separately determined in the future.

Acknowledgments

This work was supported by the Ministry of Education and Science of the Russian Federation in the frames of Competitiveness Growth Program of National Research Nuclear University MEPhI, Agreement 02.A03.21.0005.

References

[1] Ter-MikaelianM L 1972 *High-energy electromagnetic processes in condensed media* (New York: Wiley Interscience)
[2] Potylitsyn A P, Ryazanov M I, Strikhanov M N, Tishchenko A A 2010 *Diffraction radiation from relativistic particles* (Berlin: Springer)
[3] Castellano M 1997 *Nucl. Instr. Meth. A* **394** 275-280
[4] Castellano M, Chiadroni E and Cianchi A 2010 *Nucl. Instr. Meth. in Phys. Res. A* **614** 163168
[5] Karataev P et al. 2008 *Phys. Rev. ST Accel. Beams* **11** 032804
[6] Kube G, Backe H, Lauth W and Schope H 2003 *Proc. DIPAC 2003, Mainz, Germany* 40
[7] Doucas G et al. 2006 *Phys. Rev. Spec. Top. - Accel. Beams* **9** 092801
[8] Xiang D and Huang W-H 2007 *Phys. Rev. ST Accel. Beams* **10(1)** 012801
[9] Karataev P et al. 2005 *Nucl. Instr. Meth. in Phys. Res. B* **227** 158169
[10] Fiorito R B and Rule D W 2001 *Nucl. Instr. Meth. in Phys. Res. B* **173** 6782
[11] Potylitsyn A P 2005 *Nucl. Instr. Meth. in Phys. Res. B* **227** 191-197
[12] Sukhikh L G, Kube G, Potylitsyn A P and Schlott V 2011 *Proceedings of DIPAC-2011, Hamburg, Germany*
[13] Nodvick J S and Saxon D S 1954 *Phys. Rev.* **96** 180-184
[14] Schneider G et al. 1997 *Nucl. Instr. Meth. in Phys. Res. A***397** 221-231
[15] Shibata Y et al. 1994 *Phys. Rev. E* **50(2)** 1479-1484

Journal of Physics: Conference Series **517** (2014) 012024 doi:10.1088/1742-6596/517/1/012024

Feasibility of Double Diffraction Radiation Target Interferometry for Compact Linear Accelerator Micro-train Bunch Spacing Diagnostics

D A Shkitov[1*], A P Potylitsyn[1], A S Aryshev[2], J Urakawa[2]
[1] Tomsk Polytechnic University, 30, Lenin St., Tomsk, 634050, Russia
[2] High Energy Accelerator Research Organization, 1-1 Oho, Tsukuba, Ibaraki, 305-0801, Japan

E-mail: shkitovda@tpu.ru

Abstract. In this paper the simulation of the interaction between micro-train electron beams with different parameters of energy, bunch length, bunch spacing which chosen for KEK LUCX accelerator facility and a double diffraction radiation target is considered. Calculation model and several accepted assumptions on the first step of our investigations are also described. Conducted researches allow us to conclude that applying the double diffraction radiation target interferometry as a tool for non-invasive micro-train bunch spacing diagnostics for compact linear accelerator is possible.

1. Introduction

Short-pulse THz systems are used in time-domain spectroscopy to understand biological processes and to create two- and three-dimensional images [1, 2]. Much of the recent interest in terahertz (0.1 – 10 THz) radiation comes from its ability to penetrate deep into many organic materials without the damage associated with ionizing radiation such as X-rays.

Nowadays there are a few ways to generate intense beams of THz radiation: optically pumped terahertz lasers, photomixing of near-IR lasers, backward-wave oscillators, direct multiplied sources and nonlinear optical processes occurring when an intense laser beam interacts with a material [3]. Another promising technique is to generate short, high-brightness THz-frequency coherent radiation pulses using a micro-train electron beam (THz sequence of a several fs-length electron bunches) of a compact accelerator. That micro-train beam is expected on LUCX (Laser Undulator Compact X-ray) accelerator facility at KEK. In this case, development of a robust and non-invasive micro-train bunch spacing diagnostics obtains vast importance.

Recent progress in double diffraction radiation (DR) target interferometer (full abbreviation – DDRTI) development [4] gives an excellent basis for such diagnostics implementation. When bunched electron beam passes through the slit of a double DR target (two conducting screen) it emits coherent radiation with intensity being dependent on double target parts relative position (interferogram). This interferogram can be measured when one part of the target moves with respect to another along the beam trajectory.

2. Experimental setup

The LUCX accelerator facility [5] expects to generate a few tens of MeV electron bunches with 100 – 300 fs duration and variable micro-bunch spacing [6]. There are two possible locations along the beam line which can be used for the DDRTI realization: "THz section" right after the 3.6-cell RF gun and the section downstream of 12-cell accelerating structure, so-called "Compton section". The principal difference of these two locations is the electron beam energy. It was proposed to use recently tested quasi-optical THz detector (QOD) [7] for interferometer commissioning. The main electron beam parameters for both possible DDRTI locations as well as QOD parameters are summarized in table 1.

Table 1. LUCX electron beam and QOD parameters.

	"THz section"	"Compton section"
Electron energy	8.25 MeV	30 MeV
Micro-bunch length	100 fs	300 fs
Transverse size (x, y)	200 μm	
Micro-bunch spacing	0 – 10 ps	
Normalized emittance (x, y)	5π mm×mrad	
Micro-bunch charge	100 pC	
Number of micro-bunches	2 – 16	
Electron energy spread	~1%	
QOD Spectral range	0.1 – >1.0 THz	
QOD Aperture	10 mm	
Minimum distance from output window to QOD	60 cm	
Maximum distance	a few meters	

3. Model assumptions

As a first step to investigate feasibility of micro-train bunch spacing diagnostics by double DR target interferometry the simulation was performed with several reasonable assumptions. First, the pseudophoton diffraction approach [8, 9] was used but without the far-field approximation. The main idea of this approach is that the relativistic particle electromagnetic (EM) field can be described as a field of so-called pseudophoton which behaves like a real photon. That's why the expression of the single particle EM field, accounting the phase factor and knowledge of the bunch form-factor are only needed to perform a simulation. Also in the model the finite size of two rectangular flat plates (parts) in double DR target was considered. Other assumptions were as follows:

- THz detector spectral sensitivity and the influence of the output vacuum window (glass transmission) onto THz spectrum were not taken into account.
- Calculation was done for point-like detector without accounting for the detector aperture.
- Electromagnetic field of charged particles had only transverse component.
- The only longitudinal bunch size with gaussian distribution was considered.

In order to find out how the bunch length and bunch spatial distribution in micro-train influence to the interferogram shape the simulation was carried out for electron beam parameters of two energies and bunch lengths corresponding to the "THz section" and "Compton section" of the LUCX accelerator according to table 1. The simulation parameters for electron beam, target and detector location are presented in table 2.

It is important to notice that for these two different cases of electron beam energy but for the same detector location the radiation is registered in two zones: far-field and near-field.

Table 2. Total simulation parameters.

	"THz section"	"Compton section"
Electron energy	8.25 MeV (γ≈16)	30 MeV (γ≈59)
Micro-bunch length	100 fs (30 μm)	300 fs (90 μm)
Basic micro-bunch spacing	1 ps (300 μm)	
Radiation wavelength range	0.03 – 3.0 mm	
Target slit width	2 mm	
Target-to-detector distance	700 mm far-field zone	700 mm near-field zone
Target dimensions (z, x)	46×20 mm^2	
Target tilt angle to the trajectory	π/4	
Point of detector	(0,0)	
Interferograms step	20 μm	

4. Simulation results

Interferograms for different number of bunches per train are presented at the following figures (figure 1 to figure 5). On each figure the number of bunches per train were varied from 1 to 5 and corresponded interferograms were plotted. The lower curves are calculated for 1 bunch and upper curves are simulated for 2, 3, 4 and 5 bunches in the micro-train respectively. The plot sets for different micro-bunch spacing l_{ij}, i.e. the distance between "i"-th and "j"-th bunch as well as for the different RMS electron bunch lengths σ are shown at the figure 6 and figure7.

It is clear that the figure 1, 2 and 4, 5 shows some limit value of the bunch spacing, i.e. if the spacing becomes smaller than some limit value it is impossible to derive micro-train spatial distribution from the interferogram's shape. However this value depends on the combination of electron beam energy and the bunch length as can be seen from the figure 2, 3 and 5 which are obtained for the same micro-bunch spacing l_{ij}.

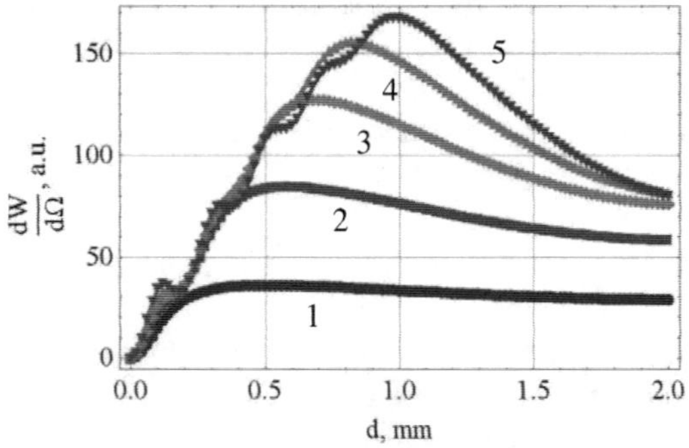

Figure 1. Simulated interferograms for different number of bunches per train: 1 to 5. l_{ij} = 200 μm, 8.25 MeV and σ = 30 μm.

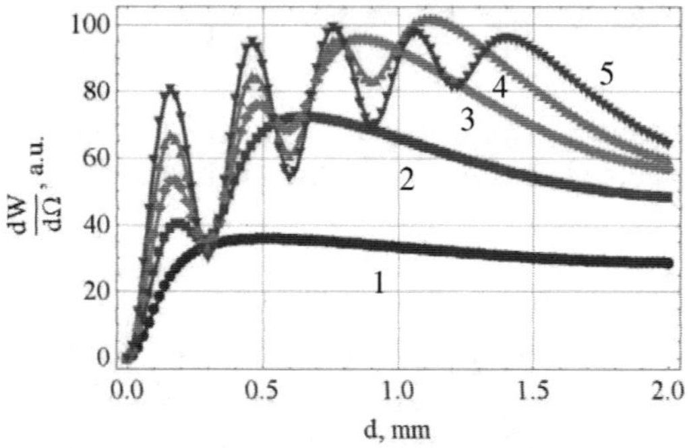

Figure 2. Simulated interferograms for different number of bunches per train: 1 to 5. l_{ij} = 300 μm, 8.25 MeV and σ = 30 μm.

It is important to notice that from the interferogram shape one can obtain the exact micro-train structure because the number of peaks in the interferograms corresponds to the number of the micro-bunches what is distinctly seen on the figure 2 and 5. Furthermore, it was observed that the position of the first minimum of the interferogram is almost equal to the distance between bunches.

Interferograms became even more clear and demonstrative for higher values of the bunch spacing.

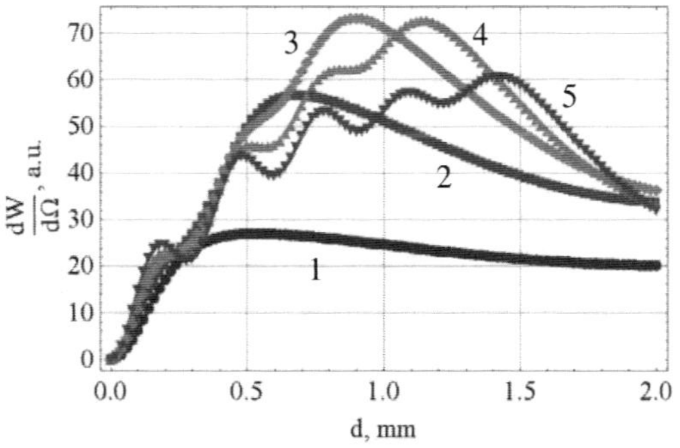

Figure 3. Simulated interferograms for different number of bunches per train: 1 to 5. l_{ij} = 300 μm, 8.25 MeV and σ = 90 μm.

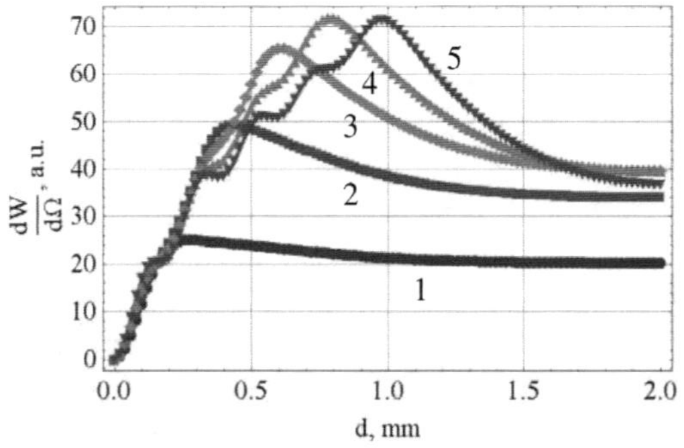

Figure 4. Simulated interferograms for different number of bunches per train: 1 to 5. l_{ij} = 200 μm, 30 MeV and σ = 90 μm.

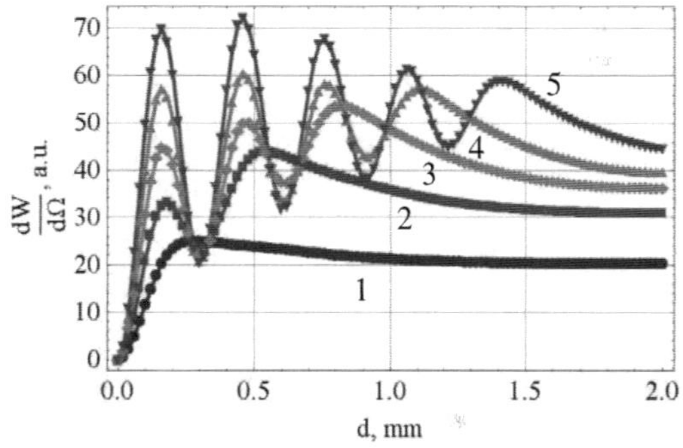

Figure 5. Simulated interferograms for different number of bunches per train: 1 to 5. l_{ij} = 300 μm, 30 MeV and σ = 90 μm.

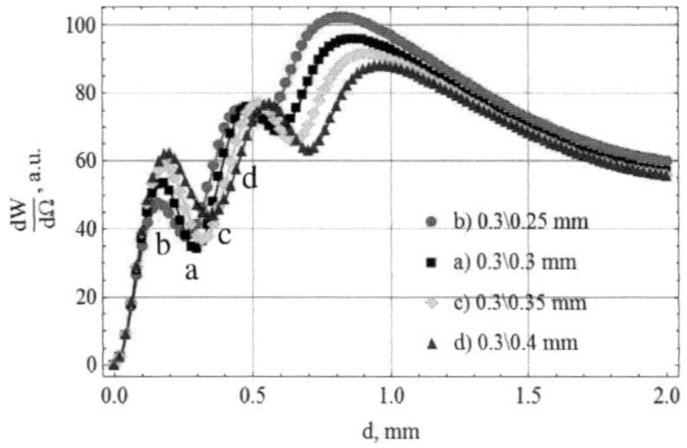

Figure 6. The interferograms simulated for different spacing between 3 bunches in the micro-train (8.25 MeV and σ = 30 μm).

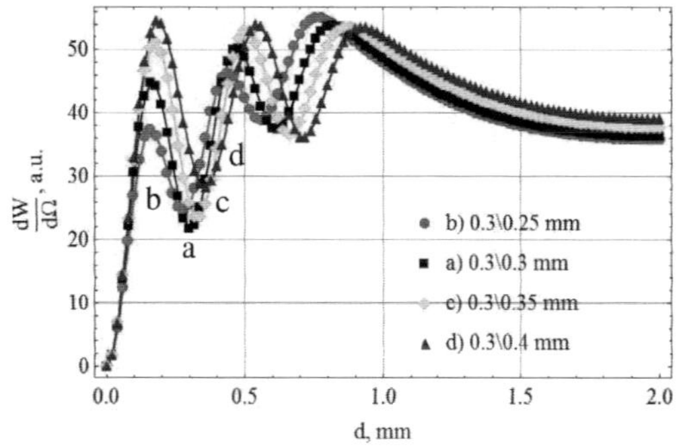

Figure 7. The interferograms simulated for different spacing between 3 bunches in the micro-train (30 MeV and σ = 90 μm).

Figure 6 and 7 shows the different bunch spacing within the micro-train influence onto DDRTI interferogram simulated for the different bunch lengths of 30 μm and 90 μm respectively. The values of l_{12} and l_{23} for corresponded plots are resented in table 3. As it was expected the maxima and minima positions of interferogram were slightly shifted when l_{23} was varied. In this case the position of the first minimum is approximately equal to the average distances between 3 bunches l_{12} and l_{23}. In addition it is noted that for reverse values of spacing ($l_{12}\backslash l_{23}$ and $l_{23}\backslash l_{12}$) the interferograms shapes are the same.

Table 3. Bunch spacing parameters for figure 6 and 7.

Figure	Bunch spacing	black (a)	red (b)	green (c)	blue (d)
6	$l_{12}\backslash l_{23}$	0.3\0.3 mm	0.3\0.25 mm	0.3\0.35 mm	0.3\0.4 mm
7	$l_{12}\backslash l_{23}$	0.3\0.3 mm	0.3\0.25 mm	0.3\0.35 mm	0.3\0.4 mm

Figure 8 shows the 3 bunch micro-train interferograms calculated for the different spacing l_{12} and l_{23} but for the same average spacing. Bunch spacings for curves from bottom to top were equal to: 0.5\0.5 (a), 0.45\0.55 (b), 0.4\0.6 (c), 0.35\0.65 (d), 0.3\0.7 (e) and 0.2\0.8 (f) mm respectively. The interferogram shape for the cases where the spacing differences were less than 40% relatively to the average value remains the same (first two curves from the bottom). And for spacing differences more than 40% the interferogram shape totally changes in comparison with the bottom curve what leads to the inability to determine the number of bunches. Furthermore this observation sets some boundary of applicability for maximum space difference between bunches in the micro-train that can be found from the interferogram.

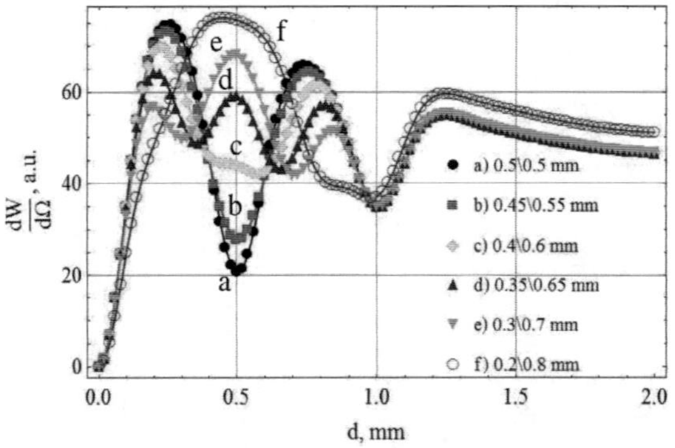

Figure 8. The interferograms simulated for different spacing between 3 bunches in the micro-train (30 MeV and $\sigma = 90$ µm).

Nevertheless the qualitative consideration of the above-mentioned assumptions in simulations (detector aperture and transversal size of bunches) allows to conclude that it will lead to the definite interferograms shape changing. Its shape will become smoother, the minimum around zero (d = 0) will be shifted towards higher values and differences between maxima and minima will become smaller. But the positions of minima will almost remain unchanged because micro-train form-factor has the determining significance and contains strongly expressed harmonics. As is well known the longitudinal electromagnetic field component of the charged particles is γ times less than transverse component and can significantly affect the results for the low electron beam energy micro-train characterization. In the relativistic case presented here the effect is negligible. However the longitudinal field component will be taken into account in the future calculation.

5. Conclusions and plans

The obtained result demonstrates a clear dependence of simulated interferograms shape on the bunch spacing in the micro-train. The possibility of the number of bunches in micro-train direct observation through the interferogram shapes was demonstrated. The average bunch spacing within the micro-train can be determined from the first minimum of the interferogram.

An additional investigation should be performed for better understanding of the micro-train beam and double DR target interferometer interaction process. As an example, the influence of the bunch form-factor shape and the beam position relative to the slit will be considered in the simulation. Accounting for the double DR target interferometer plate's adjustment inaccuracies will be implemented for better experimental setup construction and alignment. Cross-check measurements with a deflecting cavity and THz Michelson interferometer at KEK: LUCX facility is also considered.

Applying double diffraction radiation target interferometry for the non-invasive determination of the bunch spacing can lead to construction of a robust diagnostics for modern accelerators and compact THz sources.

Acknowledgments
This work was supported by JSPS KAKENHI Grant number 23226020 and 24654076 and by Grant of Russian MES program "Nauka" number 2.1799.2011.

References

[1] Mueller E R 2003 *The Industrial Physicist* (August/September)
[2] Tonouchi M. 2007 *Nature Photon.* **1** 97
[3] Gallerano G P et al. 2004 *Proceedings of the FEL Conference* 216
[4] Shkitov D A, Zhang J B, Naumenko G A, et al. 2013 *Proceedings of IPAC* **MOPME067** 583
[5] Fukuda M et. al. 2011 *Nucl. Instrum. Methods Phys. Res.* Sect. A **637** S67
[6] Aryshev A et. al. 2013 *Proceedings of the 10th Particle Accelerator Society of Japan* **SUP020**
[7] Shevelev M et. al. *Proceedings of RREPS13* to be published at *J. of Physics: Conf. Series*
[8] Ter-Mikaelian M L 1972 *High-Energy Electromagnetic Processes in Condensed Media* (New York: Wiley-Interscience)
[9] Potylitsyn A P et. al. 2010 *Diffraction Radiation from Relativistic Particles* (Berlin: Springer)

Electromagnetic fields and the radiation of a charged particle rotating along a closed trajectory around a dielectric cylinder

A S Kotanjyan and A A Saharian

Yerevan State University, 1 Alex Manoogian Street, 0025 Yerevan, Armenia
Institute of Applied Problems in Physics, 25 Nersessian Street, 0014 Yerevan, Armenia

E-mail: anna.kotanjyan@ysu.am, saharian@ysu.am

Abstract. We investigate the electromagnetic fields and the radiation intensity for a charged particle moving along an arbitrary closed orbit around a dielectric cylinder immersed into a homogeneous medium. These results generalize our previous research in the special case of a circular orbit. For the latter geometry it has been shown that under certain conditions strong narrow peaks appear in the angular distribution of the radiation intensity in the exterior medium. We discuss the influence of the trajectory shift from the circular one on the characteristics of the peaks.

1. Introduction

Due to the unique characteristics of synchrotron radiation, such as broad spectrum, high flux, high brilliance, it has wide applications (for reviews see [1]-[3]). Synchrotron light is an ideal tool for many types of research and also has industrial applications including materials science, biological and life sciences, medicine, chemistry. The wide applications of the synchrotron radiation motivate the importance of investigations for various mechanisms of control for the radiation parameters. In particular, it is of interest to consider the influence of a medium on the characteristics of the radiation. The synchrotron radiation from a charged particle circulating in a homogeneous medium was considered in [4]-[9], where it has been shown that the interference between the synchrotron and Cherenkov radiations leads to remarkable effects. New interesting features arise in the case of inhomogeneous media.

In a series of papers [10]-[16], we have investigated the influence of cylindrical boundaries between two dielectrics on the characteristics of the synchrotron radiation. It has been shown that under the Cherenkov condition for the material of the cylinder and the particle velocity, strong narrow peaks appear in the angular distribution of the radiation intensity in the exterior medium with respect to the dielectric cylinder. At these peaks the radiated energy exceeds the corresponding quantity in the case of a homogeneous medium by several orders of magnitude. The radiation propagating inside the dielectric cylinder is studied in [17, 18]. Similar features for the radiation generated by a charge moving along a helical orbit around/inside a dielectric cylinder have been investigated in [19]-[22]. The corresponding problem for a charge moving in vacuum has been widely discussed in the literature (see, for instance, [2, 3] and references given therein).

In the previous research we have assumed that the particle moves along a circular trajectory concentric with the cylinder axis or, in the case of helical motion, the projection of the helix on the plane perpendicular to the cylinder axis is a concentric circle. In realistic situations the trajectory of the particle may differ from the concentric circular one and it is of interest to investigate the influence of this shift on the parameters of the peaks in the radiation intensity. In the present paper we consider the electromagnetic fields and the radiation intensity from a charge moving around a dielectric cylinder along a trajectory the projection of which on the plane perpendicular to the cylinder axis is an arbitrary closed curve.

2. Electromagnetic fields in the exterior medium

Consider a point charge q moving along a helical trajectory around a dielectric cylinder with radius ρ_c. The permittivity for the cylinder will be denoted by ε_0 and the cylinder is immersed in a homogeneous medium with dielectric permittivity ε_1 (magnetic permeabilities will be taken to be unit). In what follows the cylindrical coordinate system (ρ, ϕ, z), with the axis z directed along the cylinder axis, will be used. We will assume that the projection of the particle trajectory on the plane perpendicular to the z-axis is an arbitrary closed curve and the corresponding motion of the charge is described by the functions $\rho = \rho_e(t)$ and $\phi = \phi_e(t)$ with $\rho_e(t) > \rho_c$. The latter condition means that the charge does not cross the cylinder surface. The projection of the charge velocity along the z-axis will be denoted $v_\parallel$ assuming that it is a constant. The components of the current density created by charge are given by the formula

$$j_l = (v_l q/\rho)\delta(\rho - \rho_e(t))\delta(\phi - \phi_e(t))\delta(z - v_\parallel t),\ l = \rho, \phi, z, \tag{1}$$

where v_ρ and v_ϕ are the corresponding components of the charge velocity.

By making use of the formula for the Green function given earlier in [10], in the Lorentz gauge for the vector potential one finds the following expression (the details will be presented elsewhere)

$$A_l(\mathbf{r},t) = -\frac{q}{\pi c} \sum_{m,n=-\infty}^{+\infty} \sum_{l'=\rho,\phi,z} \int_{-\infty}^{+\infty} dk_z\, G_{ll',n}(m, k_z, \omega_n(k_z), \rho) e^{im\phi + ik_z z - i\omega_n(k_z)t}, \tag{2}$$

where

$$\omega_n(k_z) \equiv n\omega_0 + k_z v_\parallel,\ \omega_0 = 2\pi/T, \tag{3}$$

with T being the period of the transverse motion. For the functions in the integrand of (2) we have

$$\begin{aligned} G_{ll',n}(m, k_z, \omega, \rho) &= \frac{\pi}{4i^{\sigma_l + 1 - \sigma_{l'}}} \sum_{p=\pm 1} p^{\sigma_l - \sigma_{l'}} B^{(p)}_{l',m,n} H_{m+p}(\lambda_1 \rho), \\ G_{lz,n}(m, k_z, \omega, \rho) &= \frac{i^{2-\sigma_l} k_z}{2\rho_c} H_{m,m,z}(n, \lambda_1) \frac{J_m(\lambda_0 \rho_c)}{\alpha_m V^H_m} \sum_{p=\pm 1} p^{\sigma_l - 1} \frac{J_{m+p}(\lambda_0 \rho_c)}{V^H_{m+p}} H_{m+p}(\lambda_1 \rho), \\ G_{zz,n}(m, k_z, \omega, \rho) &= -(i\pi/2) \left[J_{m,m,z}(n, \lambda_1) - H_{m,m,z}(n, \lambda_1) V^J_m / V^H_m \right] H_m(\lambda_1 \rho), \end{aligned} \tag{4}$$

where $l, l' = \rho, \phi$, $\sigma_\rho = -\sigma_\phi = 1$ and $G_{z\rho,n}(m, k_z, \omega, \rho) = G_{z\phi,n}(m, k_z, \omega, \rho) = 0$. In (4), $J_m(x)$ is the Bessel function, $H_m(x) \equiv H^{(1)}_m(x)$ is the Hankel function of the first kind. We have also defined new functions

$$F_{m',m,l}(n, \lambda_1) = \frac{1}{T} \int_0^T dt\, v_l(t) F_{m'}(\lambda_1 \rho_e(t)) e^{-im\phi_e(t) + in\omega_0 t},\ F = J, H, \tag{5}$$

with $\lambda_j^2 = \omega_n^2(k_z)\varepsilon_j/c^2 - k_z^2$, $j = 0, 1$. Other notations used in (4) are as follows:

$$V_m^F = J_m(\lambda_0\rho_c)\frac{\partial F_m(\lambda_1\rho_c)}{\partial\rho_c} - F_m(\lambda_1\rho_c)\frac{\partial J_m(\lambda_0\rho_c)}{\partial\rho_c},\ F = J, H, \tag{6}$$

and

$$\begin{aligned} B_{l,m,n}^{(p)} &= J_{m+p,m,l}(n,\lambda_1) - H_{m+p,m,l}(n,\lambda_1)\,V_{m+p}^J/V_{m+p}^H \\ &+\frac{ip^{\sigma_l-1}\lambda_0}{\pi\rho_c\alpha_m}J_m(\lambda_0\rho_c)\frac{J_{m+p}(\lambda_0\rho_c)}{V_{m+p}^H}\sum_{p'=\pm1}p'^{\sigma_l}\frac{H_{m+p',m,l}(n,\lambda_1)}{V_{m+p'}^H}. \end{aligned} \tag{7}$$

The function α_m is defined by the relation

$$\alpha_m = \frac{\varepsilon_0}{\varepsilon_1-\varepsilon_0} - \frac{\lambda_0 J_m(\lambda_0\rho_c)}{2}\sum_{p=\pm1}p\frac{H_{m+p}(\lambda_1\rho_c)}{V_{m+p}^H}. \tag{8}$$

Note that the equation $\alpha_m = 0$ determines the eigenmodes of the dielectric cylinder.

With the vector potential (2) we can find the electric and magnetic fields by using the standard formulas. In the special case of circular helix coaxial with the cylinder axis one has $F_{m',m,l}(n,\lambda_1) = v_l F_{m'}(\lambda_1\rho_e)\delta_{nm}$ and the expressions given above are reduced to the ones derived in [22].

3. Radiation intensity

Having the electromagnetic fields we can investigate the intensity of the radiation propagating in the exterior medium. For $\lambda_1^2 < 0$ the corresponding Fourier components are exponentially damped for large values ρ, and the radiation at large distances from the cylinder is present only under the condition $\lambda_1^2 > 0$. The average energy flux per unit time through the cylindrical surface of radius ρ, coaxial with the dielectric cylinder, is given by the Poynting vector:

$$I = \frac{c}{4\pi T}\int_0^T dt\int_0^{2\pi}d\phi\int_{-\infty}^{+\infty}dz\,\rho\mathbf{n}_\rho\cdot[\mathbf{E}\times\mathbf{H}]. \tag{9}$$

At large distances from the cylinder, the variable k_z in the formulas given above corresponds to the projection of the wave vector for the radiated photon on the z-axis. Instead of k_z it is convenient to introduce the angle θ between the photon wave vector and the cylinder axis in accordance with the relation

$$k_z = \frac{n\omega_0}{c}\frac{\sqrt{\varepsilon_1}\cos\theta}{1-\beta_{1\|}\cos\theta},\ \beta_{1\|} = \frac{v_\|}{c}\sqrt{\varepsilon_1}. \tag{10}$$

For the radiation frequency one has $|\omega_n|$ with

$$\omega_n = n\omega_0/(1-\beta_{1\|}\cos\theta). \tag{11}$$

Hence, n determines the harmonic number of the radiation. We also have the following expressions for λ_1 and λ_0:

$$\lambda_1 = \frac{n\omega_0}{c}\frac{\sqrt{\varepsilon_1}\sin\theta}{1-\beta_{1\|}\cos\theta},\quad \lambda_0 = \frac{n\omega_0}{c}\frac{\sqrt{\varepsilon_0-\varepsilon_1\cos^2\theta}}{1-\beta_{1\|}\cos\theta}. \tag{12}$$

The angular density of the radiation intensity at a given harmonic n, $dI_n/d\Omega$, with $d\Omega = \sin\theta d\theta d\phi$ being the solid angle element, is defined by

$$I = \sum_{n=-\infty}^{+\infty} \int d\Omega \frac{dI_n}{d\Omega}. \tag{13}$$

By using (2), at large distances from the cylinder, for the angular distribution of the radiated energy per unit time we find

$$\frac{dI_n}{d\Omega} = \frac{q^2}{4\pi^3 c} \frac{n^2\omega_0^2\sqrt{\varepsilon_1}}{\left|1-\beta_{1\|}\cos\theta\right|^3} \sum_{m=-\infty}^{+\infty} \left[\left|\sum_{p=\pm1} pD_{m,n}^{(p)}\right|^2 + \left|\sum_{p=\pm1} D_{m,n}^{(p)}\right|^2 \cos^2\theta\right], \tag{14}$$

with the notation

$$\begin{aligned} D_{m,n}^{(p)} &= \frac{\pi}{2ic}\left[J_{m+p,m}(n,\lambda_1) - H_{m+p,m}(n,\lambda_1)\frac{V_{m+p}^J}{V_{m+p}^H}\right] \\ &+\frac{i\pi\lambda_1}{2ck_z}\left[J_{m,m,z}(n,\lambda_1) - H_{m,m,z}(n,\lambda_1)\frac{V_m^J}{V_m^H}\right] \\ &+p\frac{J_m(\lambda_0\rho_c)}{c\rho_c\alpha_m}\frac{J_{m+p}(\lambda_0\rho_c)}{V_{m+p}^H}\left[\frac{k_z}{V_m^H}H_{m,m,z}(n,\lambda_1) + \frac{\lambda_0}{2}\sum_{l=\pm1}\frac{H_{m+l,m}(n,\lambda_1)}{V_{m+l}^H}\right], \end{aligned} \tag{15}$$

and

$$F_{m+p,m}(n,\lambda_1) = F_{m+p,m,\phi}(n,\lambda_1) - ipF_{m+p,m,\rho}(n,\lambda_1),\ F = J, H. \tag{16}$$

First let us consider a special case of a circular helix with $\rho_e(t) = \rho_e = \text{const}$ and $\phi_e(t) = \omega_0 t$. In this case $v_\rho = 0$ and $F_{m',m,l}(n,\lambda_1) = v_l F_{m'}(\lambda_1\rho_e)\delta_{nm}$. Hence, in (14) the term $m = n$ contributes only. The features of the radiation intensity in this special case have been discussed in [22]. By using Debye's asymptotic expansions for the Bessel and Neumann functions, it can be seen that under the condition $|\lambda_1|\rho_c < n$, at points where the real part of the function α_n, given by formula (8), is equal to zero, the contribution of the imaginary part of this function into the coefficients $D_{n,n}^{(p)}$ can be exponentially large for large values n. This leads to the appearance of strong narrow peaks in the angular distribution for the radiation intensity at a given harmonic n. The condition for the real part of the function α_n to be zero has the form:

$$\sum_{l=\pm1}\left[\frac{\lambda_1}{\lambda_0}\frac{J_{n+l}(\lambda_0\rho_c)Y_n(\lambda_1\rho_c)}{J_n(\lambda_0\rho_c)Y_{n+l}(\lambda_1\rho_c)} - 1\right]^{-1} = \frac{2\varepsilon_0}{\varepsilon_1-\varepsilon_0}, \tag{17}$$

where $Y_\nu(x)$ is the Neumann function. Note that (17) is obtained from the equation determining the eigenmodes for the dielectric cylinder by the replacement $H_n \to Y_n$. Equation (17) has no solutions for $\lambda_0^2 < 0$, which is possible only for $\varepsilon_0 < \varepsilon_1$. Hence, the peaks do not appear for the case $\lambda_0^2 < 0$ which corresponds to the angular region $\cos^2\theta > \varepsilon_0/\varepsilon_1$. As necessary conditions for the presence of the strong narrow peaks in the angular distribution for the radiation intensity one has

$$\varepsilon_0 > \varepsilon_1,\ \tilde{v}\sqrt{\varepsilon_0}/c > 1, \tag{18}$$

where $\tilde{v} = \sqrt{v_\|^2 + \omega_0^2\rho_c^2}$ is the velocity of the charge image on the cylinder surface. The second condition in (18) is the Cherenkov condition for the velocity of the charge image on the cylinder surface and dielectric permittivity of the cylinder. In figure 1, the full curve displays the dependence of the angular density for the number of the quanta radiated per unit time,

$$\frac{dN_n}{d\Omega} = \frac{1}{\hbar|\omega_n|}\frac{dI_n}{d\Omega}, \tag{19}$$

as a function of the angle θ for the parameters $\beta_{1\perp} = 0.9$, $\beta_{1\|} = 0.4$, $\rho_c/\rho_e = 0.95$, $n = 10$. We assumed that the cylinder is made of fused quartz with permittivity $\varepsilon_0 = 3.75$ and the particle moves in the vacuum ($\varepsilon_1 = 1$). The dashed curve corresponds to the radiation in the vacuum ($\varepsilon_0 = \varepsilon_1 = 1$). For the heights of the peaks one has (from left to right) $(\hbar cT/q^2)dN_n/d\Omega \approx 2.8, 17, 16.9$. For $\rho_e \sim 1$ cm the corresponding radiation starts from centimeter wavelengths and falls into the terahertz region for higher harmonics. Note that for these wavelengths the imaginary part of the quartz permittivity is small.

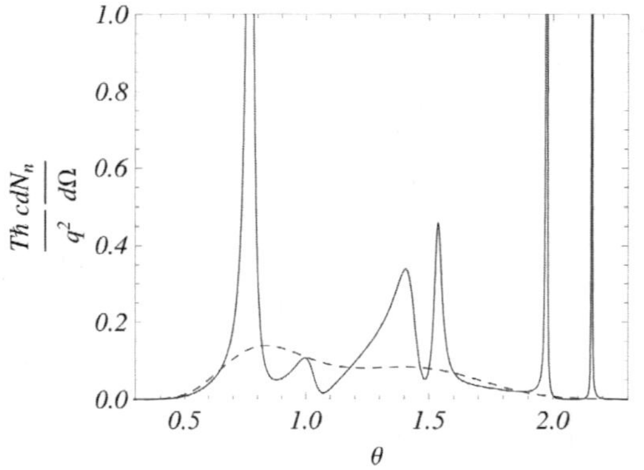

Figure 1. The dependence of the angular density for the number of radiated quanta, $(T\hbar c/q^2)dN_m/d\Omega$, per period T of the transverse motion, as a function of the angle θ for $\beta_{1\perp} = 0.9$, $\beta_{1\|} = 0.4$, $\rho_c/\rho_e = 0.95$, $n = 10$. The full curve corresponds to the radiation in the presence of a cylinder made of fused quartz ($\varepsilon_0 = 3.75$, $\varepsilon_1 = 1$) and the dashed curve is for the radiation in the vacuum ($\varepsilon_0 = \varepsilon_1 = 1$).

Analytical estimates for the heights and the widths of the peaks are given in [22]. These estimates show that at the peaks the angular density of the radiation intensity increases with increasing n. In realistic situations the growth of the radiation intensity is limited by several factors. In particular, the factors which limit the increase, are the imaginary part and the dispersion of the dielectric permittivity ε_1 (see the discussion in [22]). As it can be seen from (14), for a fixed value of the charge energy and in the case of circular motion, the product $\rho_e^2 dI_n/d\Omega$, with $n \neq 0$, is a function of the ratio ρ_c/ρ_e alone. From here it follows that for a fixed value of $\rho_e - \rho_c$ and for large values of ρ_c, the angular density of the radiation intensity, $dI_n/d\Omega$, decays as $1/\rho_e^2$. In this limit, the Cherenkov radiation remains only if the condition $\beta_{1\|} > 1$ is obeyed. The features of this radiation are discussed in [22]. Note that, in the exterior region the Cherenkov radiation is absent for the example presented in figure 1.

For general transverse motion of the charge and for a given radiation harmonic n, the angular density of the radiation intensity contains the summation over m (see (14)). Now for a given harmonic n we can have a set of peaks corresponding to different values of m and determined by the condition (17) replacing n in the indices of the cylinder functions by m. From the asymptotic formulas for the cylinder functions for large values of the order it follows that the peaks come from the part of the trajectory for which $|\lambda_1|\rho_e(t) < |m|$. At the peaks the coefficients $D_{m,n}^{(p)}$ in the expression for the radiation intensity are estimated by $D_{m,n}^{(p)} \sim Y_{m,m,z}(n, \lambda_1), Y_{m+l,m}(n, \lambda_1)$, where the functions $Y_{m,m,z}(n, \lambda_1)$ and $Y_{m+l,m}(n, \lambda_1)$ are defined by the relations (5) and (16) taking in the integrand of (5) $F_m(\lambda_1 \rho_e(t)) = Y_m(\lambda_1 \rho_e(t))$. For large values of $m > 0$ we can use the Debye's asymptotic formula

$$Y_m(my) \sim \frac{2(\operatorname{sgn} y)^{m+1} e^{m\zeta(y)}}{\sqrt{2\pi m}(1-y^2)^{1/4}}, \quad |y| < 1, \tag{20}$$

with $\zeta(y) = \ln[(1+\sqrt{1-y^2})/|y|] - \sqrt{1-y^2}$. The most strong peaks correspond to the part of the trajectory for which the function $\zeta(\lambda_1 \rho_e(t)/m)$ is maximum. For given values of n and m this corresponds to the minimum of the function $\rho_e(t)$. The parameters of the peaks depend on the form of the trajectory.

4. Conclusion

In the present paper we have considered the radiation from a charged particle moving around a dielectric cylinder along a helical trajectory the projection of which on the plane perpendicular to the cylinder axis is an arbitrary closed curve. Our aim was to study the sensitivity of the peaks appearing in the angular distribution of the radiation intensity with respect to distortions of the particle trajectory. By using the Green function from [10], the expression for the corresponding vector potential is presented in the region outside the cylinder.

We have derived a formula for the spectral-angular density of the radiation intensity in the exterior medium. The latter is given by the expression (14) with the coefficients defined by (15). In the case of a circular motion concentric with the cylinder axis, in (14) the term with $m = n$ survives only and we obtain the results previously discussed in [22]. Instead of a single peak in the angular distribution of the radiation intensity at a given harmonic n, for the case of general motion we have set of peaks corresponding to different values of m. The parameters of the peaks depend on the form of the trajectory. The most strong peaks correspond to the radiation from the part of the trajectory closest to the cylinder surface.

References

[1] Sokolov A A and Ternov I M 1986 *Radiation from Relativistic Electrons* (New York: ATP)
[2] Bordovitsyn V A (editor) 1999 *Synchrotron Radiation Theory and Its Development* (Singapore: World Scientific)
[3] Hofman A 2004 *The Physics of Synchrotron Radiation* (Cambridge: Cambridge University Press)
[4] Tsytovich V N 1951 Westnik MGU **11** 27
[5] Kitao K 1960 Progr. Theor. Phys. **23** 759
[6] Erber T White D and Latal H G 1976 Acta Phys. Austriaca **45** 29
[7] Schwinger J Tsai W-Y and Erber T 1976 Ann. Phys. **96** 303
[8] Erber T White D Tsai W-Y and Latal H G 1976 Ann. Phys. **102** 405
[9] Rynne T M Baumgartner G B and Erber T 1987 J. Appl. Phys. **49** 2233
[10] Grigoryan L Sh Kotanjyan A S and A.A. Saharian 1995 Sov. J. Contemp. Phys. **30** 1
[11] Kotanjyan A S Khachatryan H F Petrosyan A V and Saharian A A 2000 Sov. J. Contemp. Phys. **35** 1
[12] Kotanjyan A S and Saharian A A 2001 Sov. J. Contemp. Phys. **36** 7
[13] Kotanjyan A S and Saharian A A 2002 J. Contemp. Phys. **37** 263
[14] Kotanjyan A S and Saharian A A 2002 Mod. Phys. Lett. A **17** 1323
[15] Kotanjyan A S 2003 Nucl. Instrum. Methods **B201** 3
[16] Saharian A A and Kotanjyan A S 2005 J. Phys. A **38** 4275
[17] Saharian A A and Kotanjyan A S 2012 Int. J. Mod. Phys. B **26** 1250033
[18] Kotanjyan A S and Saharian A A 2012 J. Phys.: Conf. Ser. **357** 012009
[19] Saharian A A and Kotanjyan A S and Grigoryan M L 2007 J. Phys. A **40** 1405
[20] Saharian A A and Kotanjyan A S 2007 J. Phys. A **40** 10641
[21] Arzumanyan S R Grigoryan L Sh Khachatryan H F Kotanjyan A S and Saharian A A 2008 Nucl. Instrum. Methods **B266** 3703
[22] Saharian A A and Kotanjyan A S 2009 J. Phys. A **42** 135402

Zemax Simulations of Transition and Diffraction Radiation

T Aumeyr[1], M G Billing[2], L M Bobb[1,3], B Bolzon[3,4,5], P Karataev[1], T Lefevre[3] and S Mazzoni[3]

[1] John Adams Institute at Royal Holloway, Egham, Surrey, UK
[2] Cornell University, Ithaca, New York, USA
[3] CERN European Organisation for Nuclear Research, Geneva, Switzerland
[4] Cockcroft Institute, Worrington, Cheshire, UK
[5] University of Liverpool, Liverpool, Merseyside, UK

E-mail: Thomas.Aumeyr@rhul.ac.uk

Abstract.
Charged particle beam diagnostics is a key task in modern and future accelerator installations. The diagnostic tools are practically the "eyes" of the operators. Precision and resolution of the diagnostic equipment are crucial to define the performance. Transition and Diffraction Radiation (TR and DR) are widely used for electron beam parameter monitoring. However, the precision and resolution of those devices are determined by how well the production, transportation and detection of these phenomena are understood. This paper reports on initial simulations of TR and DR spatial characteristics. A good consistency with theory is demonstrated. Also, optical system alignment issues are discussed.

1. Introduction

Beam dynamic considerations demand very tight tolerances on most beam parameters and these in turn dictate most of the requirements for beam instrumentation [1]. Particle beams with extremely small emittances are generated in the damping rings and these must be conserved over several tens of km of beam lines. This requires a precise control of the beam position over a long distance. In most cases, the measurement of the beam size serves directly to compute the transverse beam emittance. For next generation linear colliders such as the Compact Linear Collider (CLIC) [1] or the ILC, transverse beam size measurements must have a resolution on the micron-scale. Currently, a laser-wire scanner [2] is the main candidate for non-invasive high resolution measurements. However, over a distance of more than 40 km many laser-wire monitors would be required. This is both costly and difficult to maintain.

1.1. Transition Radiation

Optical Transition Radiation (OTR) appearing when a charged particle crosses a boundary between two media with different dielectric properties has widely been used as a tool for transverse profile measurements of charged particle beams in various facilities worldwide. OTR monitors are simple, robust, and give a direct image of a two-dimensional beam profile. The resolution of the OTR monitors is normally defined as a root-mean-square of the so-called

OTR point spread function (PSF) [3]. The angular distribution can be used for beam angular divergence measurements [4].

Following the approach for calculating TR from a particle passing through a boundary between vacuum and an ideal conductor [5] and applying an ultra-relativistic approximation $(\theta_x, \theta_y, \gamma^{-1} \ll 1)$, the following equation is obtained for the vertical polarisation of the spectral-angular distribution of intensity [6]:

$$\frac{d^2W_y^{TR}}{d\omega d\Omega} = \frac{\alpha}{\pi^2}\frac{{\theta_y}^2}{\left(\gamma^{-2} + {\theta_x}^2 + {\theta_y}^2\right)^2} \tag{1}$$

Here θ_x and θ_y are the radiation observation angles measured either from the mirror reflection direction or from the particle trajectory, α is the fine structure constant and γ is the charged particle Lorentz factor. Eq. 1 is TR in the case of normal incidence, i.e. with no target tilt. However, in case of large target tilt angles and ultra-relativistic electrons, Eq. 1 can still be used for a tilted target.

1.2. *Diffraction Radiation*

Diffraction Radiation (DR) is produced when a relativistic charged particle moves in the vicinity of a medium [7]. The spatial-spectral properties of DR are sensitive to a range of electron beam parameters [8–10]. Furthermore, the energy loss due to DR is so small that the electron beam parameters are unchanged. Therefore DR can be used to develop non-invasive diagnostic tools.

The ODR model considers the case when a charged particle moves through a slit between two semi-planes i.e. only DR produced from the target is considered (see Fig. 1). In the case of a horizontal slit, the vertical polarisation component is sensitive to beam size [10]. Eq. 2 gives the expression for the ODR vertical polarisation component convoluted with a Gaussian distribution [10], where $t_{x,y} = \gamma\theta_{x,y}$, λ is the observation wavelength, σ_y is the rms vertical beam size, a is the target aperture size, $\overline{a_x}$ is the offset of the beam centre with respect to the centre of the slit and $\psi = \arctan\left(t_y/\sqrt{1+{t_x}^2}\right)$. This model is applicable when the transition radiation contribution from the tails of the Gaussian distribution is negligible, which means approximately $a \geq 4\sigma_y$.

$$\begin{aligned}\frac{d^2W_y^{slit}}{d\omega d\Omega} = \quad & \frac{\alpha\gamma^2}{2\pi^2}\frac{\exp\left(-\frac{2\pi a}{\gamma\lambda}\sqrt{1+{t_x}^2}\right)}{1+{t_x}^2+{t_y}^2} \\ & \times\left\{\exp\left[\frac{8\pi^2{\sigma_y}^2}{\lambda^2\gamma^2}\left(1+{t_x}^2\right)\right]\cosh\left(\frac{4\pi\overline{a_x}}{\gamma\lambda}\sqrt{1+{t_x}^2}\right) - \cos\left(\frac{2\pi a}{\gamma\lambda}t_y + 2\psi\right)\right\}\end{aligned} \tag{2}$$

This expression is valid for ultra-relativistic particles and large target tilt angles. For a slit width $a = 0$ follows considering the above approximation that $\overline{a_x} = 0$ and $\sigma_y = 0$, that Eq. 2 results in the angular distribution for OTR in Eq. 1.

2. Zemax

Optical system design is no longer a skill reserved for a few professionals. The Zemax Optical Design Program is a readily available commercial software package which integrates all the features required to conceptualise, design, optimise, analyse and tolerance virtually any optical system [11]. It is widely used in the optics industry as a standard design tool. Zemax supports two modes, geometrical ray tracing and Physical Optics Propagation (POP). Strictly speaking, geometrical ray tracing can only be used when the diffraction effect is negligible. The propagation of light is a coherent process. A wavefront traveling through free space or an optical medium, coherently interferes with itself. This is what is modelled in POP mode, using diffraction laws to propagate a wavefront through an optical system interface by interface.

In POP mode, the entire beam array must be stored in computer memory at once contrary to the ray tracing mode. Therefore, a large quantity of RAM is needed when using the POP mode for large sampling arrays and because imaging systems require large sampling to accurately model aberrations. The wavefront is modelled using an array of discretely sampled points, analogous to the discrete sampling using rays for geometric optics analysis. Each point in the array stores complex amplitude information about the beam. The entire array is then propagated in free space between optical interfaces. At each optical interface, a transfer function is computed which propagates the beam from one side of the optical surface to the other, using either Fresnel diffraction propagation or an angular spectrum propagation algorithm [12, 13]. Zemax automatically chooses the algorithm that yields the highest numerical accuracy. The diffraction propagation algorithms yield correct results for any propagation distance, for any arbitrary source and can account for any surface aperture, including User Defined Apertures (UDAs).

2.1. Physical Optics Propagation algorithm

Zemax propagates any spatial complex electric field defined by a 2D matrix. The electric field is represented in three dimensions as

$$\boldsymbol{E}(x, y, z) = E_x\hat{x} + E_y\hat{y} + E_z\hat{z}, \tag{3}$$

where all $\boldsymbol{E}$ values are complex and $\hat{x}$, $\hat{y}$ and $\hat{z}$ are the Cartesian unit vectors. Because the beam is propagating along the local z direction, the first approximation made is to neglect the E_z component. By keeping track of the electric field components along both the x and y axes, polarisation effects can be studied, such as transmission and reflection losses, polarisation aberrations and the polarisation state of the beam. If polarisation effects are not required, the x component of the field can be ignored, thus speeding up the computations.

For a beam with a radius R propagating over a distance Z, the Fresnel number N_F is defined by

$$N_F = \frac{2}{\lambda}\left(\sqrt{Z^2 + R^2} - Z\right). \tag{4}$$

Assuming $Z \gg R$, this expression reduces to

$$N_F = \frac{R^2}{\lambda Z}. \tag{5}$$

The Fresnel number is equal to the number of Fresnel rings on the source plane that can be seen from any point on the detection plane:

- Fraunhofer diffraction (far field): $N_F = 0$
- Fresnel diffraction (pre-wave zone if $R > \gamma\lambda$): $0 < N_F < 1$
- Angular Spectrum Propagation (near field): $N_F > 1$

The decision on the Fresnel numbers, and thus on the algorithms used by Zemax, is based on a pilot beam. By default, the pilot beam is an ideal Gaussian beam, with a waist, beam size, phase radius and relative z position. The initial parameters may be generated by fitting the Gaussian beam equations to the initial distribution. However, it is also possible to define a pilot beam if the electric field defined at the source is a non-Gaussian beam.

In any case, the pilot beam is then propagated from interface to interface. At each interface, new beam parameters, such as the new waist, phase radius, or position are computed. The properties of the pilot beam are then used to determine if the actual distribution is inside or outside the Rayleigh range, and what propagation algorithms are appropriate.

2.2. *Defining the source*

Several types of beam are already predefined as source in POP: Gaussian Waist, Gaussian Angle, Gaussian Size+Angle, Top Hat, File, Windows Dynamic Link Library (DLL) or Multimode. Moreover, any source of light can be provided by the user in POP. The user has to define, among others, the spatial distribution of the complex electric field of the source either in a *Beam File* or in a DLL. In both cases, the parameters of the pilot beam have to be defined. These are the pilot beam waist, the Rayleigh distance and the pilot position, both in x and y directions.

The most practical way to define a source using an external program, is to compute the initial electric field into a Windows Dynamic Link Library (DLL). Sample DLLs with source code are provided with Zemax. These examples are of great help to see how the DLL should be written to properly define the source amplitude and phase. The instructions to compile such a C source file into a DLL can be found on the Zemax support website.

With a DLL, it is possible to define between zero and eight user defined data values as input parameters in the computation of the beam properties. This is particularly useful when the electric field is defined by an equation, whose parameters have to be changed by the user from one simulation to the next (e.g. particle energy, observation wavelength, etc.).

Starting with the example C source file provided with Zemax, the only changes that had to be applied are the equation of the electric field, the user-defined data values and the pilot beam parameters which were set to 0. The OTR electric field programmed in the C source file is the approximation of the electric field for y polarisation component induced by a single electron on a target surface [14], based on the solution of Maxwell equations for a field of a charged particle moving in free space. Up to a constant it is (see [5])

$$E_{y,real} = const \cdot \frac{y}{\sqrt{x^2+y^2}} \left[\frac{2\pi}{\gamma\lambda} K_1 \left(\frac{2\pi}{\gamma\lambda} \sqrt{x^2+y^2} \right) - \frac{J_0 \left(\frac{2\pi}{\lambda} \sqrt{x^2+y^2} \right)}{\sqrt{x^2+y^2}} \right] \tag{6}$$

$$E_{y,imag} = 0.$$

Here x and y are two orthogonal coordinates of the target measured from the point of electron incidence, γ is the charged particle Lorentz factor, λ is the radiation wavelength, K_1 is the first order modified Bessel function, J_0 is the zero order Bessel function.

According to the Fresnel-Huygens principle, the field of the particles reflected off the target surface represents the radiation field. For TR the entire field is reflected and propagates towards the observation plane. In case of DR, only a part of the field is reflected, depending on the target geometry.

3. Simulations

3.1. *Comparing simulations with theory*

In Fig. 2, the ODR angular distribution for various slit sizes at the detector plane created by a single electron passing through a source with a 1 mm vertical slit ($\gamma = 4110, \lambda = 400$ nm) is shown. This corresponds to the experimental conditions at the ODR monitor at the Cornell Electron Storage Ring (CesrTA) [15]. The distance between the source and detector plane was set to 100 m to ensure the angular distribution is fully defined. Comparing with the analytical expression given in Eq. 2, a nearly perfect agreement can be observed.

At the Accelerator Test Facility (ATF), the OTR monitor uses an observation wavelength of $\lambda = 550$ nm for a beam Lorentz factor of $\gamma = 2500$ [10]. The far-field requirement for OTR is $L \gg \frac{\gamma^2\lambda}{2\pi} = 0.55$ m, therefore the distance of 100 m between source and detector plane was more than enough to achieve a perfect consistency. The size of the source was $r_{max} = 10 \cdot \frac{\gamma\lambda}{2\pi} = 2.188$ mm to fulfil an infinite target approximation. Fig. 3 shows the effect on

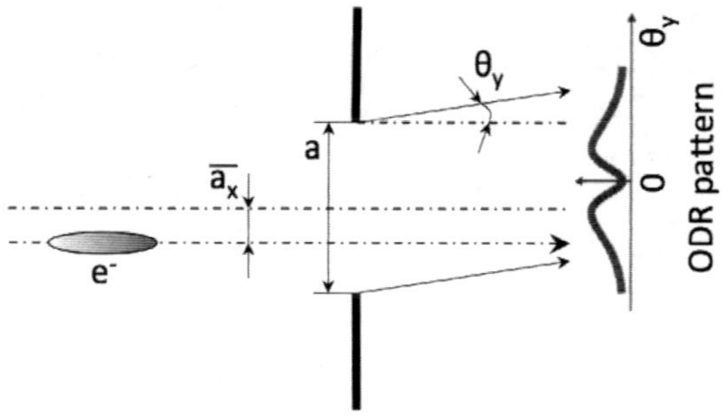

Figure 1: Geometry of the ODR production.

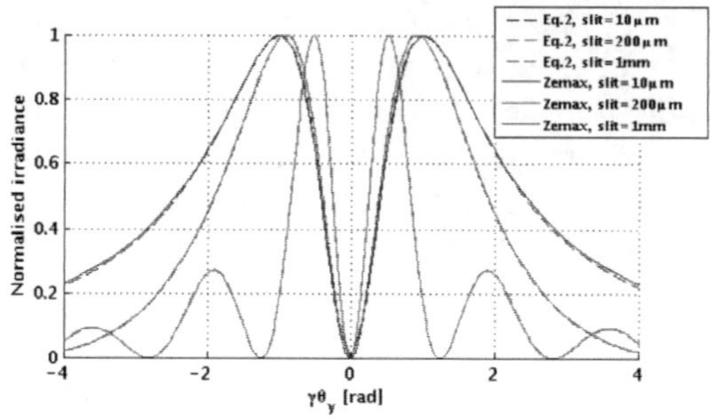

Figure 2: ODR angular distribution for various slit sizes compared with theory.

the angular distribution when moving the detector plane from the near-field into the far-field. The distribution was simulated at three different distances from the source - $\frac{\gamma^2\lambda}{2\pi}$, $2 \cdot \frac{\gamma^2\lambda}{2\pi}$ and $10 \cdot \frac{\gamma^2\lambda}{2\pi}$. This figure is in excellent agreement with analytical calculations [16].

The far-field condition for the parameters for the ODR monitor at CesrTA is fulfilled for a distance $L \gg \frac{\gamma^2\lambda}{2\pi} = 1.08$ m. The size of the simulated source was $r_{max} = 10 \cdot \frac{\gamma\lambda}{2\pi} = 2.617$ mm. As in the OTR case, Fig. 4 shows the effect on the angular distribution when moving the detector plane from the near-field into the far-field, this time using a slit width of 300 μm. The distribution was again simulated at three different distances from the source.

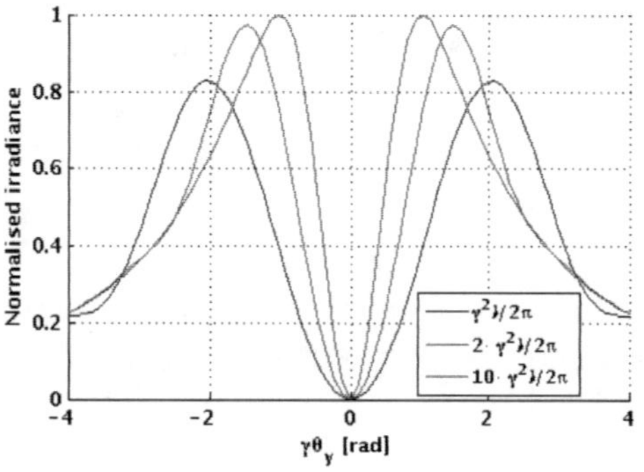

Figure 3: OTR angular distribution for various distances from source to image.

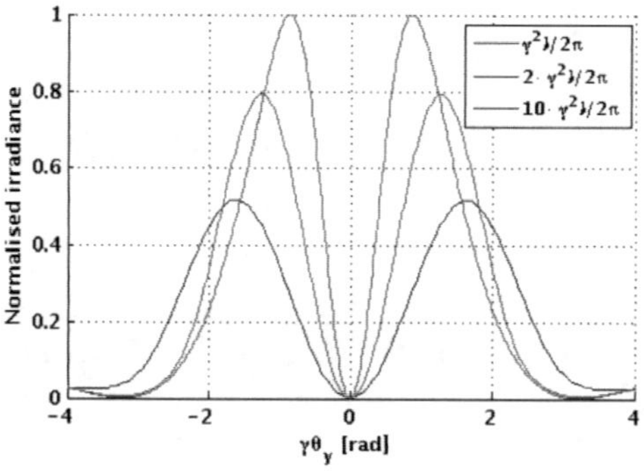

Figure 4: ODR angular distribution for various distances from source to image.

3.2. Optimisation and alignment of the optical system

After establishing agreement between simulation and theory when propagating in free space, optical systems can be optimised and studied with respect to misalignment. Inserting a biconvex lens and putting the image plane in the back focal plane (BFP) of the lens, a full angular distribution can be established in the near-field. Fig. 5 shows the ODR angular distributions for various positions of the image plane. It can be seen that the ODR angular distribution is very sensitive to distances away from the focal plane. The detector must therefore be exactly in the back focal plane (BFP). The same simulation parameters as above were used. Zemax has a built-in lens catalog, including most commercially available lenses. The position of BFP of the biconvex chosen for the ODR optical system at CesrTA is 326.63 mm.

Simulating the ODR angular distribution when tilting the biconvex lens can be seen in Fig. 6. To compare the change in the angular distribution pattern, the distributions are referenced to a reference sphere and therefore centred at $\theta_y = 0$.

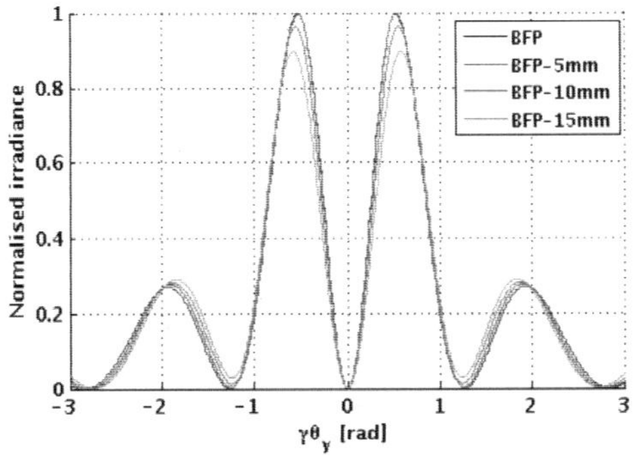

Figure 5: ODR angular distribution for various positions of the image plane (BFP is the position of the back focal plane).

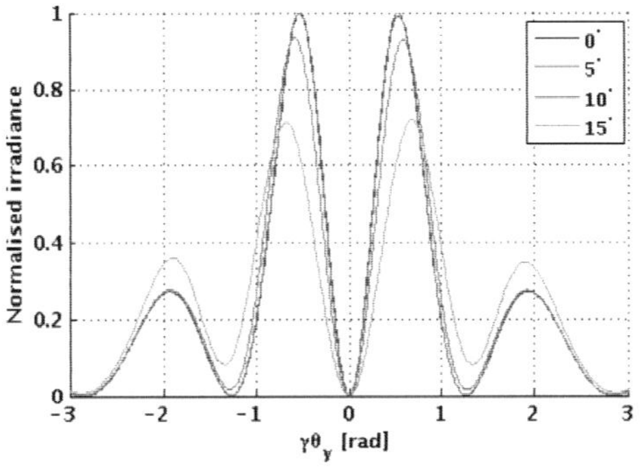

Figure 6: ODR angular distribution for various tilts about the y-axis.

4. Conclusion and future prospects

With assumptions similar to theoretical boundary conditions, Zemax simulations of OTR and ODR agree with the analytical expressions. The next step is to improve the simulation further with including a tilt angle in the incident electric fields to enable simulating a tilted target. With this, agreement can also be achieved for low energies.

Furthermore, simulations will be done applying a finite beam size. This can be achieved by displacing the single particle with respect to the optical axis across the transversal profile. The resulting angular pattern for each step can then be weighted and summed up. After comparing this with analytical equations, the software will have been proven useful for studies of any type of optical system using OTR or ODR. Introducing an off-axis incident field or even an arbitrarily shaped aperture does not slow down the Zemax simulations noticeably and is therefore the preferable method.

This will enable simulations of all misalignment errors and optimisation of a real optical system to be implemented in a real diagnostic station (including viewports, polarisers, filters, etc.). Field depth studies can then be done as well as investigations on the behaviour of the OTR PSF in real optical setups. Also, beam divergence and energy spread will be included to simulate realistic beam properties.

References

[1] CLIC Conceptual Design Report, to be published in 2012, `http://clic-study.org/accelerator/CLIC-ConceptDesignRep.php`
[2] S.T. Boogert et al., Phys. Rev. ST Accel. Beams 13 (2010) 122801.
[3] M. Castellano and V.A. Verzilov, Phys. Rev. ST Accel. Beams 6 (1998) 062801.
[4] R.B. Fiorito et al., Phys. Rev. ST Accel. Beams 9 (2006) 052802.
[5] M.L. Ter-Mikaelian, *High Energy Electromagnetic Processes in Condensed Media*, Wiley, New York (1972).
[6] A. Potylitsyn, Nucl. Inst. and Meth. B 145 (1998) 169-179.
[7] A. Potylitsyn et al., *Diffraction Radiation from Relativistic Particles*, Springer Tracts Mod.Phys. 239 (2011) 1-260
[8] M.Castellano, Nucl. Inst. and Meth. A 394 (1997) 275-280.
[9] J. Urakawa et al., Nucl. Inst. and Meth. A 472 (2001) 309-317.
[10] P. Karataev et al., Phys. Rev. Lett. 93 (2004) 244802.
[11] Zemax 13, Optical Design Program, *Users Manual*, February 20, 2013.
[12] J.W. Goodman, *Introduction to Fourier Optics*, McGraw-Hill, New York (1968).
[13] G.N. Lawrence, *Optical Modeling*, Applied Optics and Optical Engineering 11, Academic, New York (1992).
[14] M. Castellano et al., Nucl. Inst. and Meth. A 435 (1999) 297.
[15] D. Rubin et al., Proc. of PAC2009, Vancouver, Canada, WE6PFP103, p. 2751.
[16] P.V. Karataev, Phys. Lett. A 345 (2005) p. 428-438.

Journal of Physics: Conference Series **517** (2014) 012027 doi:10.1088/1742-6596/517/1/012027

Basics of multiphoton effects in coherent radiation spectra

M V Bondarenco

NSC Kharkov Institute of Physics and Technology, 1 Academic St., 61108 Kharkov, Ukraine

E-mail: bon@kipt.kharkov.ua

Abstract. Differences between classically calculated radiation spectra of coherent radiation sources and their calorimetrically measured counterparts are discussed in a nutshell. Properties of photon multiplicity spectra are highlighted. The high-intensity limit is analyzed. A simple model for quick estimates of multiphoton effects in coherent radiation spectra is given.

1. Introduction

Ultra-relativistic character of radiation from high-energy electrons usually limits its experimental observation to angle-integral spectra, which are furthermore measured by electromagnetic calorimetry. But for intense gamma-radiation sources (e.g., coherent sources based on electron transmission through periodic structures), such spectra can significantly differ from classically evaluated ones (cf. Fig. 1), because an electron has a significant chance of emitting more than one photon, and only their total energy is measured by a calorimeter. Hence the need for establishing correspondence between single-photon spectra (photon number density) and multiphoton spectra (radiative energy loss probability distributions).

A detailed theory for multiphoton effects in coherent radiation spectra was offered in [1]. Here we present an outline of its results, along with some complementary remarks. In addition, we describe a crude model suitable for quickly obtaining expectations about coherent radiation spectrum shapes at a given intensity.

2. Generalities

If $\frac{dI}{d\omega_1}$ is a classically evaluated radiation energy spectrum, the rescaled distribution $\frac{dw_1}{d\omega_1} = \frac{1}{\omega_1}\frac{dI}{d\omega_1}$ can give the corresponding photon probability density only provided the integral

$$w_1 = \int_0^\infty d\omega_1 \frac{dw_1}{d\omega_1} \tag{1}$$

(the mean photon multiplicity) is small. Otherwise, $\frac{dw_1}{d\omega_1}$ can be used as a kernel for kinetic evolution of the distribution function for the radiating electrons,

$$\frac{\partial}{\partial L}\Pi(E_e-\omega) = \int_0^E d\omega_1 \frac{\partial}{\partial L}\frac{dw_1}{d\omega_1}\left[\Pi(E_e-\omega-\omega_1)-\Pi(E_e-\omega)\right], \qquad \Pi(E_e-\omega)\big|_{L=0} = \delta(\omega). \tag{2}$$

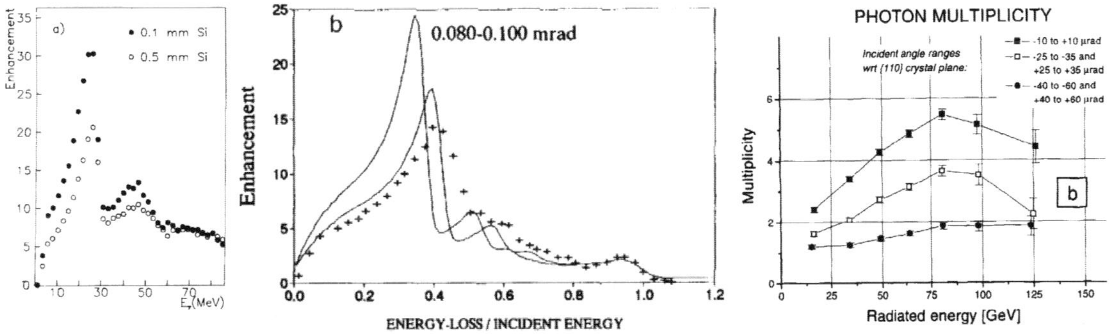

Figure 1. (a) Planar channeling radiation of 5 GeV positrons in (110) oriented Si crystal [2]. There is a difference between spectra (enhancement over the equivalent amorphous target) for crystal thickness $L = 0.1$ mm and $L = 0.5$ mm. (b) Coherent bremsstrahlung of 150 GeV electrons in a 0.6 mm thick (110) Si crystal. [3]. There is a difference between the calorimetrically measured (points) and the classically calculated radiation spectra (curves) [3]. (c) Photon multiplicities at coherent bremsstrahlung and planar channeling of 150 GeV electrons in 1.5 mm thick diamond crystal [4]. They exhibit a linear rise with ω, followed by a turnover.

or of the multiphoton spectrum,

$$\frac{\partial}{\partial L}\frac{dw}{d\omega} = \int_0^\infty d\omega_1 \left(\left.\frac{dw}{d\omega'}\right|_{\omega'=\omega-\omega_1} - \frac{dw}{d\omega} \right) \frac{\partial}{\partial L}\frac{dw_1}{d\omega_1} + W_0(L)\frac{\partial}{\partial L}\frac{dw_1}{d\omega}, \qquad \frac{dw}{d\omega}\Big|_{L=0} \equiv 0 \tag{3}$$

(inhomogeneous, but with a uniform initial condition). Quantities $\frac{dw}{d\omega}$ and Π are related via

$$\Pi(E_e - \omega) = \frac{dw}{d\omega} + W_0\delta(\omega), \qquad \int_0^{E_e} d\omega \Pi(E_e - \omega) = 1,$$

where $W_0 = 1 - \int_0^\infty d\omega \frac{dw}{d\omega} = e^{-w_1}$ is the photon non-emission probability.

The solution to the kinetic equation may be expressed as a contour integral in the complex s plane [1, 5]:

$$\frac{dw}{d\omega} = \frac{1}{2\pi i}\int_{c-i\infty}^{c+i\infty} ds e^{s\omega} \left[e^{\int_0^\infty d\omega_1 \frac{dw_1}{d\omega_1}(e^{-s\omega_1}-1)} - W_0 \right] \tag{4}$$

not involving probabilities differential in the traversed length L. The integral for $\Pi(E_e - \omega)$ is the same but without the W_0 term in the brackets. Another derivation of the contour integral representation for the resummed spectrum, not resorting to the notion of evolution with L, is based on the Poisson character of the soft photon emission process [6, 1].

Function $\Pi(E_e - \omega)$ has a δ-singularity at $\omega = 0$, whereas $dw/d\omega$ is regular everywhere. However, if integral (1) diverges on the lower limit, W_0 vanishes, and then $dw/d\omega = \Pi(E_e - \omega)$. That is the case for ionization losses, for which resummation of the type (4) was historically first carried out [5], but not always so for radiative energy losses [1].

Tips for numerical evaluation of the contour integral. At large $|\mathfrak{Im} s|$, the integrand of (4) is oscillatory due to factor $e^{s\omega}$, with the oscillation amplitude decreasing insufficiently rapidly for direct numerical computation. To improve the convergence and facilitate the numerical evaluation, one may combine $e^{s\omega}$ with the differential, whereafter integrate by parts:

$$\frac{dw}{d\omega} = \frac{1}{2\pi i\omega}\int_{c-i\infty}^{c+i\infty} ds e^{s\omega + \int_0^\infty d\omega_1 \frac{dw_1}{d\omega_1}(e^{-s\omega_1}-1)} \int_0^\infty d\omega_1 \omega_1 \frac{dw_1}{d\omega_1} e^{-s\omega_1}. \tag{5}$$

If $\frac{dw_1}{d\omega_1} \underset{\omega_1\to 0}{\sim} \frac{a}{\omega_1} + b$, at large s we have $\int_0^\infty d\omega_1 \omega_1 \frac{dw_1}{d\omega_1} e^{-s\omega_1} \underset{s\to\infty}{\sim} \frac{a}{s} + \frac{b}{s^2}$, $\int_0^\infty d\omega_1 \frac{dw_1}{d\omega_1}(e^{-s\omega_1} - 1) \underset{s\to\infty}{\sim} -a \ln sE + \frac{b}{s} - \text{const}$, hence the integrand of (5) at $|\mathfrak{Im} s| \to \infty$ decreases as s^{-1-a}. Such a rate of decrease may be sufficient, but still the integration must be carried out over a wide interval of s, and with a high working numerical precision, in order to account for significant cancellations with an oscillatory integrand. At small s, $\int_0^\infty d\omega_1 \omega_1 \frac{dw_1}{d\omega_1} e^{-s\omega_1} \underset{s\to 0}{\sim} \frac{a}{s}(1 - e^{-sE})$, which has a cusp at $s \lesssim E^{-1}$. Therefore, in case of non-zero a, it may be sensible to choose constant $c > 0$.

The role of averaging. The statistical independence of photon emissions holds only for a completely prescribed classical electromagnetic current. Therefore, any averaging procedures are to be performed *after* the resummation.

The integration over photon emission angles, though, may as well be performed before the resummation: if we do that after resummation, for angles of all the photons, we regain a resummation formula involving only the angle-integral single-photon spectrum.

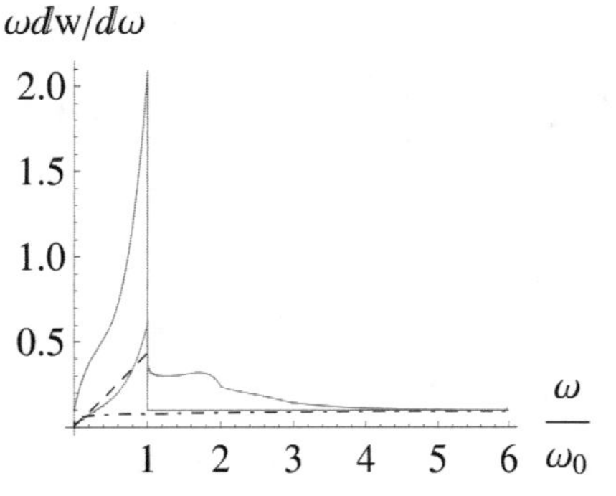

Figure 2. Red solid curve, single-photon spectrum of 'one-point' coherent bremsstrahlung, described by Eqs. (6-8), with parameters $E = 10\omega_0$, $a = 0.1$, $b\omega_0 = \frac{3}{2}w_{1c} = 2$ (moderate intensity). Green solid curve, the corresponding multiphoton spectrum. The second maximum originates entirely from multiphoton effects. Black dashed curve, low-ω asymptotics, $\omega\frac{dw}{d\omega} \simeq e^{-w_{1c}} \left(\frac{\omega}{E}\right)^a (a + b\omega)$. Black dot-dashed curve, high-ω asymptotics, $\omega\frac{dw}{d\omega} \simeq a\left(\frac{\omega}{E}\right)^a$.

3. Manifestation of multiphoton effects

In this section we will illustrate manifestation of multiphoton effects by a simple physical example. Adopt a 'one-point' dipole single-photon spectrum for the coherent radiation component:

$$\frac{dw_{1c}}{d\omega_1} = bP\left(\frac{\omega_1}{\omega_0}\right)\theta(\omega_0 - \omega_1), \qquad P(z) = 1 - 2z + 2z^2, \tag{6}$$

with $\theta(z)$ the Heaviside unit step function, whereas the incoherent component is given by[1]

$$\frac{dw_{1i}}{d\omega_1} = \frac{a}{\omega_1}\theta(E - \omega_1), \tag{7}$$

and neglecting the interference, let

$$\frac{dw_1}{d\omega_1} = \frac{dw_{1c}}{d\omega_1} + \frac{dw_{1i}}{d\omega_1}. \tag{8}$$

An exemplary graph of the multiphoton spectrum corresponding to the single-photon spectrum (6) is shown in Fig. 2. We can see that multiphoton effects cause general lowering of the coherent radiation peak (like in Figs. 1a,b). Besides that, multiphoton effects alone can give rise to a second spectral maximum at $\omega \approx 2\omega_0$, and cause suppression at low ω, below the Bethe-Heitler value. A more detailed analysis is given in [1].

[1] Here the effective ultraviolet cutoff parameter E appears to differ from the electron energy E_e, and approximately equals $E \approx 0.5E_e$ (see [1, 6]).

High-intensity limit for pure coherent radiation. If coherent radiation intensity is high, but the incoherent radiation component is negligible, the contour integral (4) can approximately evaluated by the steepest descent method, yielding a Gaussian distribution:

$$\frac{dw_{\mathrm{c}}}{d\omega} \simeq \frac{1}{\sqrt{2\pi\overline{\omega^2_{1\mathrm{c}}}}} e^{-\rho^2/2} \qquad (w_{1\mathrm{c}} \gg 1). \tag{9}$$

Here $\rho = \frac{\omega-\overline{\omega_{1\mathrm{c}}}}{\sqrt{\overline{\omega^2_{1\mathrm{c}}}}}$ is the scaling variable, and $\overline{\omega_{1\mathrm{c}}} = \int_0^\infty d\omega_1 \omega_1 \frac{dw_{1\mathrm{c}}}{d\omega_1}$, $\overline{\omega^2_{1\mathrm{c}}} = \int_0^\infty d\omega_1 \omega_1^2 \frac{dw_{1\mathrm{c}}}{d\omega_1}$ are single-photon spectral moments.

It is interesting whether the Gaussian limit is achievable at practice. For that, besides high intensity, one also needs independence of the radiation spectrum on electron impact parameters, which are to be averaged over. In oriented crystals such a condition is nontrivial to fulfill. A possible case is radiation from electrons passing through a thick axially orientated crystal, wherein fair impact parameter independence may emerge due to the dynamical chaos and dechanneling. Fig. 3 shows the measured radiation spectrum under such conditions [7], along with a fit by formula (9). The fit may be regarded as acceptable, though not perfect. A better agreement can be obtained with the account of Chebyshev's correction for skewedness [1].

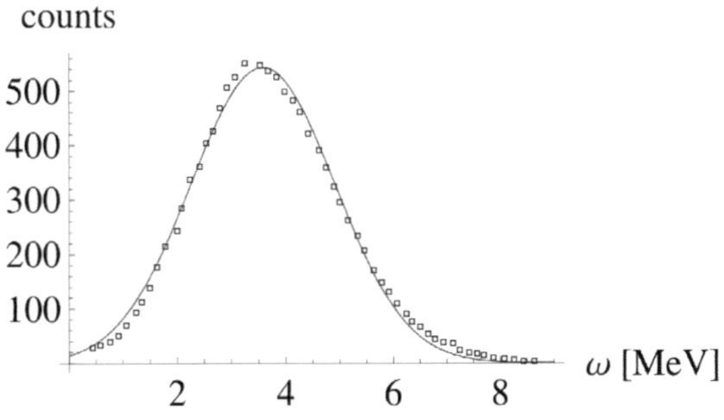

Figure 3. Energy loss probability spectrum for radiation from 40 GeV electrons traversing 2.5 cm thick Ge crystal along ⟨110⟩ axis. Points, experimental data [7]. Purple curve, the fit of Eq. (9) to the data.

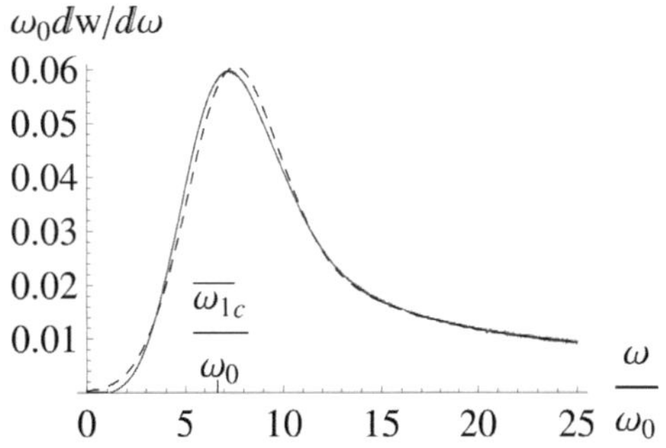

Figure 4. Multiphoton spectrum for single-photon spectrum (6-8) with parameters $b\omega_0 = 20$, $a = 0.3$. Solid purple curve, exact distribution; dashed black curve, parabolic cylinder approximation 10).

Impact of incoherent radiation on the high intensity limit. The incoherent bremsstrahlung component makes spectral moments diverge, thereby invalidating the Gaussian approximation. But still, the radiation in the Bethe-Heitler 'tail' is not intense enough for the Lévy asymptotics to become valid. Nonetheless, using convolution representations, we can attain an approximation suitable for $b\omega_0 \gg 1$, though $a < 1$ (intermediate between Gaussian and Lévy distributions) [1]:

$$\frac{dw}{d\omega} \approx \frac{1}{\sqrt{2\pi} E^a \overline{\omega^2_{1\mathrm{c}}}^{\frac{1-a}{2}}} e^{-\gamma_{\mathrm{E}} a - \rho^2/4} D_{-a}(-\rho), \tag{10}$$

$D_{-a}(z)$ being the parabolic cylinder function, and $\gamma_{\mathrm{E}} = 0.577$ Euler's constant. The behavior of distribution (10) is illustrated in Fig. 4.

4. Photon multiplicity spectrum

An observable complementary to the multiphoton spectrum is the photon multiplicity spectrum (for its definition and method of measurement see [1, 4]). For this observable, the resummation procedure gives [1]:

$$\bar{n}(\omega) = \frac{1}{dw/d\omega}\frac{1}{2\pi i}\int_{c-i\infty}^{c+i\infty} ds e^{s\omega+\int_0^\infty d\omega_1 \frac{dw_1}{d\omega_1}\left(e^{-s\omega_1}-1\right)} \int_0^\infty d\omega_1' \frac{dw_1}{d\omega_1'} e^{-s\omega_1'}. \tag{11}$$

Alternatively, the knowledge of $\frac{\partial}{\partial L}\frac{dw}{d\omega}$ from kinetic equation (3) or (2) may be used to obtain $\bar{n}(\omega)$ directly:

$$\bar{n}(\omega) = w_1 + L\frac{\partial}{\partial L}\ln\frac{dw}{d\omega}. \tag{12}$$

As an illustration, in Fig. 5 we show $\bar{n}(\omega)$ for the same conditions as in Fig. 2. There is a monotonous though stepwise rise in the region where the spectrum $\frac{dw}{d\omega}$ is well above the Bethe-Heitler value, and it changes into a falloff and a slow (logarithmic) increase where the Bethe-Heitler contribution dominates.

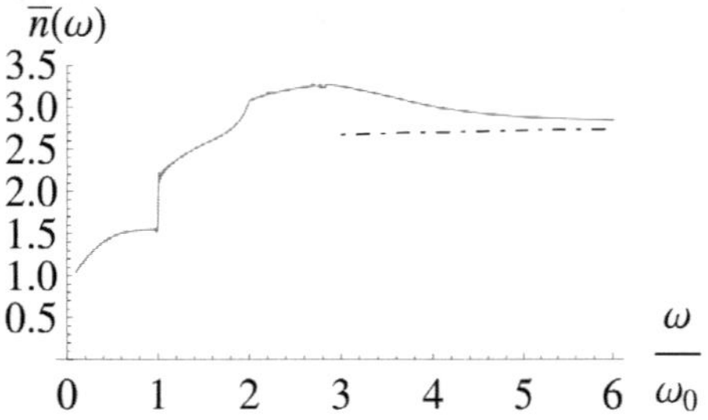

Figure 5. Solid green curve, photon multiplicity spectrum for the single-photon radiation spectrum given by Eqs. (6–8) with $a = 0.1$, $b\omega_0 = 2$, and $\epsilon = 0.1\omega_0$ (the infrared cutoff). Dot-dashed black curve, large-ω asymptotics $\bar{n}(\omega) \simeq w_{1c} + \bar{n}_i(\omega) = w_{1c} + 1 + a\ln\omega/\epsilon$.

At high radiation intensity, the Gaussian asymptotics of the spectrum $\frac{dw_c}{d\omega}$ corresponds to the linear asymptotics of the photon multiplicity spectrum:

$$\bar{n}_c(\omega) \approx w_{1c} + \frac{\overline{\omega_{1c}}}{\overline{\omega_{1c}^2}}\left(\omega - \overline{\omega_{1c}}\right) \qquad (w_{1c} \gg 1), \tag{13}$$

i. e., the step-like behavior smoothens out. Linear behavior (13) is in accord with the initial interval of experimental dependencies in Fig. 1c. With the account of incoherent radiation component, when $\frac{dw}{d\omega}$ obeys Eq. (10), $\bar{n}(\omega)$ expresses through parabolic cylinder functions, too (see [1]), and the initial linear dependence is followed by a turnover and subsequent logarithmic growth.

5. A toy model for coherent radiation

The generic features of multiphoton coherent radiation spectra may be elucidated by example of a simple model for the single photon spectrum[2]:

$$\frac{dw_1}{d\omega_1} = \frac{w_{1c}}{\omega_0}\theta(\omega_0 - \omega_1) + \frac{a}{\omega_1}\theta(E - \omega_1). \tag{14}$$

Inserting (14) to (4) gives the integral

$$\frac{dw}{d\omega} = e^{-w_{1c}-\gamma_E a}\frac{1}{2\pi i}\int_{c-i\infty}^{c+i\infty}\frac{ds}{(sE)^a}\exp\left[s\omega + \frac{w_{1c}}{\omega_0 s}\left(1 - e^{-s\omega_0}\right)\right]. \tag{15}$$

[2] The uniform spectral distribution of the coherent part (the first term) of $\frac{dw_1}{d\omega_1}$ corresponds to uniform angular distribution of radiation in electron rest frame, instead of $1 + \cos\tilde{\theta}^2$ typical for dipole radiation.

Expanding further $\exp[-\frac{w_{1c}}{\omega_0 s}e^{-s\omega_0}] = \sum_{n=0}^{\infty}\frac{1}{n!}\left(-\frac{w_{1c}}{\omega_0 s}\right)^n e^{-ns\omega_0}$ and integrating termwise, we get

$$\frac{dw}{d\omega} = \frac{e^{-w_{1c}-\gamma_E a}}{\omega_0}\left(\frac{\omega_0}{E}\right)^a \sum_{n=0}^{\lfloor\omega/\omega_0\rfloor}\frac{(-1)^n}{n!}w_{1c}^{\frac{n+1-a}{2}}\left(\frac{\omega}{\omega_0}-n\right)^{\frac{n-1+a}{2}} I_{n-1+a}\left\{2\sqrt{w_{1c}\left(\frac{\omega}{\omega_0}-n\right)}\right\}, \quad (16)$$

where I_{n-1+a} are the modified Bessel functions, and $\lfloor\omega/\omega_0\rfloor$ signifies the greatest integer smaller than ω/ω_0. Hence, for any given ω, the number of terms in the series is finite. Similarly, one can derive a closed formula for $\bar{n}(\omega)$, but it also involves derivatives of Bessel functions by index.

Eq. (16) can be used for quick generation of exemplary multiphoton distributions. Parameter w_{1c} is not in a strict correspondence with the magnitude of coherent radiation [say, b entering Eq. (6]. But since function P obeys inequality $\frac{1}{2} \leq P(z) \leq 1$ $(0 \leq z \leq 1)$, it seems reasonable to choose w_{1c} somewhere in the interval $\frac{1}{2}b\omega_0 < w_{1c} < b\omega_0$. In Fig. 6 we plot $\omega\frac{dw}{d\omega}$ for fixed a and several values of w_{1c}. The curve at moderate intensity (green band) can be compared with the green curve of Fig. 2. The correspondence is fair at $w_{1c} \approx \frac{4}{3} = \frac{2}{3}b\omega_0 = b\omega_0\int_0^1 dzP(z)$.

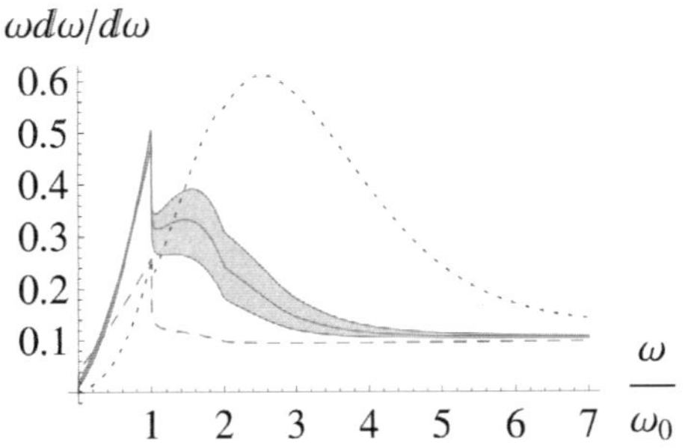

Figure 6. Energy spectra for coherent radiation described by formula (16), for parameters $E = 10\omega_0$, $a = 0.1$, and $w_{1c} = 0.3$ (red dashed curve); $\frac{4}{3} \pm \frac{1}{3}$ (green solid curves); 4 (blue dotted curve). The middle green curve is close in shape to the green curve of Fig. 2.

6. Concluding remarks

Multiphoton effects affect spectra of radiation in crystals with thicknesses $\gtrsim 1\%X_0$ (1 mm for Si). They not only bear on the overall normalization, but can also evoke a 2nd maximum, at $\omega \approx 2\omega_0$, similar to effect of secondary harmonics in the intra-crystal potential, or non-dipole effects. Besides that, they generate a suppression at low ω, superficially similar to LPM effect. Thus, one needs to be careful at interpretation of spectral features in medium-thickness crystals.

At implementation of the resummation procedure, it is essential that averaging over impact parameters must be performed at the final, not the initial stage. Sometimes that makes little difference, as for the case radiation in an amorphous medium; but it may be crucial in other cases, like for channeling radiation.

An inverse theoretical problem is the reconstruction of the single-photon spectrum from calorimetrically measured multiphoton one [8]. The complete reconstruction procedure for that (though restricted to conditions $\omega \ll E_e$ and the absence of strong averaging effects) was formulated in [1]. Its possible applications will be considered elsewhere.

References

[1] Bondarenco M V 2013 Multiphoton effects in coherent radiation spectra *Preprint* physics.acc-ph/1310.6681
[2] Bak J *et al.* 1985 *Nucl. Phys.* B **254** 491
[3] Medenwaldt R *et al.* 1991 *Phys. Lett.* B **260** 235
[4] Kirsebom K *et al.* 1996 *NIM* B **119** 79
[5] Landau L D 1944 *J. Phys. USSR* **8** 201
[6] Baier V N and Katkov V M 1999 *Phys. Rev.* D **59** 056003
[7] Medenwaldt R *et al.* 1989 *Phys. Lett.* B **227** 483
[8] Kolchuzhkin A, Potylitsyn A, Bogdanov A, and Tropin I 1999 *Phys. Lett.* A **264** 202; Kolchuzhkin A and Potylitsyn A 2001 *NIM* B **173** 126

Flux dynamics and angular distributions of relativistic electrons and positrons passing through a half-wave crystals

Yu L Pivovarov and T A Tukhfatullin

National Research Tomsk Polytechnic University

Lenin Ave 30, 634050 Tomsk, Russia

E-mail: TTA@tpu.ru

Abstract. The simulation of trajectories, angular and spatial distributions of relativistic electrons and positrons (155 MeV – 1500 MeV) penetrating through an ultra-thin crystal (half-wave crystal - HWC) has been performed taking into account initial angular beam divergence. The results show similarities and differences with non-relativistic protons channeling in a HWC.

1. Introduction

As a charged particle moves in a crystal within the ''critical angle'', with respect to atomic plane or axis, it may enter into a special regime of motion – channeling. The motion of a positive particle captured into planar channeling regime is characterized by oscillations between neighboring atomic planes, while negative particle oscillations occur in the vicinity the crystal plane.

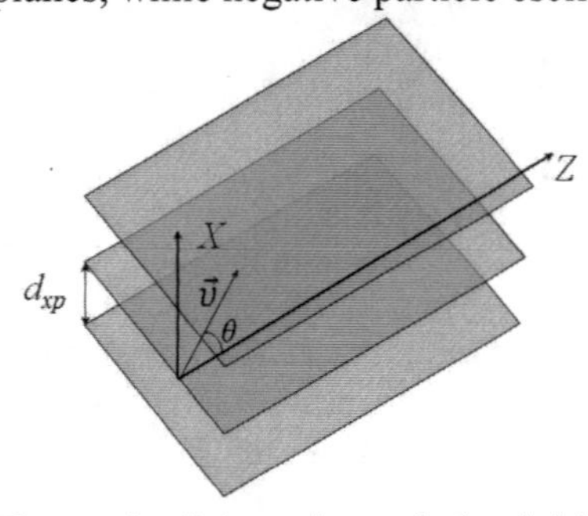

Figure 1. Orientation of the initial velocity vector of the electron/positron with respect to the channeling planes.

Recent computer simulations and experimental studies of nonrelativistic protons planar channeling in an ultra-thin crystal - "half wavelength crystal" (below – HWC) – revealed very interesting features of both flux dynamics inside a crystal and angular distributions of nonrelativistic proton beam after a crystal [1]. In particular, the authors of Ref. [1] experimentally demonstrated that channeled 2 MeV protons were mirrored by a HWC - ultra-thin silicon unbent crystal. It was also observed that over-barrier particles were deflected to the opposite direction with a dynamics similar to that of volume reflection (VR) in a bent crystal.

One can define the thickness of HWC in a following way:

$$L_{HWC} = \frac{2d_{xp}}{\theta_c}, \quad (1)$$

where d_{xp} is the interplanar distance and θ_c is the critical channeling angle:

$$\theta_C = \sqrt{\frac{2U_0}{E}} \,. (2)$$

In Eq.(2), U_0 is the maximal value of potential energy of the particle in the field of continuous potential for planar channeling. Thus defined, the L_{HWC} value is depends on the initial energy of the particle and is scaled proportional to $\sqrt{E}$. That means, in the case of relativistic electrons and positrons similar HWC effect can occur and even can be stronger (more brilliant) since the number of oscillations will be several times less compared to [1], i.e both dechanneling and non-harmonic effects should be weaker. For example, if the HWC thickness L=1 micrometer, in order to make only one oscillation (one reflection by channeling plane) the electron/positron energy should exceed at least 300 MeV in the case of Si crystal. In fact, similar effect may occur even if electron/positron makes only several oscillations during penetration through an ultra-thin crystal, at the beam energy below 300 MeV.

In our work we confirmed this by detailed computer simulations using our computer code BCM – 1.0 [2], which was successfully applied to explain (describe) recent experimental data on doughnut scattering (DS) and planar channeling scattering (PCS) of 255 MeV electrons in a 20 micrometer Si crystal, obtained at SAGA Light Source [2-3]. Simulations were carried out for several electron energies available at existing facilities: 155 MeV SPARC LNF (Italy), 255 MeV linear accelerator of SAGA Light Source synchrotron (Japan), 855 MeV and 1508 MeV - Mainz Microtron MAMI (Germany). In addition, similar calculations have been performed for positron beams with the same energies.

2. Electron/positron trajectories simulation

For the calculation of every individual charged particles trajectory we numerically solved classical equation of motion

$$\gamma m\ddot{x} = -\frac{\partial U(x)}{\partial x}, \quad \gamma m\ddot{z} = 0, \tag{1}$$

where x are the transversal coordinates, z is the longitudinal coordinate (see figure 1.), $U(x)$ here describes the potential energy of the particles in the field of periodically arranged crystal planes.

Under the channeling condition the longitudinal component of relativistic particles velocity $\upsilon_\parallel$ is nearly equal to the speed of light. The transversal component of velocity is much less than the longitudinal one: $\upsilon_\perp \equiv \dot{x} << \upsilon_\parallel$; $\upsilon_\perp / c << 1/\gamma$ (γ denotes the relativistic factor). Under these conditions, equations of motion are reduced to the non-relativistic equation (1) with the replacement of the particle mass m to the relativistic mass γm.

The initial conditions for this system contain the electron entrance point into the crystal $x(0) \equiv x_0$ and transversal components of initial velocity:

$$\upsilon_x(0) = c\sqrt{1-\frac{1}{\gamma^2}}\sin(\theta), \tag{2}$$

where θ is the angles between electron momentum and channelling plane. The point of entry of were distributed uniformly within a period of change the function $U(x)$ (within one potential well). To account for the angular divergence of the electron beam initial angle for each point of entry, 5 values of incident angle θ were generated by a random number generator with normal distribution. The dispersion of the normal distribution was 0.1 mrad≈ $\theta_c/4$). The trajectories and velocities of electrons are obtained by numerical integration of the equation of motion. For the numerical simulation we used a new computer code "Basic Channeling with Mathematica©" BCM–1.0 developed by the authors [2].

The exact periodic potentials for (220) and (111) channelling have been used in numerical solving of Eq.(1) (see, e.g. in [2]).

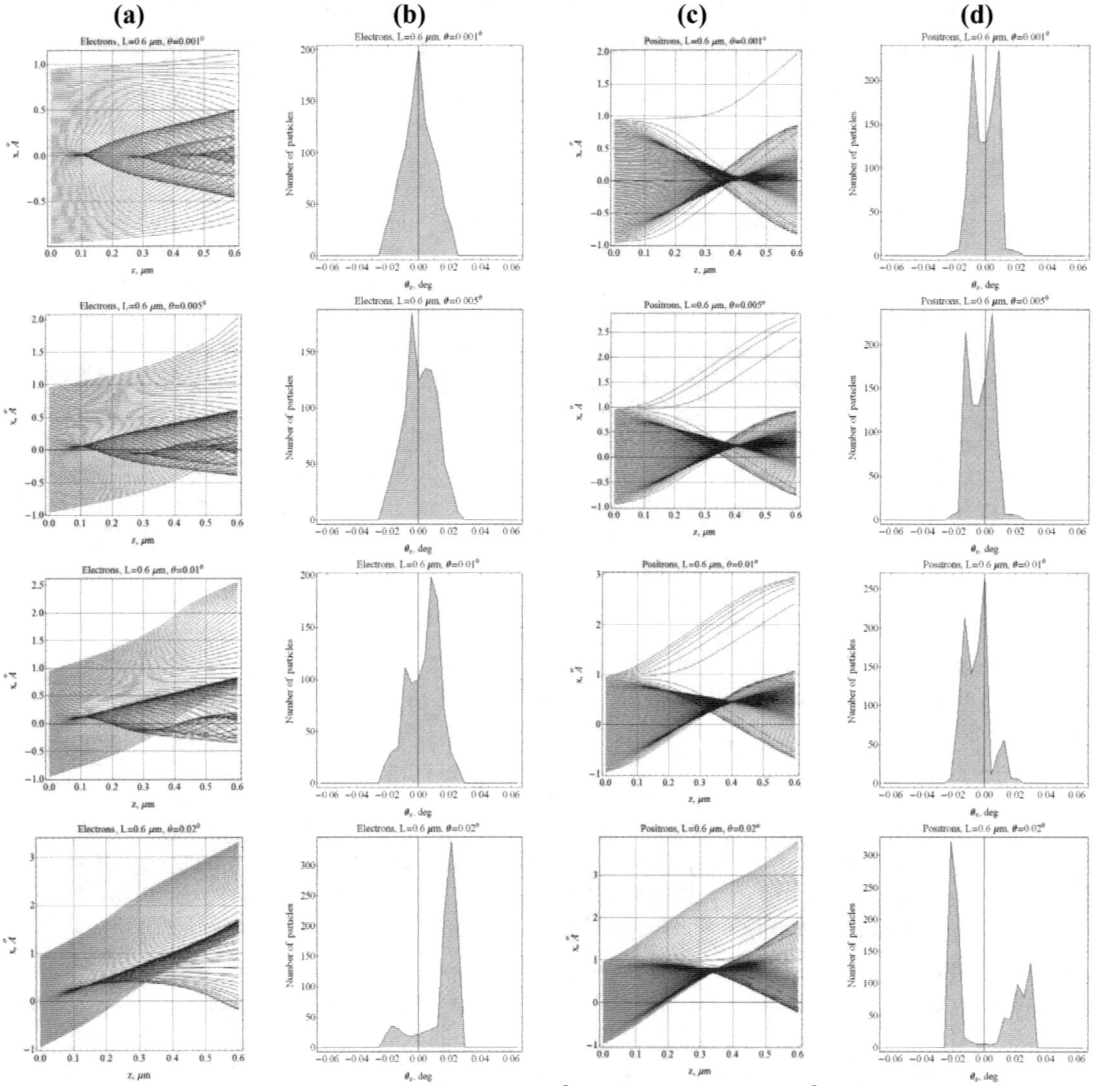

Figure 2. Simulated trajectories (statistics $N=10^2$ particles (a, c), $N=10^3$ particles (b, d),) showing flux dynamics of 255 MeV electrons (a) and positrons (c) channeled between (220) planes of a Si crystal 0.6 μm thick. Angular distribution of outgoing electrons (b) and positrons (d) versus angle $\theta_x=\upsilon_x/c$ as a result of simulation for different incident angles θ. For this case, $L_{HWC}=0.94\,\mu$m corresponding to the critical channelling angle θ_c=0.023°

3. Results of computer simulation

To obtain the spatial and angular distributions of electrons and positrons in planar channeling in HWC Si, were calculated 1000 trajectories. After that we calculated the exit angles and exit coordinate of outgoing particles. The exit angles are calculated according the formulae $\theta_x = \upsilon_x / c$, where υ_x is the transversal components of the particle velocity at the exit from the crystal.

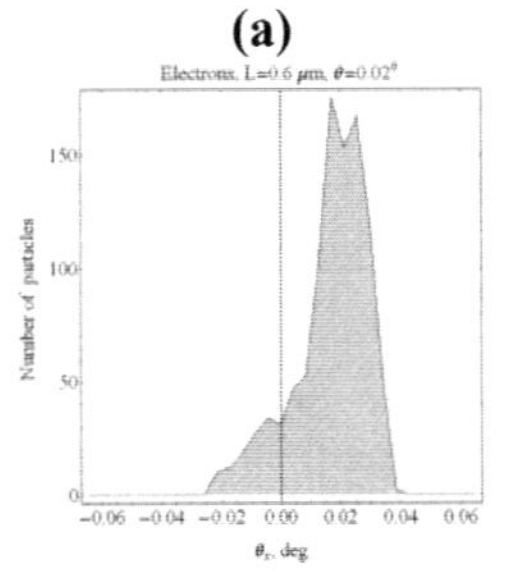

Figure 3. Angular distribution of 255 MeV electrons (a) and positrons (b) at (220) channeling in 0.6 μm Si crystal. Calculations are carried out with taking into account the angular divergence.

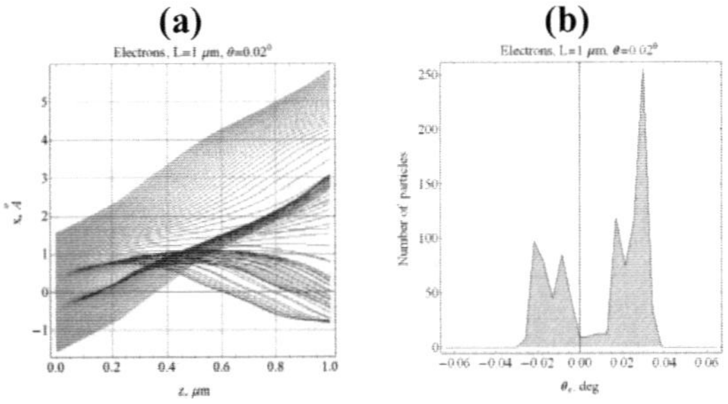

Figure 4. (a) –Simulated trajectories of 255 MeV electrons at (111) channeling in 1.0 μm Si crystal (b) – Angular distribution of outgoing electrons: appearance of two-modal structure (cf. with Fig.3a)

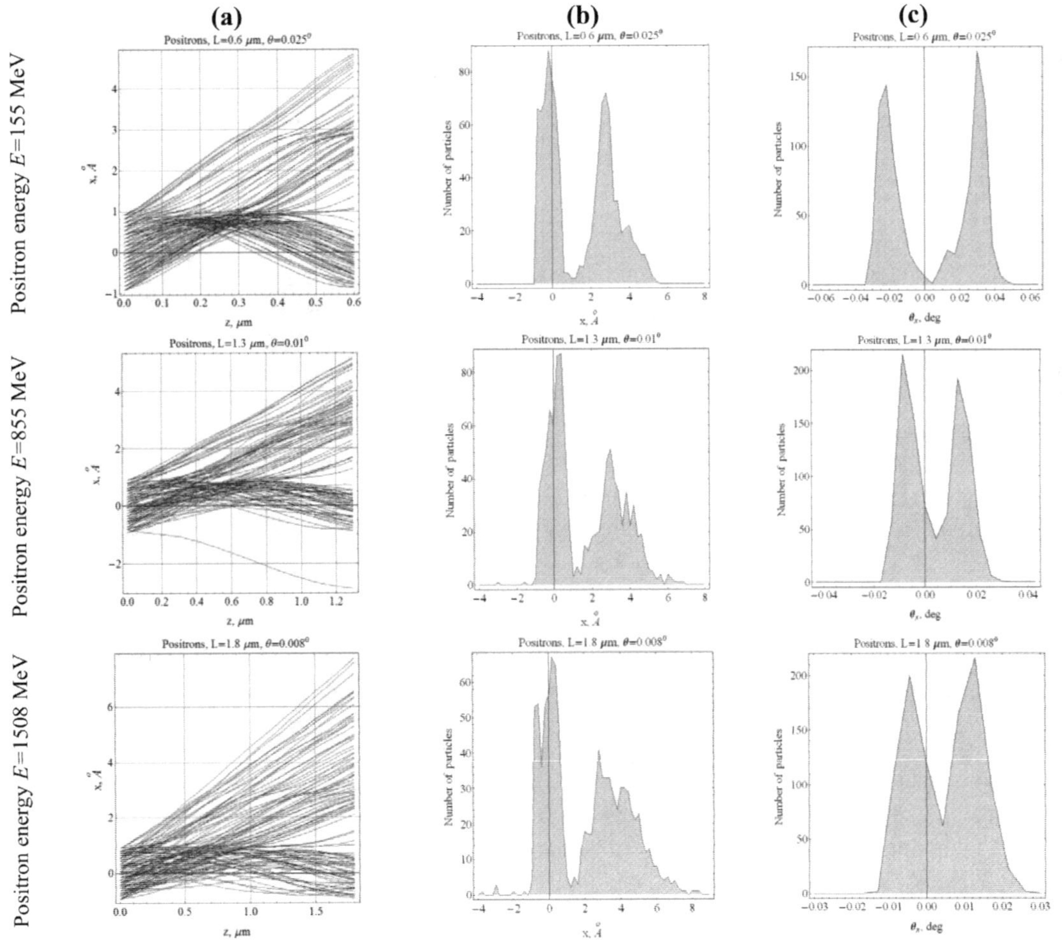

. **Figure 5.** (a) –Simulated trajectories of positrons channelled between (220) planes of a Si crystal. (b) – Spatial distribution of outgoing positrons. (c) – Angular distribution of outgoing positrons versus angle. Critical channelling angle θ_c=0.03° for positron energy E=155 MeV, θ_c=0.008° E=1508 MeV.

Results of simulation of the flux dynamics of 255 MeV electrons and positrons in a 0.6 μm Si crystal at (220) alignment is shown in figure 1. The simulation was performed without taking into account angular divergence of the incident beam. Figure 1d show that the positron beam was split into two components after interaction with the crystal. For example for angle of incidence θ=0.02°=0.87θ_c the positions of the peaks equal -0.02° and +0.025°. The first peak corresponds to channeled particles and the second peak due to over barrier particle. So, due to channeling positron beam turns by an angle approximately equal to 2 critical channeling angle. The similar effect can be observed for electrons (see in figure 1b), but the fraction of channeling electron is much less (about 10% incidence θ=0.02°). This is due to the shape of the potential well for electrons is much sharper than for the positrons.

If we take into account the angular divergence of the beam, the effect for the electrons disappears (figure 3a). While the angular divergence of positrons beam lead to the amplitude of the peaks correspond to channeling and over barrier particles become practically equal (see in figure 3b).

The effect of splitting of the electron beam into two components can be improved using of channeling along (111) planes in the silicon crystal (see in figure 3). In this case the potential well is deeper and brighter, more close to harmonic potential as for positrons

The energy dependence of the flux dynamics and angular distribution. Because the thickness of HWC is connected with initial energy of the particle (L_{HWC}=2d_{xp}/θ_c) it is possible to increase thickness energy increases the thickness of the crystal can be increased with growing of the particles energy. The results of simulation spatial and angular distribution of positrons after the HWC at (220) channeling for energies of 155, 855 and 1508 MeV are presented in figure 5. The gap between the peaks of the angular distribution are equal to 0.55° 155 MeV positrons and 0.25° for 855 MeV. This is due to the fact that critical channeling angle decreases with increasing the energy E of the channeling particle ($\theta_c = \sqrt{\frac{2U_0}{E}}$ where U_0 is the depth of potential well). We did not consider here the case of ultrarelativistic electrons/positrons (e.g. as in the recent CERN experiment [6]), since in this case to define the trajectories in a most correct way, probably one have to take into account the radiation energy loss of electrons passing through a crystal. But on the other hand the HCW thickness for this energy ~10 μm is too small to loose sufficient energy due to channeling radiation.

4. Conclusions

Computer simulations of angular and spatial distribution of relativistic electrons and positrons penetrating through a HWC using computer code BCM – 1.0 [2] was performed taking into account initial angular divergence of the particle beam. The results can be summarized:

- After penetration through a HWC, the spatial distribution of positrons has brilliant bimodal (or two-peak) structure as in the case of non-relativistic proton beam [1] (1 oscillation). The angular distribution is also split into two components. Therefore, unbent ultrathin crystal - HWC - can be used for relativistic positron beam deflection/splitting.

- After penetration through a HWC, the spatial distribution of electrons has more complex structure as in the case of positrons, and electron beam does not split into two components for (220) channelling in Si crystal. Splitting of electron beam can be observed in the case of (111) channelling in HWC Si crystal (the potential well is deeper and brighter, more close to harmonic potential as for positrons).

In addition, we want to stress that the specifics of trajectories as well as very small crystal thickness should effect on the channeling radiation spectra of relativistic electrons/positrons in a HWC. There are several reasons to measure channeling radiation in a HWC. First, we expect strong influence of trajectories "tails" first considered in [3]. Second, it is the just the case when during calculation of channeling radiation spectrum one can try the approach suggested in Ref. [4] for calculation of bremsstrahlung (BS) spectrum in the thin amorphous target. Third, it is closely connected with radiation spectrum from a particle moving in an arc [5] (never studied experimentally). The analysis of channeling radiation in a HWC is in progress and will be published elsewhere (separately).

5. Acknowledgements

The authors are grateful to N. Shul'ga and A. Mazzolari for useful discussions. This work was supported by Russian Fund for Basic Research, Grant No 12-02-01314.

6. References

[1] Guidi V, Mazzolari A, De Salvador D and Bacci L 2012 *Phys. Rev. Lett.* **108** 014801
[2] Bogdanov O V, Fiks E I, Korotchenko K B, Pivovarov Yu L and Tukhfatullin T A 2010 *J. Phys.: Conf. Ser.* **236** 012029
[3] Bogdanov O V, Pivovarov Yu L 2008 *Nucl. Instr. and Meth. in Phys. Res.* **B 266** 3852
[4] Artru X, Fomin S P, Shul'ga N F 2003 *Nucl. Instr. and Meth. in Phys. Res.* **B 66** 394
[5] Bagrov V G, Ternov I M, Fedosov N I 1982 *Journ of Exper. and Theor. Phys* **55** 835
[6] Mazzolari A, et al. Talk presented at IX International Symposium RREPS-2013 (CERN channeling radiation experiments with 120 GeV electrons/positrons)

Coherent Bremsstrahlung from planar channeled positrons

Yu P Kunashenko[1,2]
[1] National Research Tomsk Polytechnic University,
[2] Tomsk State Pedagogical University

E-mail: kunashenko@tpu.ru

Abstract. Theory of coherent bremsstrahlung from planar channeled positrons in a crystal is developed in the framework of the virtual photons method. It is shown that positron channeling results in splitting and appearance of additional structure of coherent peaks in axial coherent bremsstrahlung spectrum.

1. Introduction

It is well known that many physical processes, accompanying a passage of the high energy particles at small angles with respect to the crystal axes or planes, differ from those in an amorphous target. In fact, in this case coherent effects occur (see, e.g. [1, 2]). The coherent effect in the interaction of charged particle with a crystal arises due to periodical arrangement of the crystal atoms, when the interaction amplitudes of the particle with individual atoms are summing up in the phase. As a result of this summing, an interferential multiplier appears in the cross–section. It leads to appearance of coherent peaks when the transferred momentum coincides with one of the reciprocal lattice vectors. This happens at selected parameters (beam energy, angles of incidence into a crystal with respect to crystallographic axes or planes).

The channeling phenomenon appears when the fast charged particle passes through a crystal at small angle with respect to the crystal axis or plane, and the crystal thickness is rather large, so that the Born approximation or many–wave diffraction approach do not work. In this case, a charged particle interacts with averaged (continuous) potential of the crystal axis or plane and one can neglect the periodic part of the crystal potential.

The coherent scattering or channeling of relativistic charged particles in a crystal leads to emission of coherent bremsstrahlung (CB)[1, 2] or channeling radiation [3–5].

In fact, both channeling and coherent process can occur simultaneously – this is called a combined effect. For example, the coherent bremsstrahlung from relativistic axially channeled electrons differs from ordinary CB. In this case, the coherent peak in the energy spectrum of emitted photons is split [6]. In the case of combined effect, the CB theory should be modified.

In the present paper we investigate the combined effect in CB from planar channeled positrons. To this purpose, we consider the special geometrical conditions of the passage particle through the crystal. In our case, the positron moving between two planes of the crystal in the regime of planar channeling and simultaneously parallel to the crystallographic axis. Moreover, this axis is located between the planes and is parallel to them. These geometric conditions can be realized for example in a tungsten crystal if one select the (100) planes and <001> axis.

To study this combined effect we modified the method of virtual photons, taking into account the channelling of the radiating particles.

2. Cross–section of the photon scattering by a channeled positron

In the reference frame of the channeled positron passing through the crystal the electric field of an atom is perpendicular to its magnetic field. According to the Lorentz transformations, the fields are compressed in the longitudinal direction of the particle motion. Therefore, in a moving reference frame the electromagnetic field of the crystal is similar to the fields of a plane wave (if channeled positron relativistic factor $\gamma > 1$). Therefore, the electromagnetic field of the crystal can be replaced by a flux of virtual photons. Further, a process of bremsstrahlung can be described in terms of the virtual photons scattering by a channeled positron [1].

It is convenient to consider the scattering of photons by channeled positron in a coordinate system moving with the longitudinal velocity of channeled positron. In the moving reference frame one can use the non relativistic approach to the scattering process.

The cross–section of the photon scattering by a channeled positron was found in [7] :

$$d\sigma = \frac{\alpha^2\hbar^2}{m^2c^2}\frac{\hbar\omega_2}{\hbar\omega_1}\left|\sum_{pol}\sum_n\left\{\frac{(\mathbf{e}_2\mathbf{R}_{fn})(\mathbf{e}_1\mathbf{A}_{ni})}{(E_i+\hbar\omega_1)-E_n}+\frac{(\mathbf{e}_1\mathbf{A}_{fn})(\mathbf{e}_2\mathbf{R}_{ni})}{E_i-E_n-\hbar\omega}\right\}\right|^2 \times \tag{1}$$
$$\delta\left[(\mathbf{k}_{||_i}+\mathbf{k}_{||_1})-(\mathbf{k}_{||_f}+\mathbf{k}_{||})\right]\delta\left[(E_i+\hbar\omega_1)-(E_f+\hbar\omega)\right]d^2\mathbf{k}_{||_f}d\Omega\, d\hbar\omega_2 .$$

Here $\sum_{pol}$ is the sum over incident and scattered photon polarization vectors $\mathbf{e}_1$ and $\mathbf{e}_2$ and

$$M_{fi} = \sum_n\left\{\frac{(\mathbf{e}_2\mathbf{R}_{fn})(\mathbf{e}_1\mathbf{A}_{ni})}{(E_i+\hbar\omega_1)-E_n}+\frac{(\mathbf{e}_1\mathbf{A}_{fn})(\mathbf{e}_2\mathbf{R}_{ni})}{E_i-E_n-\hbar\omega}\right\}, \tag{2}$$

is the process matrix element, with

$$\mathbf{A}_{kl} = -\frac{e}{mc}\int\Psi_k^*(\mathbf{r})\hat{\mathbf{p}}\sqrt{\frac{2\pi\hbar}{\omega_1}}\exp[-i\mathbf{k}_1\mathbf{r}]\Psi_l(\mathbf{r})d\mathbf{r}, \tag{3}$$

being the matrix element of the photon absorption and

$$\mathbf{R}_{kl} = -\frac{e}{m}\int\Psi_k^*(\mathbf{r})\hat{\mathbf{p}}\sqrt{\frac{2\pi\hbar}{\omega_2}}\exp[-i\mathbf{k}_2\mathbf{r}]\Psi_l(\mathbf{r})d\mathbf{r}, \tag{4}$$

being the matrix element of photon emission from the channeled positron.

In equations (13), $\mathbf{k}_1$ is the photon wave vector, $\mathbf{e}_1$ is the polarization vector and ω_1 is the frequency of the incident photon while $\mathbf{k}_2$, $\mathbf{e}_2$ ω_2 are the wave vector, polarization vector and frequency of the scattered photon, respectively, $\hat{\mathbf{p}}$ is a positron momentum operator and $\Psi_l(\mathbf{r})$ is a channeled positron wave function.

If the incident photon moves along the OZ axis then its wave vector is:

$$\boldsymbol{k} = \{0,0,k\}, \quad k = \omega / c,$$

while the wave vector of scattered photon can be written as:

$$\mathbf{k}' = \{k'\sin\Theta\cos\Phi, k'\sin\Theta\sin\Phi, k'\cos\Theta\}, \quad k' = \omega'/c .$$

Here, Θ and Φ are the photon scattering angles. The scattering angles are given in cylindrical coordinates.

The channeled positron wave function has a form

$$\Psi_{i(f)}(\mathbf{r}) = \varphi_{i(f)}(x)\exp(i\mathbf{k}_{||i(f)}\mathbf{r}_{||}),$$

here the exponential $\exp(i\mathbf{k}_{||}\mathbf{r}_{||})$ describes the positron free motion along the crystal plane with $\mathbf{k}_{||}$ being its longitudinal wave vector while $\varphi(x)$ is a channeled positron transverse wave function. We chose the continuous planar potential for positrons in the form [4]:

$$V(x)=\frac{4V_0}{d^2}x^2.$$

The transverse wave functions corresponding to this potential are well known [3, 8].

The energies of the incident and scattered photons are related by the formula [7]:

$$\hbar\omega_2=\hbar\omega_1\cos\Theta-mc^2+ \\ +mc^2\sqrt{\left(1-\frac{(\hbar\omega_1)^2+4mc^2(\varepsilon_{\perp i}+\hbar\omega_1-\varepsilon_{\perp f})+\hbar\omega_1(\hbar\omega_1\cos 2\Theta-4mc^2\sin\Theta)}{2m^2c^4}\right)\mathrm{ctg}^2\Theta}. \tag{5}$$

3. The cross section of coherent bremsstrahlung from a planar channeled positron

When the particle is moving parallel to the axis of the crystal, it is enough to consider account its interaction with single crystallographic axis. The electrostatic potential of the crystal axis is a sum of the Coulomb potentials of the individual atoms forming the axis and is equal to:

$$V(\vec{r})=\sum_{i=1}^{N}V_1(|\vec{r}-\vec{r}_i|),\quad V_1(r)=\frac{Ze}{r}\exp\left(-\frac{r}{R}\right). \tag{6}$$

Here V_1 is the potential of a single atom, R is the screening radius, N is the number of atoms in the axis, Z is atomic number.

The spectrum of virtual photons of crystal axis as it seen in the rest frame of the positron moving parallel to the axis is given by [9]:

$$n(\omega)d\omega=\frac{Z^2e^2}{\pi^2}N\times \\ \left\{\left[L-B\exp\left[-\left(\frac{\hbar\omega}{\gamma\hbar c}\bar{u}\right)^2\right]\right]+\frac{2\pi}{d}\sum_{g_n}B\exp\left[-\left(\frac{\hbar\omega}{\gamma\hbar c}\bar{u}\right)^2\right]\delta(k_1-g_n)|S|^2\right\}\frac{d\omega}{\omega}, \tag{7}$$

$$L=\pi\ln\left[\frac{a\lambda^{-2}}{(\hbar\omega/\gamma\hbar c)^2+R^{-2}}\right],\quad a\approx 1,$$

$$B(x)=\pi\{-(1+x)e^xEi(-x)-1\},\quad x=\left[\left(\frac{\hbar\omega}{\gamma\hbar c}\right)^2+R^{-2}\right]\bar{u}^2.$$

Here d is the lattice constant, ω is the frequency of virtual photon, $\lambda=mc/\hbar$ is the Compton wave length of electron , $\bar{u}^2$ is the mean–square displacement of the crystal atom from equilibrium position, $Ei(-x)$ is the exponential integral function, $g_n=2\pi n/d$ is the 1D reciprocal lattice vector, $|S|$ is the structure factor of a crystal axis, γ is the positron relativistic factor. To compare, the spectrum of virtual photons of a separate atom is:

$$n_{\mathrm{am}}(\omega)d\omega=\frac{Z^2e^2}{\pi^2}L\frac{d\omega}{\omega}. \tag{8}$$

If we neglect the terms of order of $(\hbar\omega/\gamma\hbar c)$ in the functions L and B in equations. (7,8) than we obtain the Ter–Mikaelian result for virtual photons spectrum [1].

The photons scattering cross–section by a channeled positron (1) should be transformed into the laboratory coordinate system. For this purpose we use the formula (5) and the Lorentz transformation for the photon energy (the Doppler shift):

$$\hbar\omega_r = \gamma\, \hbar\omega'(1-\beta\cos\Theta'). \tag{8}$$

Here we denote $\beta = v/c$, and v is the longitudinal velocity of channeled positron in the laboratory coordinate system, $\hbar\omega_r$ is the energy of emitted photon.

In accordance with the virtual photons method, the cross–section of CB by channeled positron is:

$$\frac{\sigma^{CR}(\omega_r)}{d\omega_r} = \int\limits_{\omega_{MIN}}^{\omega_{MAX}} \frac{\sigma\,(\omega_r,\omega)}{d\omega_r d\omega} n(\omega)d\omega, \tag{9}$$

and the cross–section of positron bremsstrahlung in amorphous target is

$$\frac{\sigma^{AM}\,(\omega_r)}{d\omega_r} = N \int\limits_{\omega_{MIN}}^{\omega_{MAX}} \frac{\sigma_1(\omega_r,\omega)}{d\omega_r d\omega} n_{\mathrm{am}}(\omega)d\omega, \tag{10}$$

here $\sigma\,(\omega_r,\omega)$ is cross–section of the virtual photon scattering by the channeled positron in the laboratory reference frame and $\sigma_1(\omega_r,\omega)$ is cross–section of the virtual photon scattering by the free positron (positron) in the laboratory reference frame. The limits of integration are determined by the condition $-1 \le \cos\Theta \le 1$:

$$\omega_{\mathrm{MIN}} = \frac{\hbar\omega + \gamma(1+\beta)(\varepsilon_{\perp f} - \varepsilon_{\perp i})}{\gamma(1+\beta)\hbar}, \quad \omega_{\mathrm{MAX}} = \frac{\hbar\omega + \gamma(1-\beta)(\varepsilon_{\perp f} - \varepsilon_{\perp i})}{\gamma(1-\beta)\hbar}.$$

Here $\varepsilon_{\perp i(f)}$ is the transverse energy of initial (final) state of channeled positron.

The obtained cross–section for CB from channeled positron should be averaged over all possible transition of channeled

$$\frac{\sigma^{CR}(\omega_r)}{d\omega_r} = \frac{1}{\mathrm{N}} \sum_{fi} \frac{\sigma_{fi}^{CR}(\omega_r)}{d\omega_r}. \tag{11}$$

Here N is a number of transitions .

The results of calculation of the CB cross–section are shown in the figure 1: figure 1 a – contribution to the CB cross-section only from the channeled positrons (summed over all transitions of positron), figure 1 b – ordinary CB cross-section from free (not channeled) positrons in a crystal (calculated according to Ter-Mikaelian book [1]) and figure 1 c – total CB cross-section from free and channeled positrons. The target is (100) tungsten crystal, and positron emits CB interacting with the <001> axis.

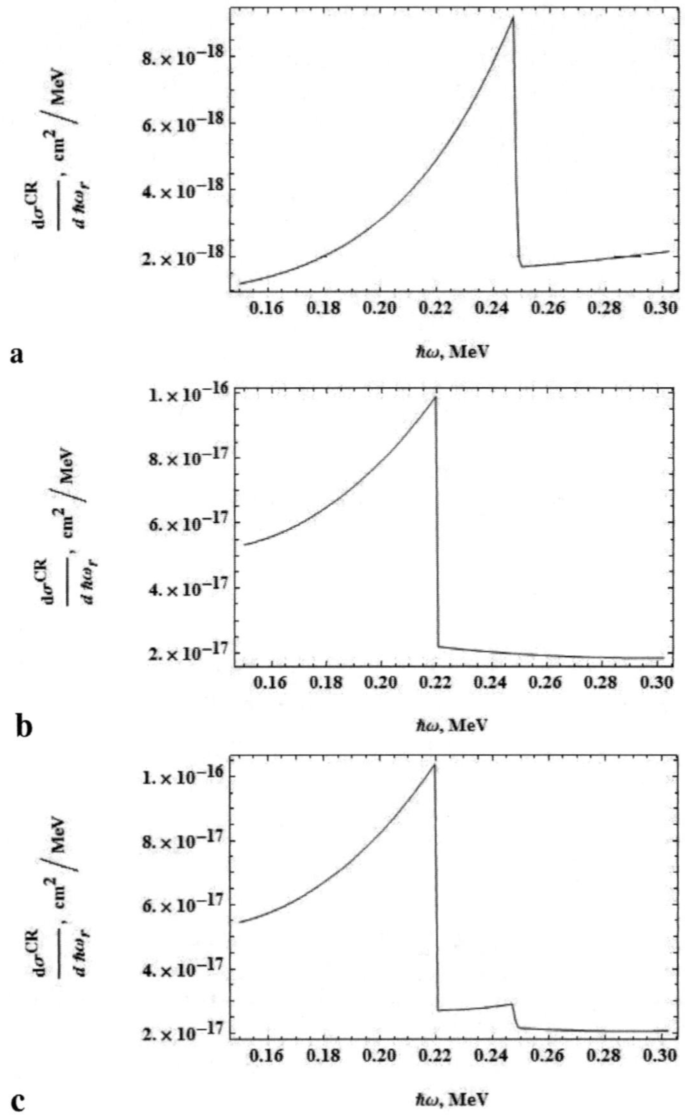

Figure 1. The cross–section of CB from positrons.

4. Conclusions

The results of calculation show that channeled positrons give small contribution to the total CB from positrons in a crystal. This correction is ten times less than ordinary radiation. As a result, one has appearance the additional structure of the coherent peak.

The combined effect in CB from relativistic axially channeled electrons was experimentally detected in [6].

In the present paper we use simple approximation for the continuous crystal potential. In a future we plan to use more exact transverse wave function (see, e.g. [10]).

The similar combined effect for coherent e+e- pair photoproduction in a field of the crystal axis was considered in [11]. In that case a positron was created in a bound (channeled) state with a crystal axis.

Acknowledgments

The author is grateful to Prof. Yu.L. Piviovarov for the fruitful discussion.

The work is partially supported by the Russian Education Federal Agency program «Support of Scientific Schools», contract 224.2012.2.

References

[1] Ter-Mikalian M L 1972 *High Energy Electromagnetic processes in Condensed Media* (New-York: Wiley Interscience)

[2] Uberall H 1956 *Phys. Rev.* **103** 1055

[3] Bazylev V A, Zhevago N K 1987 *Radiation of Fast Particles in Matter and External Fields* (Moscow: Nauka, in Russian)

[4] Baier V N, Katkov V M and Strakhovenko V M 1998 *Electromagnetic Processes at High Energies in Oriented Single Crystals* (Singapore: World Scientific Publishing Co)

[5] Akhiezer A I, Shul'ga N F 1996 *High energy Electrodynamics in matter* (Luxembourg: Gordon and Breach)

[6] Amosov K Yu, Vnukov I E , Naumenko G F, Potulitsin A P, Saruchev V P 1992 *JETP Lett* **55** 612

[7] Schtep U V and Kunashenko Yu P 2012 *Bulletin of TSPU* **7** 56

[8] Landau L D and E M Lifshiys E M 1975 *Quantum Mechanics* (Oxford: Pergamon)

[9] Kunashenko Yu P and Pivovarov Yu L 1996 *Nucl. Ins. and Meth.* **B119** 137.

[10] Korotchenko K B, Pivovarov Yu L, Tukhfatullin T A 2008 *Nucl. Instrum. Methods in Phys. Res.* **B 266** 3753

[11] Kunashenko Yu P 1997 *Poverhnost' Rentgenovskie sinchrotronuee i neitronnue issledovaniya* **8** 107

Quantum chaos manifestation in the axial channeling

N F Shul'ga[1], V V Syshchenko[2], V S Neryabova[2] and A Yu Isupov[3]

[1] Akhiezer Institute for Theoretical Physics of the NSC "KIPT", Akademicheskaya Street, 1, Kharkov 61108, Ukraine

[2] Belgorod State University, Pobedy Street, 85, Belgorod 308015, Russian Federation

[3] Laboratory of High Energy Physics (LHEP), Joint Institute for Nuclear Research (JINR), Dubna 141980, Russian Federation

E-mail: shulga@kipt.kharkov.ua, syshch@bsu.edu.ru, syshch@yandex.ru

Abstract. Energy levels of classically chaotic system demonstrate mutual repulsion described by Wigner distribution of distances between neighboring levels. The statistical properties of energy levels for the axially channeling electron in the system of [110] atomic strings of a silicon crystal are studied in the present report. The inter-level spacing distributions obtained by numerical calculation demonstrate satisfactory agreement with a theoretical predictions for chaotic systems.

1. Introduction

Quantum chaos means the study of semiclassical behavior of systems whose classical motion exhibits chaos [1]. The most prominent characteristic of that behavior is the statistics of energy levels. It was demonstrated in many cases [2] that the spacing s between adjacent (neighboring) levels satisfies Wigner distribution

$$p(s) = (\pi s/2D^2)\exp(-\pi s^2/4D^2) \tag{1}$$

(where D is the mean level spacing) for chaotic systems, while exponential (sometimes called as Poisson) distribution

$$p(s) = (1/D)\exp(-s/D) \tag{2}$$

for regular ones. The last case manifests itself as a tendency to group the levels into shells.

The hypothesis of universality of the distribution (1) for classically chaotic systems is tested in the present article for the case of axial channeling of fast electrons in crystals.

2. Transverse energy levels of axially channeling particles

When an electron is incident to the atomic string of the crystal under small angle ψ, it can be captured by the attractive potential of the string. The finite motion in such potential is called the axial channeling (see, e.g., [3, 4, 5]). This motion could be (with good accuracy) described as a motion in the uniform string potential $U(x, y)$ (i.e. the potential of the atomic string averaged along its axis z). The longitudinal (i.e. parallel to the string axis) component of the particle momentum $p_{\parallel}$ is conserved in such field. The motion in the transverse plane will be described in this case by the two-dimensional Schrödinger equation [3]

$$\hat{H}\Psi(x, y, t) = i\hbar\frac{\partial}{\partial t}\Psi(x, y, t) \tag{3}$$

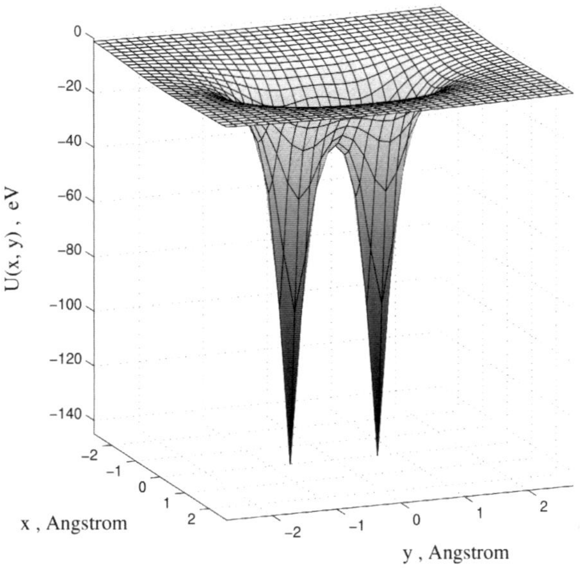

Figure 1. Electron potential energy (6) in the field of two neighboring atomic strings [110] of a silicon crystal.

with the Hamiltonian

$$\hat{H} = -\frac{\hbar^2}{2E_{\parallel}/c^2}\left(\frac{\partial^2}{\partial x^2} + \frac{\partial^2}{\partial y^2}\right) + U(x,y) \ , \tag{4}$$

in which the value $E_{\parallel}/c^2$ (where $E_{\parallel} = \sqrt{m^2c^4 + p_{\parallel}^2c^2}$) plays the role of the particle mass.

The finite motion in the (x, y) plane would be characterized by the discrete set of the Hamiltonian (4) eigenvalues $E_{\perp}$. Such quantization of the transverse motion energy had been observed in many experiments for moderately fast electrons ($E_{\parallel} \sim 1$ MeV), when the total number of transverse energy levels is small (see, e.g., [4] and references therein).

The uniform string potential could be approximated by the formula [3]

$$U_1(x,y) = -U_0 \ln\left(1 + \frac{\beta R^2}{x^2 + y^2 + \alpha R^2}\right), \tag{5}$$

where for the [110] string of silicon $U_0 = 60$ eV, $\alpha = 0.37$, $\beta = 3.5$, $R = 0.194$ Å (Thomas-Fermi radius); the least distance between two parallel strings $a/4 = 5.431/4$ Å (where a is the lattice period). So, the uniform potential, in which the electron transverse motion takes place, will be described by two-well function (Figure 1)

$$U(x,y) = U_1(x, y + a/8) + U_1(x, y - a/8) \tag{6}$$

(neglecting the influence of far-away strings). The electron finite motion in such potential (corresponding to negative values of the transverse motion energy $E_{\perp}$) is known as axial channeling [3].

The potential (6) is suitable for quantum chaos investigations because the motion in such potential is classically chaotic if the transverse energy exceeds the saddle point of the potential well [5]. However, Schrödinger equation with such complicated potential as Eq. (6) requires numerical methods for its solution. In the present article we have found the transverse energy levels for axially channeling electrons in two-string potential (6) using the so-called spectral method [6], first applied to the channeling in [7, 8].

In brief, the method is based on the numerical simulation of the time evolution of initial wave function according to Schrödinger equation (3). The correlation function between wave functions at the initial and current time momenta, $\Psi(x, y, 0)$ and $\Psi(x, y, t)$,

$$P(t) = \int_{-\infty}^{\infty} \int_{-\infty}^{\infty} \Psi^*(x, y, 0)\Psi(x, y, t)\, dxdy, \tag{7}$$

contains information about the energy eigenvalues. Indeed, every solution of the time-dependent Schrödinger equation (3) could be expressed as the superposition

$$\Psi(x, y, t) = \sum_{n,j} A_{n,j} u_{n,j}(x, y) \exp(-iE_n t/\hbar) \tag{8}$$

of the Hamiltonian eigenfunctions $u_{n,j}(x, y)$,

$$\hat{H} u_{n,j}(x, y) = E_n u_{n,j}(x, y),$$

where the index j is used to distinguish the degenerate states corresponding to the energy E_n. Computation of the correlation function (7) for the wave function of the form (8) gives

$$P(t) = \sum_{n,n',j,j'} \exp(-iE_{n'} t/\hbar) A^*_{n,j} A_{n',j'} \int_{-\infty}^{\infty} \int_{-\infty}^{\infty} u^*_{n,j}(x, y) u_{n',j'}(x, y) dxdy =$$

$$= \sum_{n,n',j,j'} \exp(-iE_{n'} t/\hbar) A^*_{n,j} A_{n',j'} \delta_{nn'} \delta_{jj'} = \sum_{n,j} |A_{n,j}|^2 \exp(-iE_n t/\hbar). \tag{9}$$

Fourier transformation of (9) leads to the expression

$$P_E = \int_{-\infty}^{\infty} P(t) \exp(iEt/\hbar)\, dt = 2\pi\hbar \sum_{n,j} |A_{n,j}|^2 \delta(E - E_n). \tag{10}$$

We see that the Fourier transformation of the correlation function looks like a series of δ-form peaks, positions of which indicate the energy eigenvalues.

However, integration of the numerically obtained correlation function $P(t)$ with the exponent in (10) could be carried out only over the finite time interval:

$$P_E = \int_0^T P(t) \exp(iEt/\hbar)\, dt = T \sum_{n,j} |A_{n,j}|^2 \exp(i(E - E_n)T/2\hbar) \,\mathrm{sinc}((E - E_n)T/2\hbar), \tag{11}$$

where $\mathrm{sinc}(x) = \sin(x)/x$. As a result, we obtain a series of peaks of finite width (inverse proportional to T) with sidelobes instead of infinitely narrow δ-like peaks (10). Employing normalized Hanning window function

$$w(t) = (1/T)(1 - \cos(2\pi t/T)) \tag{12}$$

in the integrand (11),

$$P_E = \int_0^T w(t) P(t) \exp(iEt/\hbar)\, dt, \tag{13}$$

substantially suppresses these sidelobes [6]. Typical correlation function shape in this case is presented in Figure 2.

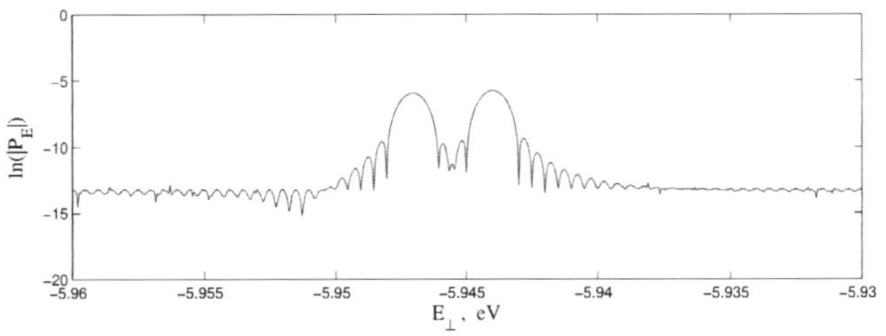

Figure 2. Correlation function (13) for electron of energy $E_{\|} = 400$ MeV with initial wave function (15) in the potential well (6) indicates two eigenvalues in the presented interval of transverse motion energy $E_{\perp}$.

Let two neighboring eigenvalues of $E_{\perp}$ are well-resolved in our numerical computation if not less than four side maxima are situated between two main ones. This leads to the criterion

$$T = 16\pi\hbar/\Delta E_{\perp} \tag{14}$$

for the total evolution time T necessary for the desirable energy levels resolution $\Delta E_{\perp}$. The resolution achieved in our computations is not worse than $\Delta E_{\perp} = 0.004$ eV. Further computational details are described in our previous papers [9, 10].

Note that the statistics of energy levels have to be studied separately for every type of eigenstate symmetry of the quantum system [2]. The potential well (6) has two planes of mirror symmetry: $x = 0$ and $y = 0$. So, every eigenstate of the electron in the potential (6) belongs to one of four classes of symmetry:

$$\Psi_{++}(-x, y) = \Psi_{++}(x, y), \quad \Psi_{++}(x, -y) = \Psi_{++}(x, y), \tag{15}$$

$$\Psi_{+-}(-x, y) = \Psi_{+-}(x, y), \quad \Psi_{+-}(x, -y) = -\Psi_{+-}(x, y), \tag{16}$$

$$\Psi_{-+}(-x, y) = -\Psi_{-+}(x, y), \quad \Psi_{-+}(x, -y) = \Psi_{-+}(x, y), \tag{17}$$

$$\Psi_{--}(-x, y) = -\Psi_{--}(x, y), \quad \Psi_{--}(x, -y) = -\Psi_{--}(x, y). \tag{18}$$

The level spacing distributions obtained for the semiclassical domain (close to the top of the potential well) demonstrate quite good agreement with the prediction (1) for every of these four symmetry classes. The distributions computed for the transverse energy domain

$$-10 \text{ eV} \leq E_{\perp} \leq -2.6 \text{ eV} \tag{19}$$

are presented in Figure 3. Small discrepancy could be caused by the limited statistics (the arrays of ~ 250 energy levels have been analyzed whereas the typical ones in [2] contain some thousands of levels).

Another possible reason of the mentioned discrepancy is a specific shape of the potential (6): rapid narrowing of the potential well with the depth increase leads to as much rapid increase of the average inter-level distance. Indeed, in the lower half of the energy interval (19)

$$-10 \text{ eV} \leq E_{\perp} \leq -6.3 \text{ eV} \tag{20}$$

the average inter-level distance D is about twice larger than in the upper one

$$-6.3 \text{ eV} \leq E_{\perp} \leq -2.6 \text{ eV}. \tag{21}$$

As a result, we can see better agreement of the distributions calculated for the intervals (21) and (20) with Wigner function (1) (Figures 4 and 5) than for the total interval (19).

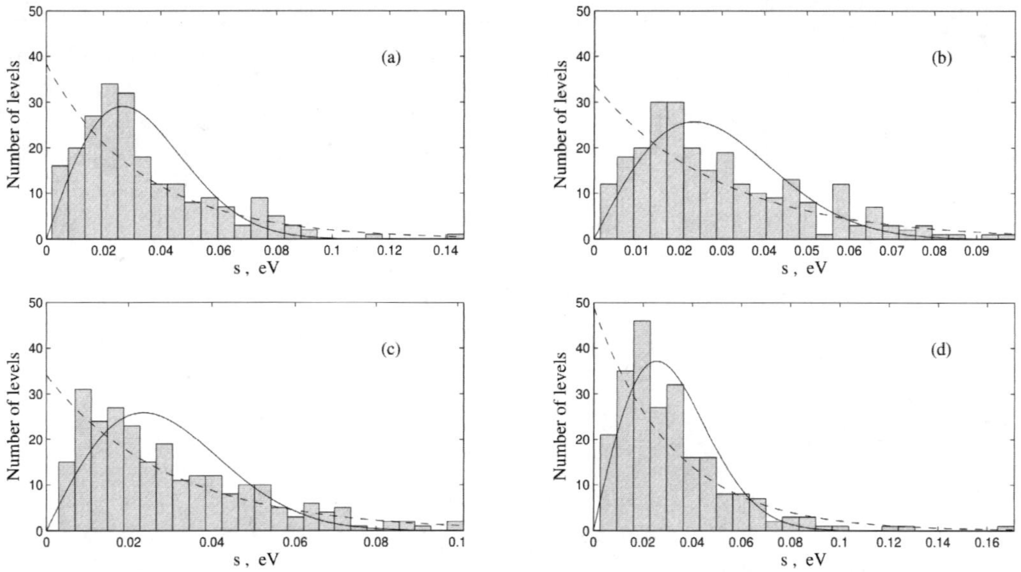

Figure 3. Transverse energy levels spacing distribution in the range (19) for $E_{\|} = 400$ MeV electrons channeled in the field of two parallel [110] strings of silicon crystal (histograms) as well as theoretical predictions (1) for chaotic motion (solid line) and (2) for regular motion (dashed line). The histograms are plotted for odd eigenstates (16), (17) (panels (a) and (b), respectively) and for even eigenstates (15), (18) (panels (c) and (d), respectively).

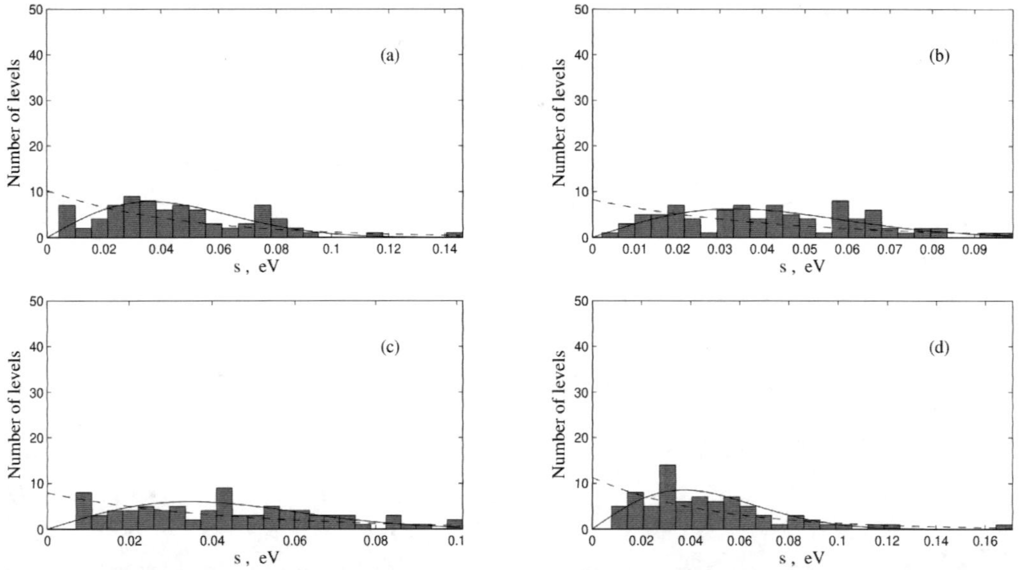

Figure 4. The same as in Figure 3 for the transverse energy domain (20).

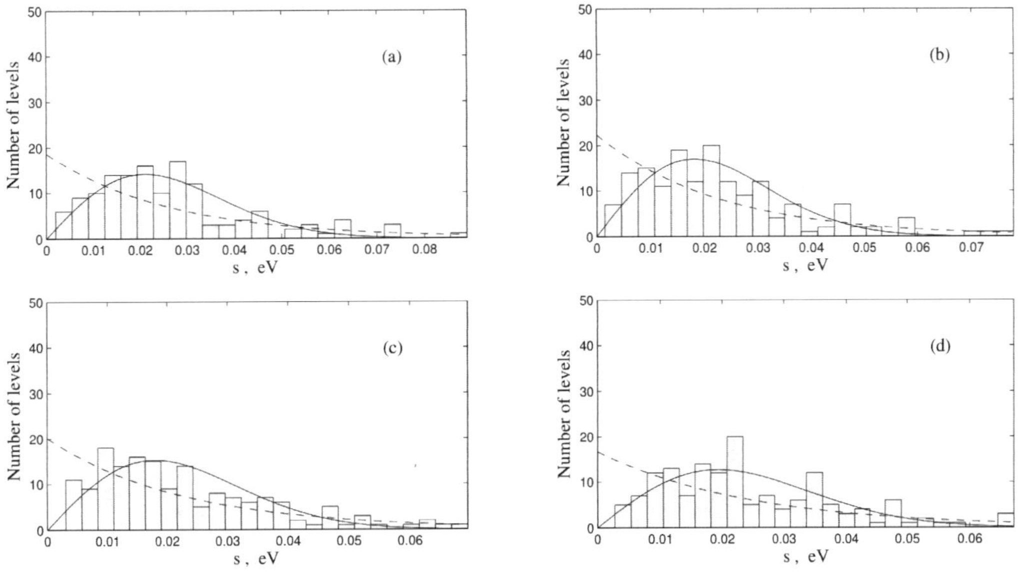

Figure 5. The same as in Figure 3 for the transverse energy domain (21).

3. Conclusion

Wigner distribution (1) of distances between adjacent energy levels of quantum system is the pronounced manifestation of quantum chaos phenomenon. Building of the inter-level distances distribution requires computational methods which allow one to find densely lying eigen-energies in the potential wells with sophisticated configuration. This problem naturally arises under consideration of the axial channeling in the silicon crystal near [110] axis of electrons with energy of some hundreds of MeV and higher. As we demonstrate in the present paper, the spectral method is sufficient and effective for this problem.

The obtained results demonstrate a different degree of agreement between the Wigner distribution and the calculated inter-level distance ones for the different transverse energy ranges. This fact could have some causes like the difference of average inter-level distance for these ranges and presence of regular (in the classic limit) motion domains among the chaotic motion. Anyway, this problem requires further more detailed investigations.

References

[1] Berry M V 1987 *Proc. R. Soc. Lond.* A **413** 183
[2] Stöckmann H-J 2000 *Quantum Chaos. An Introduction* (Cambridge: Cambridge University Press)
[3] Akhiezer A I and Shul'ga N F 1996 *High-Energy Electrodynamics in Matter* (Amsterdam: Gordon and Breach)
[4] Uggerhøj U I 2005 *Rev. Mod. Phys.* **77** 1131
[5] Akhiezer A I, Shul'ga N F, Truten' V I, Grinenko A A and Syshchenko V V 1995 *Physics-Uspekhi* **38** 1119
[6] Feit M D, Fleck J A Jr and Steiger A 1982 *J. Comput. Phys.* **47** 412
[7] Dabagov S, Ognev L I 1988 *Sov. Phys. Tekh. Phys.* **33** 158
[8] Dabagov S, Ognev L I 1988 *Nucl. Instrum. and Methods* B **30** 185
[9] Shul'ga N F, Syshchenko V V and Neryabova V S 2013 *Journal of Surface Investigation. X-ray, Synchrotron and Neutron Techniques* **7** 279
[10] Shul'ga N F, Syshchenko V V and Neryabova V S 2013 *Nucl. Instrum. and Methods* B **309** 153

Controlling of the channeling radiation by ultra-short laser pulses

V Yaralov
Alikhanian National Scientific Laboratory (Yerevan Physics Institute)
Br Alikhanians 2, Yerevan 0036, Armenia

E-mail: yaralov@mail.yerphi.am

Abstract. The interaction of an electron channelled in a crystal with laser pulses was investigated within the frameworks of semi-classical theory by means of numerical methods using the density matrix approach. It was shown that for the given pulse shape the radiation yield from an electron depends on the phase of carrier frequency in the peak of laser pulse and on the phase of non-diagonal element of density matrix at the start of laser pulse action. The dependences of populations and channeling radiation densities on the electron transit time in the crystal, as well as the corresponding spectra of channeling radiation under the influence of laser pulses (with different parameters) on the channelled electron have been calculated. The possibility of an increase in the yield of channeling radiation by an appropriate choice of parameters of ultra-short laser pulses in principle was demonstrated.

1. Introduction.

The radiation from plane channelled electron of the energy of several ten MeV in a single crystal in an external field is dealt with, the transverse energies (of electron) being of the order of several eV. The treatment of the problem under these energies shall be necessarily in the quantum mechanical approach. Now specify the required channeling theory terms of the problem. The motion of electron in transverse direction is under periodical potential and according to the solid-state theory its transverse stationary states are grouped in discrete energy zones and are identified by the zone number n and Bloch momentum k. The transitions between the transverse levels (n, k) besides the radiative ones, occur due to the scattering as well as under the influence of the external field. In electromagnetic interactions the Bloch momentum of electron is conserved in the dipole approximation. Various values of the Bloch momentum in transitions above may be treated as independent ones. In numerical calculations the continuous distribution of the Bloch momentum is replaced with a discrete one. The calculation of emitted radiation at the transition between two zones under consideration is carried out for each Bloch momentum for following summation. Thus, the problem is reduced to the radiation from a two-level electron in an external field.

2. The calculation method.

The electron is considered as a quantum object in the classical external field. Interacting with the external field and subjecting to scattering in a crystal the channelled electron has to be described by the density matrix. Corresponding equations for density matrix elements are called Bloch optics equations and contain the channeling parameters as well as the external field parameters and phenomenological damping constants. Such an approach was applied by Rabi in the works on magnetic resonance [1]. In optics these methods were used by Mollow [2]. The calculation scheme from this work was further developed and applied to channeling [3, 4]. In these works the calculations of channeling radiation in the constant amplitude monochromatic field with averaging of the initial phase of driving field have been made. In the present work the processes of channeling radiation with a fixed initial phase of the driving field with a variable amplitude are examined. The change of the external field in a crystal and the time structure of the electron beam are not taken into account.

At the incidence of electron on a crystal the transverse electron levels became populated with corresponding probabilities. Let us consider two levels. Their populations form the initial diagonal elements of the density matrix of the two-level electron. The initial electron state is described by wave function (plane wave expanded into series in transverse levels) and therefore the initial non-diagonal elements of the two-dimensional density matrix in question are also known and contain the real parts only. At the motion of electron in the crystal they become complex. The scattering of electron in the crystal leads to damping of the matrix elements. Now consider the two-dimensional linear electronic space. The lower and upper levels form the basis of this space. Any transverse electron state may be expanded in terms of the basis states. Each operator in the configuration space is assigned to an operator acting in the electronic space (second quantization) [2, 5]. The matrix elements of the dipole moment $\boldsymbol{d}$ in configuration space between the upper and lower levels are considered to be the matrix elements in the electronic space. The density matrix elements become the matrix elements of the density operator in the electronic space. After substitution of the dipole moment $\boldsymbol{d}$ with the second quantized operator the delayed radiation of the electron $\boldsymbol{E}(r,t) = \frac{1}{c^2 r}\left[\left[\frac{d^2}{dt^2}\boldsymbol{d}\left(t-\frac{r}{c}\right),\boldsymbol{n}\right]\boldsymbol{n}\right]$ also becomes an operator in the electronic space. The channeling radiation is considered to be a quantum object and is calculated by using the correlation function: $G(r,t,\tau) = < E^-(r,t)\, E^+(r,\ t+\tau) >$, where E^-, E^+ – are negative- and positive-frequency parts of the expansion of radiation into series at t and $t+\tau$ moments in the detection point $\boldsymbol{r}$ [6]. The angular brackets $< >$ mean the statistical averaging. The spectral-angular distribution of radiation density is expressed via the Fourier transformation of the correlation function with respect to parameter τ: $\frac{d^3E}{d\omega\, dt\, d\Omega} = \frac{cr^2}{2\pi}\tilde{G}(r,t,\omega)$, where E is the radiated energy, ω is the spectral frequency, $d\Omega$ is the space angle element, c is the light velocity. In the calculations we use system of units ħ = c = 1: $\omega\ (eV), t\ (\frac{1}{eV})$. The derivation of this formula as well as of the density matrix equations and the mathematical manipulations is developed in [4] in detail. The complexity of the calculation algorithm is justified by the generality of approach. The time-dependent density matrix elements contain all the required information about the transverse motion and spectral-angular radiation density at any time moment. This algorithm makes it possible to answer any questions connected with the external controlling field.

The density matrix equations for a constant amplitude monochromatic field are solved analytically. For the variable amplitude we must divide the path of the electron in the crystal into small intervals, where amplitude is regarded as constant and then calculate each interval separately. An ultra-short laser pulse in time domain can be expressed by monochromatic wave $E(t) = A(t)\cos[\varphi(0) + \omega_0 t]$ with variable amplitude. Such a representation is correct for an arbitrary time signal under the condition that the width of its spectral distribution is much less than the central frequency ω_0 called the carrier frequency [7]. In this case the amplitude (envelope) $A(t)$ is a slowly varying function in comparison with the harmonic function cos. The calculation method using the stepping of the electron path enables us to calculate the action of any pulse shapes and their combinations as well as action of the chirped pulses (with phase modulation) $E(t) = A(t)\cos[\varphi(t) + \omega_0 t]$.

3. The results of the calculations.

Let us consider 9 MeV electron channelled in the planes (110) of the 13 μm diamond crystal and the radiation produced by the transition between two levels 1, 2 in the planar channel. The plane-polarized laser pulses are directed against the electron beam. This case is not realistic with respect to some parameters. It has been chosen for comparison with already existing results both theoretical and experimental [4, 8]. The aim of the present work is to investigate the features of the channeling and channeling radiation under the influence of ultra-short laser pulses and the demonstration of the possibility of the increase of radiation in principle.

The interaction of the channeled electron with an external field depends on the initial sign (plus or minus) of the entrance angle of the electron with respect to channel planes and on the initial phase of the driving field. To investigate the phase dependence one needs to fix the transverse component of the electron momentum at entrance moment. Rotating the crystal with respect to the electron beam direction we change the comparative parts of positive and negative projections. The crystal rotation leads to an increase or decrease of the radiation yield. Thus the peak splitting of angular distribution is possible. It depends on the beam divergence. In figures 1, 2 the calculated spectra for various beam divergences and rotation angles are shown. Further we deal with a crystal rotation by 0.8 mrad and the beam divergence equal to 0.3 mrad. The part of the positive angles in a beam is equal 0.996.

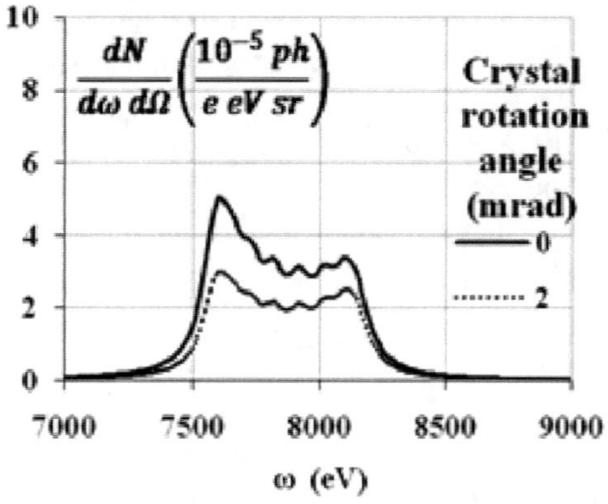

Figure 1. Beam divergence = 1 mrad.

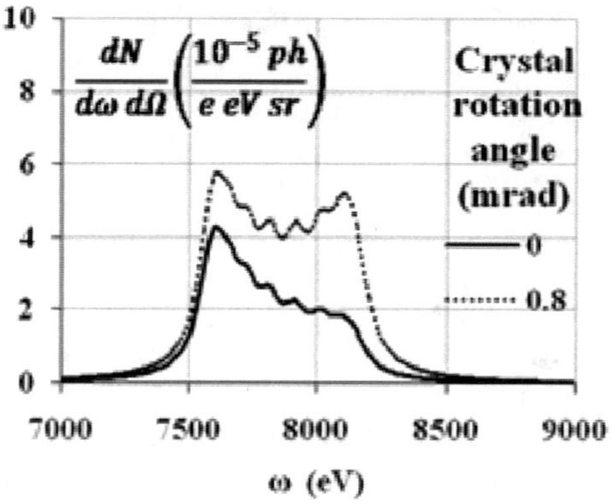

Figure 2. Beam divergence = 0.3 mrad.

Let us consider the channelled electron in the external monochromatic electric field with a constant amplitude E(t) = Ecos [φ(0) + ω_0t], where ω_0 is the frequency of the driving field near the transition frequency of the considered levels (close to a resonance). The calculated time-dependent populations of the upper level and the radiation densities with respect to time for the central frequency of the spectrum (7830 eV) are shown in figures 3, 4 for initial phases π, 0.

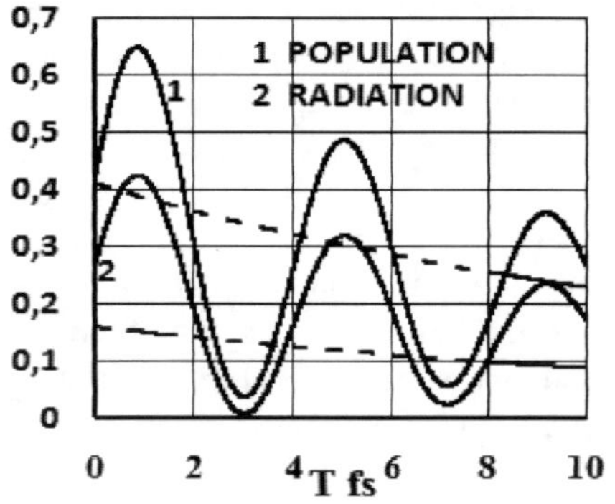

Figure 3. The initial phase = π.

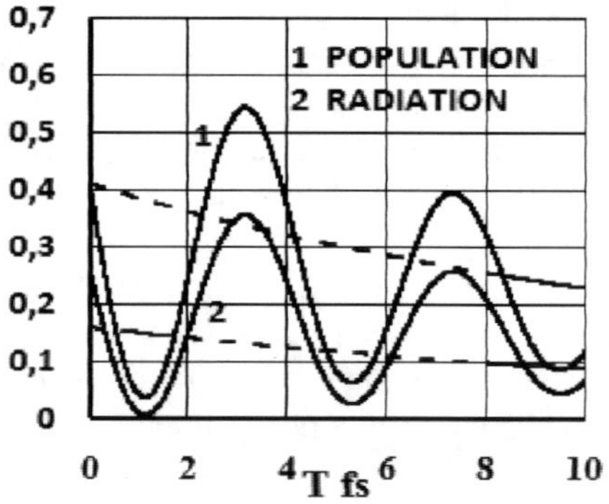

Figure 4. The initial phase = 0.

The damping curves (dashed) correspond to usual channeling whereas the oscillating ones correspond to the channeling in the external field. The upper oscillating line corresponds to upper level population. The lower oscillating line corresponds to radiation density $\frac{dN}{d\omega\, dt\, d\Omega}\left(\frac{10^{-5}ph}{e\, eV\, sr}\right)$. The field amplitude divided by the crystal field is equal to $\frac{E}{E(crystal)} = \frac{E}{10^9\left(\frac{V}{sm}\right)} = 0.1$. Let us note the features seen on the figures. First, the oscillating radiation density is proportional to the upper level population. Second, the change of the initial phase by π changes the direction of population variation to the opposite. Accordingly the radiation density is also changed. Finally, the damping and oscillating populations at the initial point have the same values. The radiation densities with and without the field at the initial point have different values. Such a jump of the radiation density is due to stimulated radiation in the presence of the external field.

In figures 5, 6 the spectral distributions of the channeling radiation in the monochromatic external fields with different initial phases and amplitudes are shown. The central peak height does not depends on a wave phase. The passage time of the electron in the 13 μm crystal is equal to 43 fs. The laser pulse with the duration of 200-300 fs in its medium range can be represented as monochromatic field with the constant amplitude.

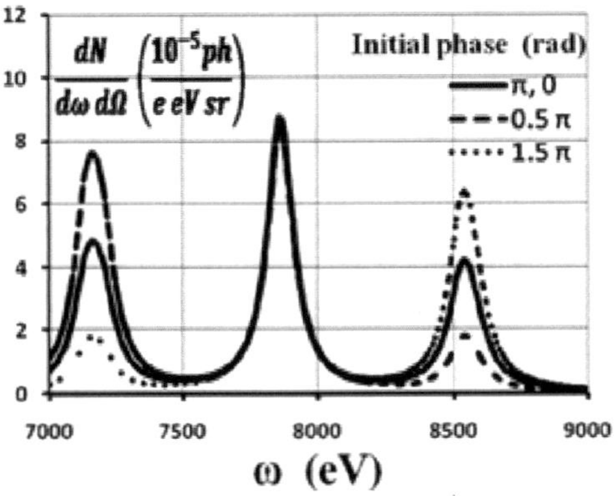

Figure 5. The spectral distribution in the monochromatic field. E / Ecrystal = 0.1.

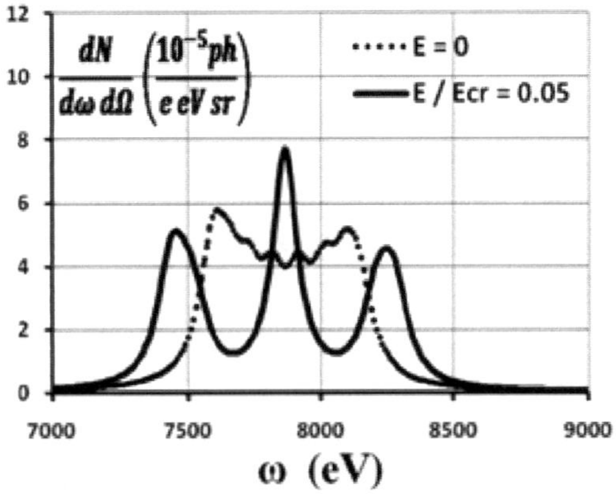

Figure 6. The spectral distribution in the monochromatic field. E/Ecrystal = 0, 0.05.

To control the population and radiation density one needs to take a driving field with a duration less than a period of the population (Rabi period). Let us consider laser pulse represented by a monochromatic wave with amplitude modulation $E(t) = A(t)\cos[\varphi(0) + \omega_0 t]$. The envelope amplitude can be taken in the form of truncated Gaussian $A(t) = A_{max} EXP\left(-\frac{(t-T)^2}{T^2}\ln\frac{A_{max}}{A_{min}}\right)$, where A_{min}= 0.005, A_{max}= 0.1 are the minimal and maximal values, T is the Gaussian full half-width and t = (0-2T). The values less than A_{min} are negligible. The laser pulse is characterized by the phase value in the envelope peak (CEP, Carrier Envelope Phase). Now it is possible to create a train of identical ultra-short laser pulses with required CEP and fixed pulse repetition frequency [7]. For the given pulse shape the initial phase and the CEP depend of each other $\varphi(T) = \varphi(0) + \omega_0 T$. It is more suitable to use the initial phase for the calculations.The calculated increase of the radiation yield provided by a short-time pump of upper level population with help of a single laser pulse is shown in figures 7, 8. The Gaussian is shown by lower line.

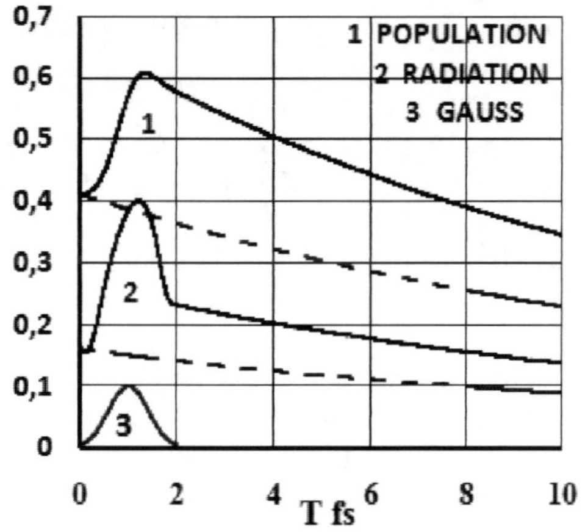

Figure 7. The populations and radiation densities under single pulse action.

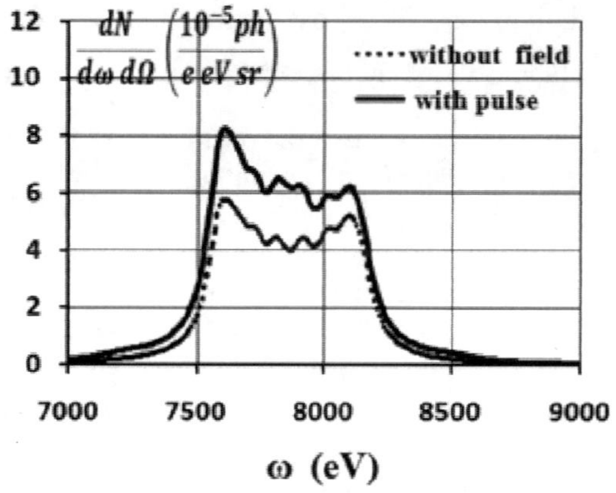

Figure 8. The corresponding spectra.

The initial value of the phase of complex off-diagonal element of the density matrix at electron entrance in crystal is equal to 0 (or π). It is changed periodically during the motion of the electron in the crystal. The result of the influence of the laser pulse with fixed initial phase and fixed shape on the channelled electron depends on the correlation between the pulse initial phase and the off-diagonal element phase. That is why the start time of the laser pulse is important. This is shown in figures 9, 10 for the pulses with time delay ΔT.

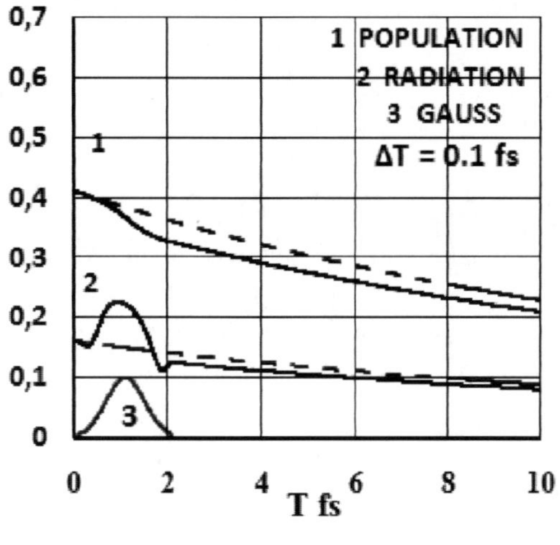

Figure 9. Pulse-delay time = 0.1 fs.

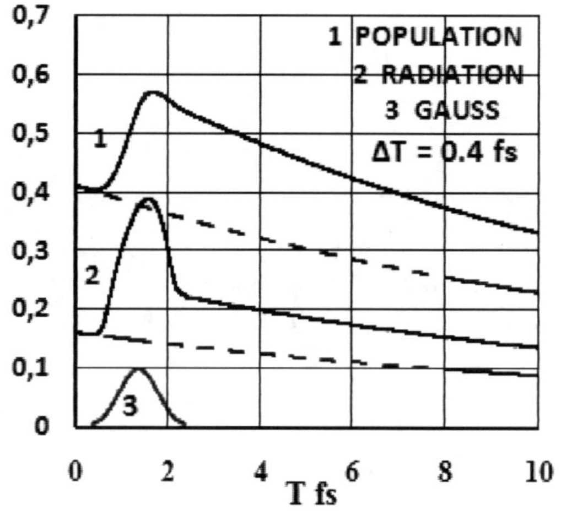

Figure 10. Pulse-delay time = 0.4 fs.

4. Conclusion.

The carried out calculations show a possibility of controlling of the channeling radiation by ultra-short laser pulses. The field of the unchirped laser pulse with a sufficient long duration can be considered as monochromatic and provides the coherent stimulated radiation and the sharp peak in the spectral distribution. In addition, the action of a monochromatic field does not depend on the initial phase. It is necessary to investigate the interaction of the channelled electron with laser pulses in greater detail and to choose the appropriate parameters of the process (the inter-planar spacing, the electron energy, the laser pulse parameters) to have for it a practical application.

References

[1] Allen L and Eberly J H 1975 Optical resonance and two-level atoms (Wiley-Interscience Publication JohnWiley and Sons)
[2] Mollow B R 1970 *Phys. Rev.* A **2** 76-80
[3] Yaralov V 2004 *Proc. of Conf. "Laser Physics-2004* 93
[4] Avakian R O, Ispirian K A and Yaralov V 2006 *Nucl. Instr. Meth.* B **252** 20-25
[5] Lax M 1963 *Phys. Rev.* **129** 2342-48
[6] Glauber R 1963 *Phys.Rev.* **130** 2529-39
[7] Diels J C and Rudolph W 2006 Ultrashort laser pulse phenomena
[8] Genz H, Groening L, Hoffmann-Staschek, Richter A, Hofer M, Hormes J, Nething U, Sellschop J P F, Toepffer C and Weber M 1996 *Phys. Rev.* B **53** 8922-36

The new principle for intensity calculation of X-rays dynamically diffracted in single crystals with defects

A M Yeghiazaryan, K M Gevorgyan, A K Atanesyan

Institute of Applied Problems of Physics of NAS RA, 25 Hr.Nersisyan str., 0014, Yerevan, Armenia

E-mail: `Atanesyan@iapp.sci.am`

Abstract. A new principle is introduced for X-ray intensity calculation dynamically scattered in single crystals with defects. The principle is based on Bragg diffraction dynamic theory for X-rays in ideal single crystals and small angle scattering theory. The principle is substantiated by juxtaposition of the corresponding formula with the particular solution of the integral-differential equation for a dynamically scattered X-ray wave in a weakly distorted single crystal.

1. Introduction

The further development of the dynamic theory for X-ray scattering in deformed crystals and crystals with defects is most important for the structural study of solid materials from the aspect of X-ray scattering phase problem solution. This leads to an unambiguous interpretation and decoding of X-ray diffraction and interference images on the purpose of precise determination of the structural distortions of the studied materials. This theory is quite developed and the integral-differential equation of the studied direct problem is obtained [1]. Nevertheless the integration of this equation is an uneasy task and the reverse problem formulation and solution for the determination of precise structures of materials with these equations is an unsolved problem. The direct problem solution, the calculation of complex amplitude of X-ray wave dynamically scattered in a distorted crystal for given analytic distortions is a difficult problem. Usually the problem is not solved analytically.

The numerical integration of the obtained equations strongly restricts the capabilities of developed strict dynamic theory of X-ray scattering in distorted crystals, in respect of new predictions, for experiment development in this spear. The uncertainty of the theoretical interpretation of experimental results, to some extent, is simplified when the crystal is distorted by definite external influences (a thermal gradient, ultrasound field). Therefore the further development of X-ray diffraction theory and experiment from the point of mutual supplementation, of course, is connected to the study of crystal distorted by definite external analytical influences [2, 3]. For the purpose of overcoming of above mentioned problems, in work [4] we suggest a new principle of calculation of X-ray intensities (or thermal neutrons) dynamically scattered in crystals with defects. This principle ensues from the juxtaposition of the dynamic theory of X-ray Bragg diffraction with the theory of their small angle scattering. The fairness of the new principle is substantiated by the juxtaposing of the corresponding formula with the particular solution of the integral-differential equation for a dynamically scattered X-wave in a weakly distorted crystal. .

2. The Essence of the New Principle

Let's assume that a beam of X-ray falls at the Bragg angle on a single crystal with a defect (figure 1). We will consider the amplitude of the diffracted wave as a result of double-phase diffraction.

In the first phase a beam of X-ray dynamically scatters on an perfect single crystal (Bragg diffraction). In the second phase the waves diffracted in the first phase undergo a small angle scattering on a defect. In the first phase the dynamic Bragg diffraction of an X-ray beam takes place on an perfect single crystal, for what the amplitudes of the diffracted waves are known. In the second phase the waves diffracted in the first phase undergo Fraunhofer or Fresnel diffraction on a defect depending upon the defect size and the parameters of intensity registration geometry of the diffracted radiation. The essence of the new principle is that the amplitude of the X-wave diffracted in a crystal with a defect can be calculated by adding the scattered amplitudes of the two phases.

Let's assume that a monochromatic X-ray beam with amplitude $\mathbf{\Psi}_0^{(i)}(\mathbf{r})$ falls at the Bragg angle on a non absorbing single crystal with a static defect (figure 1). According to the new

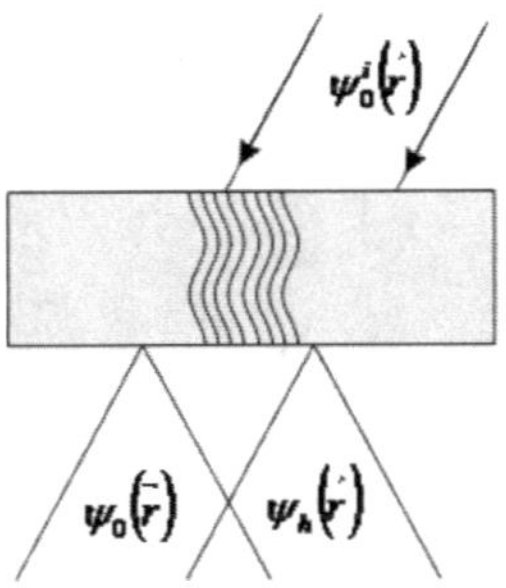

Figure 1. The scheme of the incidence of a ribbon-like beam on the crystal

principle the amplitudes of waves $\mathbf{\Psi_0}(\mathbf{r})$ and $\mathbf{\Psi_h}(\mathbf{r})$ in the first approximation, can be represented in the form;

$$\mathbf{\Psi_{0,h}}(\mathbf{r}) = \mathbf{\Psi^0_{0,h}}(\mathbf{r}) \pm \frac{\mathbf{v}}{4\pi} \int \mathbf{\Psi^0_{0,h}}(\mathbf{r}') \frac{\exp(\mathbf{ik}\,|\mathbf{r}-\mathbf{r}'|)}{|\mathbf{r}-\mathbf{r}'|} \mathbf{U}(\mathbf{r}')\mathbf{dr}' \qquad (1)$$

where $\mathbf{\Psi^0_{0,h}}(\mathbf{r})$ – correspondingly, the amplitudes of the waves dynamically diffracted on a perfect crystal v– the parameter characterizing the strength of interaction of wave $\mathbf{\Psi^0_{0,h}}(\mathbf{r})$ - with the defect, $\mathbf{U}(\mathbf{r}')$ the local replacement of the atoms from their equilibrium position in the perfect crystal, bfr the radius-vector of the observation point, the integration is made within the volume of the defect distribution.

The plus and minus signs difference follows form the energy conservation law as the observed strengthening of wave $\mathbf{\Psi_h}(\mathbf{r})$ scattered on a defect occurs at the expense of weakened wave $\mathbf{\Psi_0}(\mathbf{r})$

Expression (1) submits the solution of the volume problem. From the analysis of this solution it is possible, particularly, to predict the phenomenon of X-beam pumping form the passage direction to the reflection direction, the phenomenon of the beam focusing, diffraction on a defect in a crystal. In some cases, it is possible to set the reverse problem of determination of the local shift function $\mathbf{U}(\mathbf{r}')$ of atoms form their equilibrium position in the prefect crystal, etc.

Particularly, let's consider the phenomenon of modulation of X-rays multiple times scattered on a perfect crystal of quartz where an acoustic standing wave exists. According to the experiment scheme [5] (figure 2) a ribbon-like monochromatic beam of X-ray falls on a single crystal of quartz at the exact Bragg angle when an acoustic standing wave exists along the

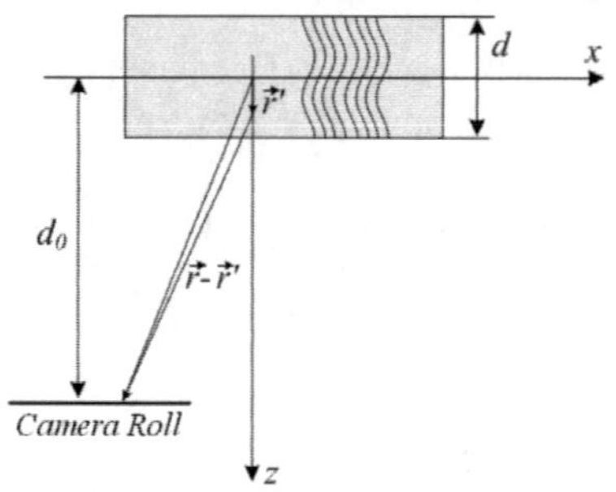

Figure 2. The scheme of the beam diffraction on the crystal in the presence of standing acoustic wave

crystal width. Under the conditions of the submitted experiment it is possible to admit that $U = U_0 \cos(2\pi k_{ac} z)$, where $k_{ac})$ – the wave number of the acoustic wave.

Expression (1) is brought to the form;

$$\mathbf{\Psi_{0,h}(x,y) = \Psi^0_{0,h}(x,y) \mp \frac{vU_0}{4\pi} \int_{-\frac{d}{2}}^{\frac{d}{2}} \Psi^0_{0,h}(0,z) \frac{\exp(ik\,|r-r'|)}{|r-r'|} \cos(2\pi k_{ac} z) dz} \tag{2}$$

where $\Psi^0_{0,h}$ - the plane wave solution for the perfect crystal, d the width of the single crystal. Taking into account the numerical values of the parameters of geometry and radiation of the experiment, the integral in expression (2) is brought to a Fresnel integral, which real part, with the precision of the constant factor has the form:

$$\int_{-\frac{d}{2}}^{\frac{d}{2}} \cos[-kz \sin\theta_B \frac{\Delta x}{2d_0} + k\frac{z^2 \cos\theta_B}{4d_0}] \cos(2\pi k_{ac} z) dz, \tag{3}$$

where θ_B – the precise Bragg angle, Δx – the distance of the viewed point on the photo plate from its center; k – the wave number of X-wave inside the crystal. The results of the numerical integration are submitted on figure 3.

From the numerical integration (9) follows that in the geometrical cross sections of both the passing and reflected beams occur certain amplitude oscillations, what precisely characterize the experimental results [5].

Expression (1) presents the general expression for the amplitudes of neutrons and X-ray beams multiple times scattered on single crystals with defects of various forms. From this expression analysis, it is possible to predict and quantitatively evaluate different known and new dynamic phenomena while the diffraction of these beams.

Let's reveal the physical essence of parameter v, characterizing the interaction power of the radiation with the defect. For this purpose, let's consider the problem of the dynamic scattering of a plane X-ray wave $\Psi_0(\mathbf{r}) = \mathbf{\Psi^j_0(r)} \exp(-\mathbf{ik^j_0 r})$ in weakly deformed crystal with electric polarization $\chi(\mathbf{r})$. The fundamental equation of this scattering has the following form:

$$(\Delta^2 + k_0^2)\mathbf{\Psi} = -\nabla \times \nabla \times (\chi\mathbf{\Psi}) \tag{4}$$

where $\mathbf{\Psi}(\mathbf{r})$ – the vector of electric induction of the wave scattered in the crystal, $k_0 = \frac{2\pi}{\lambda}$ - the wave number of the X-wave.

Differential equation (4) is equivalent to the following integral equation [1]:

$$\Psi(\mathbf{r}) = \mathbf{\Psi_0(r)} + \frac{\mathbf{1}}{\mathbf{4\pi}} \int \mathbf{dr'} \frac{\exp(-\mathbf{ik}\,|\mathbf{r}-\mathbf{r'}|)}{|\mathbf{r}-\mathbf{r'}|} \nabla' \times \nabla' \times [\chi(\mathbf{r'})\mathbf{\Psi}(\mathbf{r})] \tag{5}$$

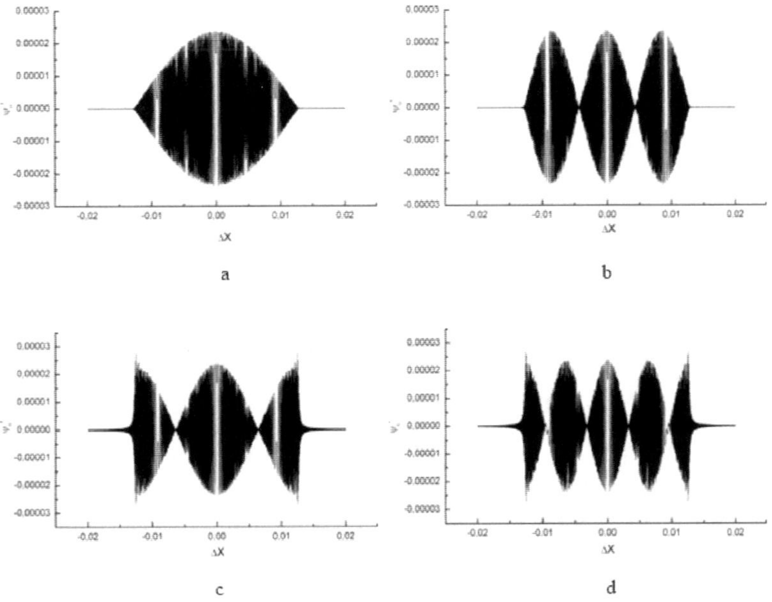

Figure 3. The results of numerical integration for the first (a), third (b), fifth (c) and seventh (d) harmonicas

We search the solution of equation (5) by applying the method of sequential approximation. In the first approximation of the solution, in the integrand expression, we put, instead $\mathbf{\Psi}(\mathbf{r}')$, the amplitude $\mathbf{\Psi}(\mathbf{r}')$ of the wave scattered in the perfect crystal. In weakly deformed crystal, the electric polarization $\chi(\mathbf{r})$ can be presented in the form:

$$\chi(\mathbf{r}') \approx \chi_0(\mathbf{r}' - \mathbf{U}(\mathbf{r}')) \tag{6}$$

where $\chi_0(\mathbf{r}')$ – the electric polarization of non-deformed crystal. According to Lagrange's formula, it is possible to write as:

$$\chi_0(\mathbf{r}' - \mathbf{U}(\mathbf{r}')) = \chi_0(\mathbf{r}') - \nabla\chi_0\mathbf{U}(\mathbf{r}'). \tag{7}$$

Taking into account expressions (4–7), the solution of equation (5) can be presented in the form:

$$\Psi(\mathbf{r}) = \mathbf{\Psi}^0(\mathbf{r}) - \frac{1}{4\pi}\int d\mathbf{r}'\frac{\exp(-i\mathbf{k_0}\,|\mathbf{r}-\mathbf{r}'|)}{|\mathbf{r}-\mathbf{r}'|}\nabla'\times\nabla'\times(\nabla\chi_0\mathbf{U}(\mathbf{r}')\mathbf{\Psi}^0(\mathbf{r}')). \tag{8}$$

As $\frac{\nabla\chi_0\mathbf{U}(\mathbf{r}')}{\chi_0} << 1$ so for the double-wave approximation of scattering, from the corresponding to χ equation (4), we'll obtain:

$$\nabla\times\nabla\times(\nabla\chi_0\mathbf{U}(\mathbf{r}')\mathbf{\Psi}^0(\mathbf{r}')) = \frac{\nabla\chi_0\mathbf{U}(\mathbf{r}')}{\chi_0}2\mathbf{k}_0^2\mathbf{\Psi}^0(\mathbf{r}'). \tag{9}$$

By inserting expression (9) into (8), we'll obtain the final solution of the formulated problem.

$$\mathbf{\Psi}(\mathbf{r}) = \mathbf{\Psi}^0(\mathbf{r}) - \frac{1}{4\pi}\int d\mathbf{r}'\frac{\exp(-i\mathbf{k_0}\,|\mathbf{r}-\mathbf{r}'|)}{|\mathbf{r}-\mathbf{r}'|}\frac{\nabla\chi_0\mathbf{U}(\mathbf{r}')}{\chi_0}2\mathbf{k}_0^2\mathbf{\Psi}^0(\mathbf{r}') \tag{10}$$

From the juxtaposition of solution (10) and expression (1), the physical essence of parameter v is reveal, which characterize the interaction strength between a wave and a defect:

$$v = \frac{2k_0^2}{\chi_0} |\nabla \chi_{\mathbf{0}}| \cos\theta \tag{11}$$

where θ - the angle between the bftors $\nabla \chi_{\mathbf{0}}$ and $\mathbf{U}(\mathbf{r}')$.

Taking into account the obtained expression for parameter (11), according to the new principle, the amplitude of the wave diffracted on a defect in a single crystal can be calculated with the formula:

$$\mathbf{\Psi_{0,h}}(\mathbf{r}) = \mathbf{\Psi^0_{0,h}}(\mathbf{r}) \pm \frac{\mathbf{2k_0^2}}{\chi_{\mathbf{0}}} |\nabla \chi_{\mathbf{0}}| \cos\theta \int \mathbf{\Psi^0_{0,h}}(\mathbf{r}') \frac{\exp(\mathbf{ik_0}\,|\mathbf{r}-\mathbf{r}'|)}{|\mathbf{r}-\mathbf{r}'|} \mathbf{U}(\mathbf{r}')\mathbf{dr}' \tag{12}$$

Particularly, in the far region of the diffraction of a plane monochromatic wave on a defect (Fraunhofer diffraction), the second member presents the Furrier-image of the distribution of the local displacement of the atoms $\mathbf{U}(\mathbf{r}')$. Consequently, if the amplitude $\mathbf{\Psi_{0,h}}(\mathbf{r})$ of the diffracted on a defect wave is known, then it possible to solve the reverse problem of the determination of the field $\mathbf{U}(\mathbf{r}')$ by calculating of the reverse Furrier-image of the amplitude $\mathbf{\Psi_{0,h}}(\mathbf{r})$.

3. Conclusion

In the end it is possible, particularly, to study the phenomenon of the pumping of X-rays into the reflection direction and to predict new phenomena, it is possible to study the focusing of X-ray beams diffracted on defects in crystals. In some cases it possible to formulate the reverse problem of the determination of the function of local replacement of atoms $\mathbf{U}(\mathbf{r}')$ from their equilibrium location in the perfect crystal, etc.

References

[1] Shen Q 1999 *Phys. Lett.* **83** 4784
[2] Mkrtchyan A R, Navasardyan M A, Mirzoyan V Gh 1982 *Pisma v JTF.* **8** 677
[3] Gabrielyan R G, Aslanyan A A, 1986, *Izv. AN. Arm. SSR, Phizika* **21** (No 6) 337
[4] Egiazaryan A M, Gevorgyan K M, Atanesyan A K 2011 *Crystallografy* **56** (No.3) 439
[5] Mirzoyan V Gh, Egiazaryan A M, Baghdasaryan E G, Mirzoyan P V 2007 *Izv. NAN. Armenia, Phizika* **42** (No.6) 355

About zone structure of a stack of a cholesteric liquid crystal and isotropic medium layers

A H Gevorgyan[1,2]*, G K Matinyan [3], M Z Harutyunyan[2] and E M Harutyunyan[1]

[1]Institute of Applied Problems of Physics NAS RA, 25 H. Nersisyan Str., 0014, Armenia
[2] Yerevan State University, 1 A. Manukyan Str., 0025, Yerevan, Armenia
[3]The State Agrarian University of Armenia, 74 Teryan Str., 009, Yerevan, Armenian

*E-mail: agevorgyan@ysu.am

Abstract. The optical properties of a stack of metamaterial-based cholesteric liquid crystal (CLC) layers and isotropic medium layers are investigated. CLCs with two types of chiral nihility are defined. The peculiarities of the reflection spectra of this system are investigated and it is shown that the reflection spectra of the stacks of CLC layers of these two types differ from each other. The influence of: the CLC sublayer thicknesses; incidence angle; local dielectric (magnetic) anisotropy of the CLC layers; refraction indices and thicknesses of the isotropic media layers on the reflection spectra and other optical characteristics of the system is investigated.

1. Introduction

Material science has energetically been developing recently, and its part, concerning the optical materials, has been developing more vigorously. In particular, metamaterials are of great interest [1]. Investigations of photonic crystals (PCs) still are of great interest both for their wide application in science and techniques, and for developing of the modern technology of creating new media. They have a photonic band gap (PBG) in their transmittance spectrum that can be changed either by external fields, or by changes of the crystal internal structure [2]. The optical devices made on the base of PCs possess such properties, as: multifunctionality and tunability; compactness and low energetic losses; high reliability and good compatibility with other optical devices. Cholesteric liquid crystals (CLCs) are known as PCs with easily tunable parameters (their parameters can be tuned by external electric, magnetic and strong light fields, thermal gradients, or UV radiation, etc.). Then it was reported in some works [3, 4] about multiple PBGs of one dimensional structures containing CLC and isotropic layers. Recently the chiral nihility media exhibit great interest. The concept of chiral nihility in electromagnetism was introduced by Lakhtakia [5], as a medium in which both dielectric and magnetic permittivities are zero. Then Tretyakov et al applied the nihility concept to the isotropic chiral metamaterials [6].

In the present work, the concept of nihility is generalized for the structurally chiral media (such as CLCs) and, using it, the peculiarities of the stack formed of CLC layers with nihility and isotropic media layers are investigated.

2. Reflection Spectra. Spectra of the Polarization Characteristics

The problem was solved by the Ambartsumian's *layer addition modified method* [4]. We analyze the spectra peculiarities of the multilayer structure that is a stack of CLC layers with nihility and an isotropic medium layers. As it is known (for instance, see [7]), the solutions of Maxwell's equations (for the normal light incidence) for the CLC with the dielectric and magnetic tensors having the forms:

$$\hat{\varepsilon}(z)=\varepsilon_m\begin{pmatrix}1+\delta_\varepsilon\cos2az & \delta_\varepsilon\sin2az & 0\\ \delta_\varepsilon\sin2az & 1-\delta_\varepsilon\cos2az & 0\\ 0 & 0 & 1-\delta_\varepsilon\end{pmatrix},\hat{\mu}(z)=\mu_m\begin{pmatrix}1+\delta_\mu\cos2az & \delta_\mu\sin2az & 0\\ \delta_\mu\sin2az & 1-\delta_\mu\cos2az & 0\\ 0 & 0 & 1-\delta_\mu\end{pmatrix}, \quad (1)$$

are as follows:

$$\mathbf{E}(z,t)=\sum_{j=1}^{4}[E_j^+\mathbf{n}_+\exp(ik_j^+z)+E_j^-\mathbf{n}_-\exp(ik_j^-z)]\exp(-i\omega t), \quad (2)$$

where: $\varepsilon_m=(\varepsilon_1+\varepsilon_2)/2$, $\mu_m=(\mu_1+\mu_2)/2$, $\delta_\varepsilon=(\varepsilon_1-\varepsilon_2)/(\varepsilon_1+\varepsilon_2)$, $\delta_\mu=(\mu_1-\mu_2)/(\mu_1+\mu_2)$; ε_1 and ε_2 are the principal values of the dielectric tensor; μ_1 and μ_2 are those of the magnetic tensor and $a=2\pi/p$, and p is the helix pitch; $\mathbf{n}_\pm=(\mathbf{x}\pm i\mathbf{y})/\sqrt{2}$ are the circular polarization orts; ω is the light frequency in vacuum; k_j^+ and k_j^- are the z components of the wave vectors ($k_j^+-k_j^-=2a$), which are defined from the dispersion equation and have the following form:

$$k_j^+=a\pm K_{1,2},\ k_j^-=-a\pm K_{1,2},\quad j=1,2,3,4, \quad (3)$$

where

$$K_{1,2}=\sqrt{\frac{\omega^2}{c^2}\frac{\varepsilon_1\mu_2+\varepsilon_2\mu_1}{2}+a^2\pm\sqrt{\left(\frac{\omega^2}{c^2}\frac{\varepsilon_1\mu_2-\varepsilon_2\mu_1}{2}\right)^2+4a^2\frac{\omega^2}{c^2}\varepsilon_m\mu_m}}. \quad (4)$$

As it is well known, (see [6]) isotropic chiral nihility metamaterials are the media with the parameters, $\varepsilon=0$, $\mu=0$, and $\rho\neq0$ (ρ is the chirality parameter). As the CLC is locally anisotropic, we define *the CLC with chiral nihility* as the CLC with $\varepsilon_m=\mu_m=0$ and $p\neq0$. There can be two types of helicoidal structures with such parameters, namely: 1. $\frac{\varepsilon_1}{\mu_1}=\frac{\varepsilon_2}{\mu_2}<0$ (the first type chiral nihility); 2. $\frac{\varepsilon_1}{\mu_1}=\frac{\varepsilon_2}{\mu_2}>0$ (the second type chiral nihility).

Both types differ from each other only by the phase of the dielectric and magnetic permittivities modulation; in the first case these phases coincide, and in the second case the phase difference is $\pi/2$. For these conditions, $\sqrt{\left(\frac{\omega^2}{c^2}\frac{\varepsilon_1\mu_2-\varepsilon_2\mu_1}{2}\right)^2+4a^2\frac{\omega^2}{c^2}\varepsilon_m\mu_m}=0$ and we have: $K_{1,2}=\sqrt{\frac{\omega^2}{c^2}\frac{\varepsilon_1\mu_2+\varepsilon_2\mu_1}{2}+a^2}$. In the first case, $\frac{\omega^2}{c^2}\frac{\varepsilon_1\mu_2+\varepsilon_2\mu_1}{2}+a^2>0$ and, therefore, $K_{1,2}$ are always real (of course, in the absence of absorption or amplification); meanwhile, in the second case, the values of $K_{1,2}$ are purely imaginary at the wavelengths, $\lambda\leq p\sqrt{\left|\frac{\varepsilon_1\mu_2+\varepsilon_2\mu_1}{2}\right|}$. According to these, in the first case, the CLC layer is transparent, and in the second case, there appears a PBG of a new type, namely, it comes out in the spectrum region, $\lambda \le p\sqrt{\left|\frac{\varepsilon_1\mu_2+\varepsilon_2\mu_1}{2}\right|}$, where the CLC completely reflects the incidence light with any polarization.

First, let it be noticed once more that the optical property peculiarities of the stack composed of CLC and isotropic medium layers at low dielectric and magnetic anisotropies of CLC layers were discussed in [8]. Below we consider the simplest case if the isotropic layers refraction coefficient, $n=1$, that is, we investigate the peculiarities of the equidistant layers band gap of the CLC with nihility. Then, we consider the case, $n_0=1$, that is, we assume that the stack is in vacuum. Besides, if it is not stipulated beforehand, we assume the first sublayer of the system isotropic. As our numerical calculations show, for the first type nihility stack, reflection is completely absent $(R\equiv 0)$ for any polarization of the incident light, as the case is for the single CLC layer. In figure 1 the reflection spectrum of the normally incident light for the stack with CLC layers with the second type nihility is presented. The incident light has left (the solid curve) and right (the dashed curve) circular polarization. The CLC layers helix is a right one. It is seen from the figure that in the shortwave region of the spectrum, $\lambda \le p\sqrt{\left|\frac{\varepsilon_1\mu_2+\varepsilon_2\mu_1}{2}\right|}$, there is a PBG of the same type as in the case of the single CLC layer, but new PBGs are formed, too. Comparison of the reflection spectra with those for the single CLC layer shows:

1. In the case of the stack, polarization sensitivity emerges, despite the non-sensitivity of the isotropic layers and single CLC layers of first type nihility.
2. In contrast to the second type nihility single CLC layer, this system has multiple PBGs of both two types, namely, the ones independent of the incident light polarization and the others selective in regard to the polarization.

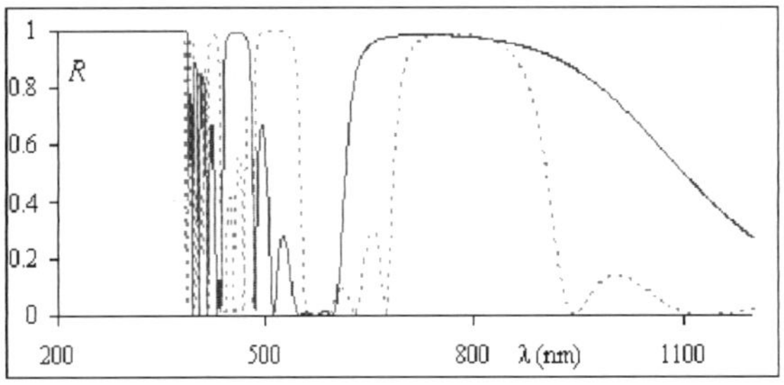

Figure 1. The reflection spectra at normal light incidence. The incident light has left circular polarization (the solid line) and the right circular polarization (the dashed line). CLC sublayer thicknesses $d=2p$, p=420 nm, isotropic sublayer thicknesses $d_1=200\,nm$, the sublayers number in the system is, s=10; $\varepsilon_1=-\varepsilon_2=-0.9$, $\mu_1=-\mu_2=-0.9$.

In figure 2, the polarization plane rotation and polarization ellipticity spectra (the first row) and those for the azimuth and the ellipticity of the first eigen polarization (the second row) for the first type nihility CLC stack (the left column) and those for the second type nihility CLC stack (the right column) are presented for the normal light incident. In figure 3*a, b*, the incident light is polarized along the x axis. As it is seen from figure 3, in both cases the system possesses optical activity, and the transmitted light has elliptic polarization, and its ellipticity has been changed strongly with the change of λ. The eigen polarizations (EPs) of the system are orthogonal elliptic polarizations for both types of nihilities. In the first case the ellipticity of the EP monotonously increases (with respect to its module) if λ increases, and in the second case the EPs are orthogonal circular polarizations.

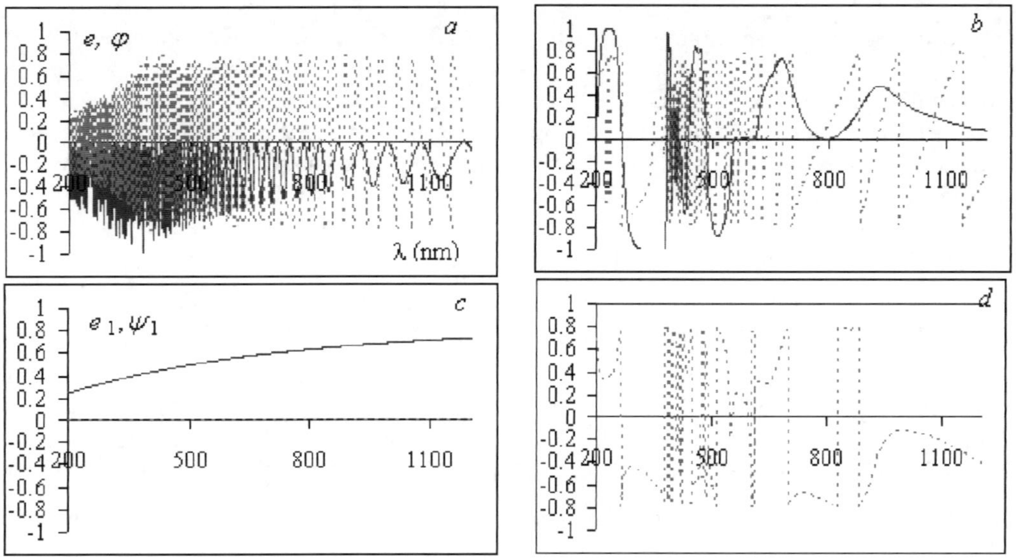

Figure 2. The polarization plane rotation (the dashed line) and polarization ellipticity (the solid line, the first row) and the spectra of the azimuth (the dashed line) and the ellipticity (the solid line) of the first eigen polarization (the second row) for the stack with CLC layers with the first type nihility (the left column) and the second type nihility (the right column). s = 50 (the left column), s = 10 (the right column). The other parameters are the same as in figure 1.

3. Isotropic Layers Thicknesses Influence.

Now we study the influence of the isotropic layers thicknesses on the band gap structure of the subject system. As it was mentioned above, in the case of the first nihility CLC stack, reflection is completely absent independent of the incident light polarization and of the thicknesses of the isotropic layers. In figure 3 the evolution of the reflection spectra for the second type nihility CLC layers stack, if the isotropic layers thickness, d_1, is changed. The incident light has left (*a*) and right (*b*) circular polarization. The bright regions correspond to stronger reflection. As it is seen from the figure, the short wavelength PBG is non-sensitive to the thickness changes of the isotropic layers. It is also seen from the figure, that, changing the isotropic layers thicknesses, one can change the number, the frequency width and frequency location of the PBG, consequently, one can tune the reflection/transmission of the subject system.

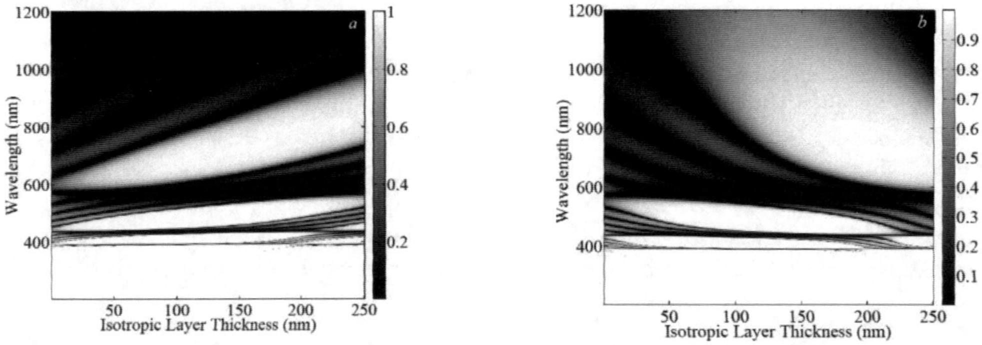

Figure 3. The reflection spectra evolution, if the isotropic layers thicknesses change. The incident light has left (the right column) and right (the left column) circular polarizations. The other parameters are the same as in figure 1.

4. CLC Layers Thicknesses Influence.

Now we pass to the investigation of the influence of the CLC layers thickness changes on the reflection spectra. We consider the case – as above – when the first sub-layer is isotropic. In figure 4, the reflection spectra evolution if the CLC layers thickness, *d*, is changing, is presented for system with CLC layers with the second type nihility. As it is seen from the figure, the frequency locations and frequency widths of the PBGs are functions of the CLC layers thicknesses. The short wavelength PBG practically does not change if the layers thickness, *d,* is changed. When the CLC layers thicknesses increase, the PBGs number and their frequency widths are changed. The PBGs frequency widths increase periodically; then they decrease, and all these take place in significantly large intervals, especially in the long-wave region. Also, it is seen from this figure, that the bright lines are practically vertical in the long-wave region for the right circularly polarized incident light. This means that the CLC layers thickness changes can either give rise PBGs with a very large frequency width, or lead to their vanishing.

5. Influence of the CLC local dielectric anisotropy.

The dielectric anisotropy, $\Delta=(\varepsilon_1-\varepsilon_2)/2$, is of the order of 0,5 or less, for the ordinary CLCs, but recently some artificial crystals (metamaterials) have been created having dielectric anisotropy varying within large intervals. It seems that there can be made such CLC-like helical periodic media on their base which possess huge local anisotropy. Such media with comparatively weaker anisotropy have been made long ago [9]. On the other hand, recently the interest in CLCs doped with nano-particles (or ferro-electric, or ferro-magnetic particles) is very great (see [10] and the citations in there). The presence of these nano-particles in the CLC structure leads to: significant increase of local (dielectric and magnetic) anisotropy; a significant change of the temperature of the phase transition from the isotropic phase to the liquid crystalline one; significant change of the frequency width and frequency location of the PBG; a change of the CLC elasticity coefficients; significant increase of tunability of CLCs, etc.

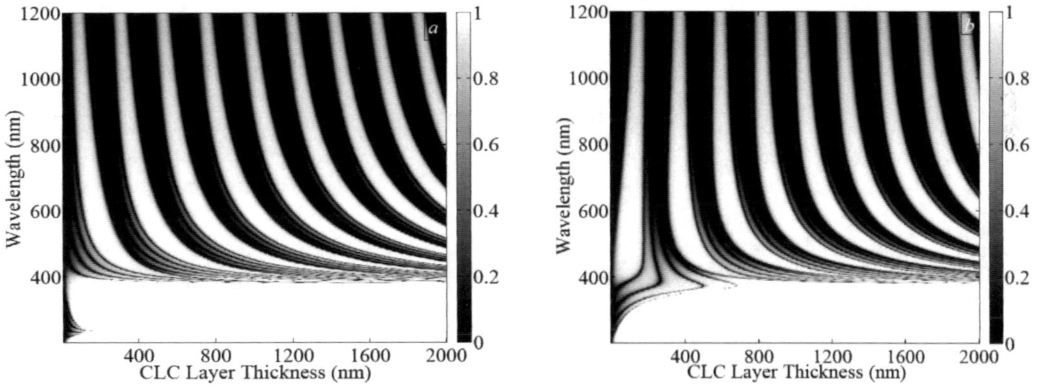

Figure 4. The reflection spectra evolution, if the CLC layers thicknesses change. The incident light has left (the left column) and right (the right column) circular polarizations. The other parameters are the same as in figure 1.

It follows from the above-said that the investigation of the optical properties of the stack of CLC layers with chiral nihility and isotropic medium layers, having different values of local dielectric and magnetic anisotropy, can have great interest. Presenting the principal values of the local dielectric and magnetic tensors of the CLC sub-layers in the form: $\varepsilon_{1,2}=\mp(\varepsilon_0-x)$ and $\mu_{1,2}=\pm(\mu_0-x)$, we investigate the influence of x on the reflection spectra.

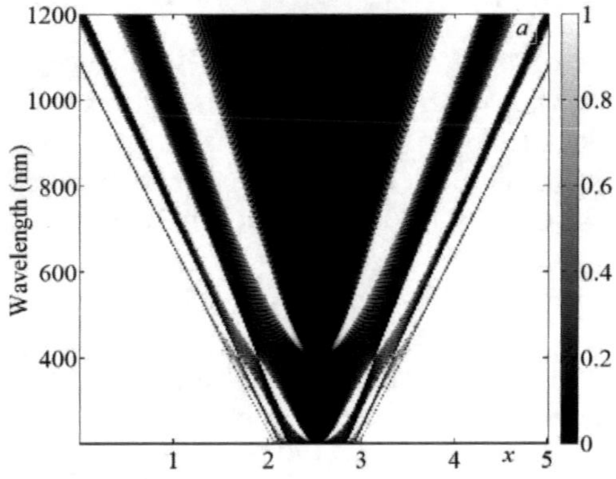

Figure 5. The evolution of the reflection spectra, when x (characterizing the anisotropy) is changing. The incident light has right (the left column) and left (the right column) circular polarizations. The other parameters are the same as in figure 1.

In figure 5, the evolution of the reflection spectra if x (characterizing the local dielectric and magnetic anisotropies) is changed are presented. As it is seen from the figure, no PBG is formed near the values, $x = \varepsilon_0 = \mu_0$ ($\varepsilon_0 = \mu_0 = 2.5$), that is, at $\varepsilon_1 = \varepsilon_2 = \mu_1 = \mu_2 = 0$, which is natural. In the case of the stack with CLC layers with chiral nihility of the second type, if $|\varepsilon_0 - x|$ (or $|\mu_0 - x|$) increases, the frequency width of the PBGs that are not selective to the incident light polarization is significantly increased and, for the larger values of this parameter, total reflection takes place practically in the whole subject wavelength range. This is natural too, because PBG is the above-said shortwave one with its longwave border shifted closer to the longwave region with the increase, $|\varepsilon_0 - x|$ (or $|\mu_0 - x|$).

4. Conclusion

Concluding, we have to note that we investigated the reflection spectra peculiarities of the stack of CLC layers with chiral nihility and isotropic medium layers. These investigations give much information about new possibilities of application of chiral PCs, in optics and photonics. Two types of CLCs with chiral nihilities were defined.

The main peculiarities of the reflection spectra of the usual single CLC layer are their polarization sensitivity (the reflection spectra are not identical, for the two orthogonal circular polarizations of the incident wave). In the case of the single CLC layer with the second type chiral nihility, a new type PBG appears which does not possess polarization sensitivity. In the stack case, this new type PBGs appears in any type of isotropic layers. Moreover, this system possesses multiple PBGs. This property of the subject system can find wide application, in particular, in display manufacturing.

It was shown that changing the system parameters (the CLC layers thicknesses, their dielectric anisotropy, the isotropic layer thicknesses, their refraction indices, etc.) one can change the PBGs' number, their frequency width and frequency distance (and in an essentially wide range) their character (as they are selective or non-selective with respect to the incident light polarization) their gyrotropic properties, etc.

Taking into account the possibility of tuning of local parameters by the changes of the external fields (electric, magnetic, mechanical, thermal, light, etc.), especially in the case if the system structure is soft, or the possibility of changing the internal structure of the system, the subject system has great perspectives, in the sense of its applications in photonics.

In particular, the subject system can find application as a tunable optical filter or mirror, or as a system allowing to obtain 100% polarized radiation from a non-polarized light (without any loss), or as a anti-reflecting systems, etc.

References

[1] Shalaev V M 2007 *Nature Photonics* **1** 41
[2] Sakoda K 2001 *Optical Properties of Photonic Crystals* (Berlin: Springer)
[3] Ha N Y, Ohtsuka Y, et al 2008 *Nature Mat.* **7** 43
[4] Gevorgyan A H 2012 *Phys. Rev. E.* **85** 021704
[5] Lakhtakia A 2002 *Int. J. of Infrared and Millimeter Waves* **23** 813
[6] Tretyakov S, Nefedov I, et al 2003 *J. of Electromag. Waves and Appl.* **17** 695
[7] de Vries H 1951 *Acta Crystallogr.* 4 219
[8] Harutyunyan M Z, Gevorgyan A H and Matinyan G K. 2013 *Opt. Spectrosc.* **114** 601.
[9] Robbie K, Brett M J and Lakhtakia A 1996 *Nature* **384** 616.
[10] Jeng S-C, Hwang S-J, Hung Y-H and Chen S-C. 2010 *Opt. Express.* **18** 22572.

Photonic Density of States of a Stack of Cholesteric Liquid Crystal and Isotropic Medium Layers

A H Gevorgyan[1,2], M Z Harutyunyan[2], G K Matinyan[3] and S A Mkhitaryan[2]
[1] The Institute of Applied Problems of NAS, RA; 25, Hr. Nersisyan St, Yerevan
[2]Yerevan State University, 1 Al. Manookian St., 025, Yerevan,Armenia
[3]Armenian National Agrarian University, Terian St., 74, 009, Yerevan, Armenia

E-mail:agevorgyan@.ysu.am

Abstract. We calculated the photonic density of states (PDS) of the eigen polarizations in the system composed of a stack of layers of a cholesteric liquid crystal and an isotropic medium. The reflection spectra and PDS peculiarities, as well as the peculiarities of absorption and emittance were investigated. We obtained the dependences of the PDS as a function of parameters characterizing absorption and gain. It was shown that the subject system can be used in lasers for obtaining low threshold lasers with the emittance wavelength tunable in wide ranges. The system also can work as a multi-position trigger.

1. Introduction

Recently the science of optical and laser materials has been in great interest. The photonic crystals (PCs) and metamaterials have been in wide application as laser cavities. The lasing in cholesteric liquid crystals (CLCs) and PCs goes on being intensely investigated (see, for instance, [1, 2] and the wide literature cited therein). The possibilities of decreasing the lasing threshold in the CLC and multilayer systems with CLC layers have been studied theoretically [3-5] and experimentally [6]. Reference [7] experimentally investigated some novel hybrid structures composed of a dye-doped low-molecular-weight CLC sandwiched into a multilayered polymer CLC films with right- and left-hand polymer CLC layers and they evaluated the lasing characteristics of this system. It was shown that a lasing with an extremely reduced threshold can be observed. In the papers [8, 9] it was reported the observation of continuous wave lasing in the dye doped CLC, and the effect of losses and gain on the photonic density of states (PDS) was calculated. On the other hand, recently the PCs with multi-photonic band gap (PBG) have been of great interest.

In this paper we investigate the influence of absorption, amplification, layer thickness, dielectric borders, local dielectric anisotropy etc on the PDS of the stack of the CLC and an isotropic medium layers. The urgency of the problem is the fact that the lasing threshold is significantly lowered in the multi-layer systems with CLC layer(s).

2. Results and discussion

The problem was solved by the Ambartsumian's layer addition modified method [4]. The ordinary and extraordinary refractive indices of the CLC sublayers (with thicknesses d_1) are taken to be: $n_o = \sqrt{\varepsilon_2} = 1.4639$ and $n_e = \sqrt{\varepsilon_1} = 1.5133$ (ε_1, ε_2 are the principal values of the local dielectric permittivity tensor of the CLC); the CLC sub-layer helix is right handed and its pitch is: p = 420 nm.

Thus, if the right circularly polarized (RCP) light is normally incident onto a single CLC layer, it has a PBG (which is in the range of $\lambda = 614.8 \div 635.6$ nm), and the light with the left circular polarization (LCP) does not have any PBG. The refraction coefficient, n, of the isotropic layers (with thicknesses d_2) is chosen as: $n = \sqrt{\varepsilon} = 1.7$.

In figure 1, the reflection spectrum for $s = 100$ (s is the sub-sublayers number in the system) in the normal incident case is presented. The incident light polarizations coincide with the polarization of the first (the solid curve) and the second (the dotted curve) eigen polarizations (EPs). The EPs of the system are circular and the modules of the ellipticities of the EPs differ from the unit mainly in the PBGs. Outside of these ranges the modules of the ellipticities are practically equal to the unit.

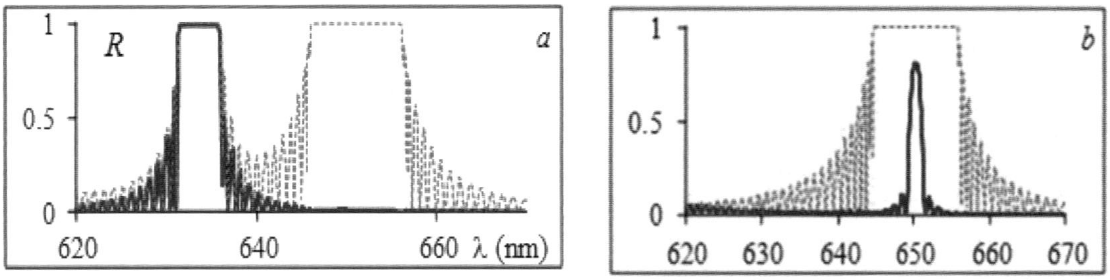

Figure 1. The spectrum of reflectance at normal light incidence. The incident light has (solid line) the first EP and (dashed line) the second EP. d_2=1000 nm. *a*: d_1=1200 nm. *b*: d_1= 3p.

As it is seen from figure 1, in contrast to the case of the single CLC layer, this system has multi-PBGs. There are PBGs of two types, namely, the one that does not depend on the incident light polarization and the other that is selective with respect to the incident light polarization. Moreover, if the CLC sublayer thicknesses are multiples of the helix pitch, then the selective reflection regions overlap the non-selective regions (figure 1*b*), otherwise these regions are separate (figure 1*a*).

Now we consider the absorption/radiation peculiarities and the PDS of the stack of the CLC and the isotropic medium layers. The divergence in the PDS at the band edges in the single CLC layer and at defect modes in the CLC with defects in its structure was shown theoretically and experimentally in [4, 5, 10]. Some new and important results for the PDS of CLC layer, as well as for the CLC-NLC structure were obtained in [11–13]. The investigation of the PDS peculiarities is important because of the following. For the laser emission it has been shown that, for instance, analyzing the case of the Fabry-Perot resonator, the threshold gain, g_{th}, can be related directly to the PDS, that is, ρ, as in [12]: $g_{th} \propto n/(d\rho)$, where n is the refractive index inside the resonator of the length d, and ρ is the maximum PDS. Furthermore, according to the space-independent rate equations, the slope efficiency of lasers can be shown to be inversely proportional to the threshold energy and, therefore, directly proportional to ρ [14].

For the isotropic medium case we have: $\rho_{iso} = n_s / c$, where n_s is the refractive index of the system surrounding medium, and c is the speed of light in vacuum. Both the losses and the gain can be incorporated through introducing an imaginary part to the CLC dielectric permittivity tensor components and to the isotropic medium dielectric permittivity.

Let the CLC and the isotropic medium layers be doped with dye molecules. Then this system is an amplifier in the presence of a pumping wave, i.e. we are discussing a planar resonator with active elements. The presence of the dye leads to the change of local refraction indices of the system sub-layers. In this case, in the presence of the pumping wave, the effective imaginary parts of the local refraction indices in the CLC ($n''_{o,e}$) and those of the isotropic layers (n'') are negative ($n_{o,e} = \sqrt{\varepsilon_{1,2}} = n'_{o,e} + in''_{o,e}$, $n = \sqrt{\varepsilon} = n' + in''$), and all they characterize the gain. In the case of the absence of the pumping wave (when the imaginary parts of the local CLC indices, $n''_{o,e}$, and those of the isotropic layers, n'', are positive), the quantity, $A=1-(R+T)$, characterizes the light energy absorbed by the system, while, in the case of amplification, the value of $|A|$ characterizes the radiation emitted by the system (we assumed that the incident light intensity is unit, $I_0 = 1$).

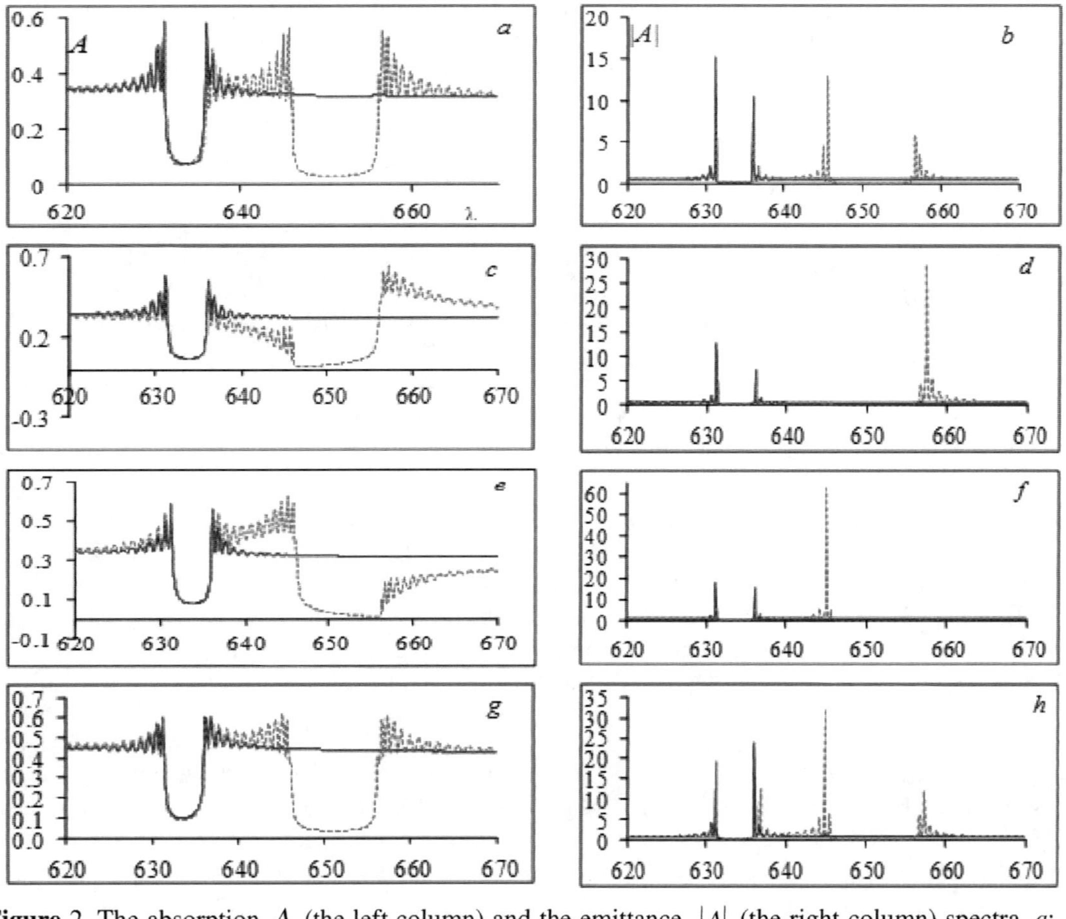

Figure 2. The absorption A (the left column) and the emittance, $|A|$ (the right column) spectra. *a*: $\varepsilon''_1 = \varepsilon''_2 = 0.001$, $\varepsilon'' = 0$. *b*: $\varepsilon''_1 = \varepsilon''_2 = -0.001$, $\varepsilon'' = 0$. *c*: $\varepsilon''_1 = 0.002$, $\varepsilon''_2 = 0$, $\varepsilon'' = 0$. *d*: $\varepsilon''_1 = -0.002$, $\varepsilon''_2 = 0$, $\varepsilon'' = 0$. *e*: $\varepsilon''_1 = 0.$, $\varepsilon''_2 = 0.002$, $\varepsilon'' = 0$. *f*: $\varepsilon''_1 = 0.$, $\varepsilon''_2 = -0.002$, $\varepsilon'' = 0$. *g*: $\varepsilon''_1 = \varepsilon''_2 = 0$, $\varepsilon'' = 0.002$. *h*: $\varepsilon''_1 = \varepsilon''_2 = 0$, $\varepsilon'' = -0.002$. The other parameters are the same as in figure 1*a*.

We characterize the degree of the arrangement of the dipole transition momentums of the guest molecules by the order parameter, S_d, defining it through the average of $\cos\vartheta$: $S_d = 3\langle\cos\vartheta\rangle/2 - 1/2$, where ϑ is the angle between the local optical axis and the dipole transition momentum of the guest molecules. The possible maximum value of this order parameter is the unit, $S_d = 1$, and it corresponds to the ideal orientation of the transition dipole momentums along the local optical axis; and the isotropic orientation of the transition dipole momentums corresponds to the value, $S_d = 0$, while the minimum value, $S_d = -0.5$, corresponds to the distribution of those momentums in the plane that are perpendicular to the local optical axis.

Journal of Physics: Conference Series **517** (2014) 012034 doi:10.1088/1742-6596/517/1/012034

In figure 2, the spectra of absorption, A (the left column) and those of the emittance, $|A|$ (the right column) are presented. The first row corresponds to the isotropic absorption and gain in the CLC layers, and the second and third ones are for the anisotropic absorption and gain in the CLC layers (assuming that the absorption and gain in the isotropic layers are absent); finally, the fourth row corresponds to the absorption and gain in the isotropic layers if they are absent in the CLC layers. The solid lines correspond to the first EP, and the dashed ones are for the second EP. As it is seen from figure 2, in the diffraction reflection regions suppression of absorption and emittance takes place. In the presence of anisotropic absorption and gain in the CLC layers, complete suppression of absorption and emittance take place at one of the PBG border, and there is an anomalously strong absorption and emittance at the other border. These indicate that at these parameters of the system, in the PBGs selective with respect to the polarization of incident light and nearby the region of the single CLC layer PBG, the modulation of the total field in the CLC layers is the same as that in the single CLC layer.

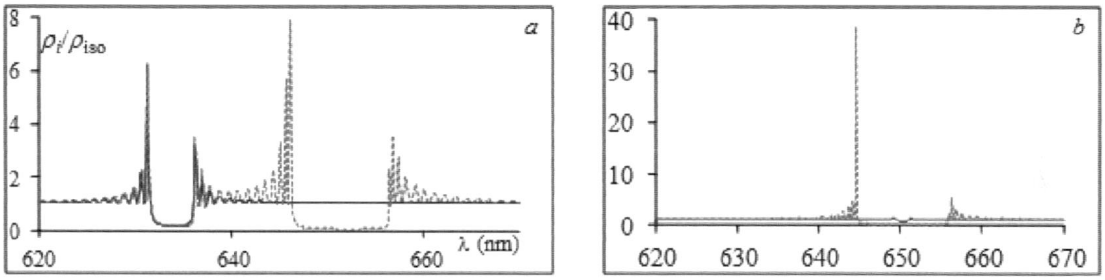

Figure 3. The spectra of the relative PDS ρ_i / ρ_{iso}. The incident light has the first EP (i=1, solid line) and the second EP (i=2, dashed line). The parameters are the same as in figure 2.

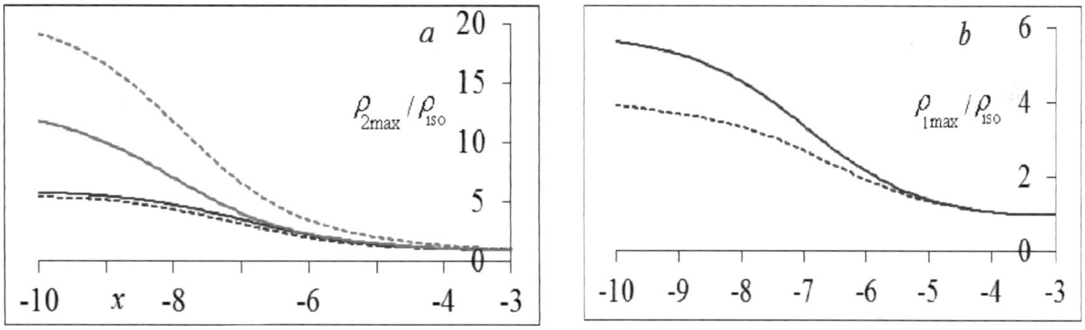

Figure 4. The dependence of the maximum, $\rho_{1,2\max} / \rho_{iso}$, of the PDS of the first (b) and the second (a) EPs at the short (the solid lines) and long (the dashed lines) wavelength edges on the parameter, $x = \ln(\mathrm{Im}\,\varepsilon_m)$, characterizing absorption. The isotropic absorption in the CLC layers is considered and the isotropic layers are assumed non-absorbing. The blue (dark) lines correspond to the nonselective PBG, and the red (light) lines are for the selective one.

Now we pass to the PDS. In figure 3, the spectra of ρ_i / ρ_{iso} are presented in the above said two cases considered in figure 1. As it is known, in the single CLC layer the value of ρ_i / ρ_{iso} for the diffracting EP has two principal maxima nearby the PBG borders (the short wavelength edge (SWE) maximum and the long wavelength edge (LWE) one), meanwhile, it is constant for the non-diffracting

EPs. As it is seen from figure 3, for the incident wave with the first EP, the value of ρ_i / ρ_{iso} has two principal maxima nearby the non-selective PBG borders, and they are essentially weaker in the second case, that is in the case when the CLC sublayer thicknesses are multiples of the helix pitch. For the incident wave with the second EP, the value of ρ_i / ρ_{iso} has four principal maxima both nearby the selective and non-selective PBG borders in the first case (that is, if the CLC layers thicknesses are not multiples of the helix pitch) and there are only two of them in the second case. Note that in the last case the PDS near the SWE reaches essentially larger values.

Now we investigate the influence of the absorption and gain on the PDS. We consider only the first case when the CLC layers thicknesses are not multiples of the helix pitch. In figure 4, the dependences of the maximum $\rho_{1,2\max} / \rho_{iso}$ of the PDS of the first (*b*) and the second (*a*) EPs at the short (the solid lines) and long (the dashed lines) wavelength edges on the parameter, $x = \ln(\operatorname{Im}\varepsilon_m)$, characterizing the absorption, are presented. The case of the isotropic absorption in the CLC layers is considered and the isotropic layers are assumed to be not absorbing. The blue (dark) lines correspond to the shortwave PBG border, and the red (light) lines are for the long wavelength one.

As it is seen from figure 4, the PDS decreases dramatically with the increment of the parameter, $x = \ln(\operatorname{Im}\varepsilon_m)$, characterizing the absorption.

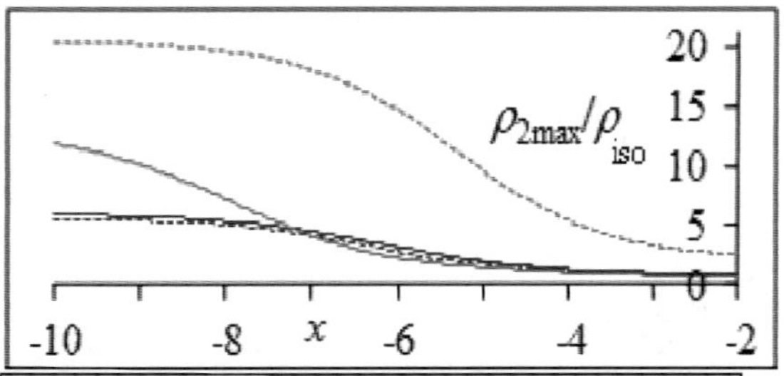

Fgure 5. The same as in figure 4*a*; the dependences for the anisotropic absorption if an imaginary term is included only in the dielectric constant parallel to the CLC local director.

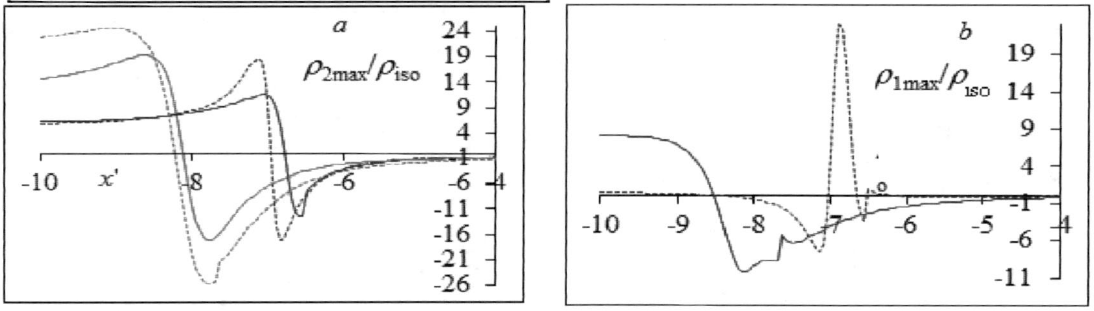

Figure 6. The same as in figure 5. The dependences of the maximum, $\rho_{1,2\max} / \rho_{iso}$, of the PDS of the first (*b*) and the second (*a*) EPs at the short (the solid lines) and long (the dashed lines) wavelength edges on the parameter, $x' = \ln(-\operatorname{Im}\varepsilon_m)$, characterizing the gain. The blue (dark) lines correspond to the nonselective PBG, and the red (light) lines are for the selective one.

An interesting situation is observed in the case of anisotropic absorption in the CLC sublayers. In figure 5, the same is presented as in figure 4*a*, that is, the dependences of the maximum, $\rho_{2\max} / \rho_{iso}$, on the parameter, *x*, for the anisotropic absorption if an imaginary term is included only in the dielectric constant parallel to the CLC sublayers local director. An interesting effect is observed here. If the absorption is weak, then the PDS for the selective with respect to the polarization PBG in its SWE is practically unaffected at first, when the loss is increased, while the PDS in the LWE is diminished sharply. Meanwhile, if an imaginary term is included only in the dielectric constant perpendicular to the local director, then the reverse takes place.

Now about the gain. In figure 6 the dependences of the maximum $\rho_{1,2\max} / \rho_{iso}$ of the PDS of the first (*b*) and the second (*a*) EPs at the short (the solid lines) and long (the dashed lines) wavelength edges on the parameter, $x' = \ln(-\mathrm{Im}\varepsilon_m)$, characterizing the gain are presented. If the gain (that is, the parameter, $x' = \ln(-\mathrm{Im}\varepsilon_m)$) increases, the maximum of the PDS increases too. The further increment of the gain leads to a resonance-like change of the maximum of the PDS; then the PDS diminishes. There exists a critical value of x' beyond witch the lasing mode is quenching and the feedback vanishes. The critical values of x' for SWE and LWE both for selective and non-selective PBGs are different. The existence of the critical value for the gain indicates the possibility of using the system as a trigger. Tuning the pumping intensity (for the corresponding density of the dye molecules), one can pass from the lasing regime to that of quenching.

3. Conclusions

We investigated the PDS peculiarities of the EPs for the stack of layers of a CLC and an isotropic medium layers both in the absence of absorption and gain and in their presence. It was shown that the subject system possesses multiple PBGs. There are PBGs of two types: the ones selective with respect to the incident light polarization and those non-selective to that. It can be shown that the change of the CLC sublayer thicknesses and the isotropic sublayer thicknesses leads both to the frequency overlapping of the PBGs and the width changes.

Taking into account the possibility of tuning the width, the number and the frequency location of these regions by the external fields (electric, magnetic, light, mechanical, heat, etc) or the possibility of changing the internal structure of the system, the system we offer seems having perspectives also for other fields of science and technologies.

The investigation of the influence of the gain on the PDS showed that if the gain is increased then the maximum of the PDS increases too; meanwhile, it decreases if absorption is present. Then, it was shown that in the presence of gain there exists a critical value of x' (the parameter characterizing the gain) beyond which the lasing mode is quenched and the feedback vanishes. Besides, the critical values for x' at the SWE and LWE are different.

The present work was carried out with partial financial support of the Armenian National Science and Education Fund (ANSEF Grant No. Opt- 3517).

References

[1] Blinov L M and Bartolino R (ed). 2010 *Liquid Crystal Microlasers* (Kerala: Transworld Research Network).
[2] Coles H and Morris S 2010 *Nature Photon.* **4** 676.
[3] Belyakov V A 2006 *Mol. Cryst. Liq. Cryst.* **453** 43.
[4] Gevorgyan A H and Harutyunyan M Z 2009 *J. Mod. Opt.* **56** 1163.
[5] Gevorgyan A H, Oganesyan K B and *et al* 2010 *Opt. Communn.* **283** 3707
[6] Matsuhita Y, Huang Y and *et al* 2007 *Appl. Phys. Lett.* **90** 091114
[7] Takanishi Y, Ohtsuka Y and *et al* 2010 *Opt. Express.* **18** 12909
[8] Munoz A, McConney M E and *et al* 2012 *Opt. Lett.* **37** 2904
[9] Mavrogordantos Th K, Morris S and *et al* 2008 *Nature Mat.* **7** 43
[10] Schmidtke J and Stille W 2003 *Eur. Phys. J. B.* **31** 179
[11] Gevorgyan A H and Kocharian A N 2012 *Opt. Communn.* **285** 2854
[12] Moreira M F, Relaix S and *et al* 2010 *Liquid Crystal Microlasers* (Kerala: Transworld Research Network)

[13] Gevorgyan A H, Oganesyan K B and *et al* 2013 *Laser Phys. Lett.* **10** 125802
[14] Bendickson J M, Dowling J P and Scalora M 1996 *Phys. Rev. E.* **53** 4107

The Optics of Gyrotropic Crystals in the Field of Two Counter-Propagating Ultrasound Waves

A H Gevorgyan[1,2], G K Matinyan [3], E M Harutyunyan[1] and M A Hovhannisyan[1]

[1]Institute of Applied Problems of Physics NAS RA, 25 H. Nersisyan Str., 0014, Armenia
[2] Yerevan State University, 1 A. Manukyan Str., 0025, Yerevan, Armenia
[3]The State Agrarian University of Armenia, 74 Teryan Str., 009, Yerevan, Armenian

E-mail: agevorgyan@ysu.am

Abstract. We consider oblique light propagation through a layer of a gyrotropic crystal in the field of two counter-propagating ultrasound waves. The problem is solved by Ambartsumyan's layer addition modified method. The results of the reflection spectra for different values of the problem parameters are presented. The possibilities of such system applications are discussed.

1. Introduction

Recently the photonic crystals (PCs) which enable to tune light propagation are of great interest [1, 2]. Chiral PCs have rich optical properties [3-10]. Of particular interest are the self-organizing structural-chiral PCs (cholesteric liquid crystals (CLCs), cholesteric smectic liquid crystals, etc.) because of their rich optical properties [11]. For these crystals, the photonic band gap (PBG) exists only for the light with one circular polarization (for normal light incidence) coinciding with the sign of the medium chirality. Nevertheless, another periodic change in the chiral media parameters is possible; namely, it is possible to create one dimensional (1D) chiral PCs in other ways as well. Recently, the chiral layer media made of layers of usual gyrotropic and non-gyrotropic media or of the layers of gyrotropic media of different types [3-10] are of great interest. A gyrotropic layer placed inside an external ultrasound field becomes a 1D chiral PC [12-14]. Below we name such media as *chiral periodic media* (CPM). As it is possible to tune the parameters of the ultrasound, hence, we can obtain a chiral PC with tunable parameters. Currently, for manufacturing acousto-optical devices of information processing (modulators, deflectors, filters, processors, detectors, etc) the very crystals with both good photo-elastic and gyrotropic properties and are widely used. To such crystals, first of all, the para-tellurite (TeO_2), tellurium (Te), quarts (α-SiO_2) etc are related.

Light diffraction on elastic waves is used in practical devices, mainly, because of the significant difference between the light speed and the sound speed. Light diffraction on acoustic waves is used to obtain optical radiation modulation.

In this work oblique light propagation through a layer of a gyrotropic crystal located in the field of two counter-propagating ultrasound waves is considered.

2. Theory

Let us consider light reflection/transmission through a finite layer from an isotropic-gyrotropic crystal placed in an ultrasound field. Let the medium layer occupy the space between the planes, z=0 and $z=d$ (d is the layer thickness). A plane ultrasound wave propagates along the axis, z. The ultrasound wave converts the parameters, ε, μ and $\bar{\gamma}$, into functions of the coordinate, z (ε and μ are the layer dielectric permittivity and magnetic permeability, and $\bar{\gamma}$ is the parameter of the natural gyrotropy of this layer). Here we assume the following dependences for these parameters' changes:

$$\begin{pmatrix} \varepsilon(z) \\ \mu(z) \\ \bar{\gamma}(z) \end{pmatrix} = \begin{pmatrix} \varepsilon \\ \mu \\ \bar{\gamma} \end{pmatrix} \left[1 + \begin{pmatrix} \Delta\varepsilon_1 \\ \Delta\mu_1 \\ \Delta\bar{\gamma}_1 \end{pmatrix} \sin(K_1 z) - \begin{pmatrix} \Delta\varepsilon_2 \\ \Delta\mu_2 \\ \Delta\bar{\gamma}_2 \end{pmatrix} \sin(K_2 z) \right], \tag{1}$$

where the modulation depths, $\Delta\varepsilon_1$, $\Delta\mu_1$, $\Delta\bar{\gamma}_1$, and $\Delta\varepsilon_2$, $\Delta\mu_2$ and $\Delta\bar{\gamma}_2$ are assumed less than the unit; $K_{1,2} = 2\pi / \Lambda_{1,2}$, and $\Lambda_{1,2}$ are the ultrasound wavelengths. These periodic excitations are changed both in space and time. Particularly, if the ultrasound is a traveling wave, then the periodic excitation is moving with the ultrasound speed. As the ultrasound speed is much less than the light speed, the periodic excitation caused by the ultrasound can be considered stationary and in Maxwell's equations it is possible to neglect the dependences of the media parameters upon time, that is, while calculating the terms, $\frac{1}{c}\frac{\partial \vec{D}}{\partial t}$ and $\frac{1}{c}\frac{\partial \vec{B}}{\partial t}$, we neglect the medium parameter time derivatives, though we take their time dependences into account in the final results. As it is known, if the optical activity is absent ($\bar{\gamma}$ =0), this can be done if Ω/ω<<1 [15]. For $\bar{\gamma}$ ≠0, we must also require that:

$$\left|\frac{\Omega}{\omega}\Delta\varepsilon_{1,2}\right|,\ \left|\frac{\Omega}{\omega}\Delta\mu_{1,2}\right|,\ \left|\frac{\Omega}{\omega}\Delta\bar{\gamma}_{1,2}\right| << \left|\frac{\omega}{c}\bar{\gamma}\right|, \tag{2}$$

to rightfully retain $\bar{\gamma}$ in Maxwell's equation – simultaneously disregarding the time derivatives of the medium parameters. These conditions can be easily carried out. We assume that the incidence plane coincides with the plane, (x, z), and that the wave is incident at the angle, α, to the normal of the layer border (coinciding with the (x, z) plane). Let us expand the components of the electric field amplitudes of the incident, reflected and transmitted waves into the projections parallel (p-polarization) and perpendicular (s-polarization) to the incident planes:

$$\vec{E}_{i,r,t} = E^p_{i,r,t}\vec{n}_p + E^s_{i,r,t}\vec{n}_s = \begin{pmatrix} E^p_{i,r,t} \\ E^s_{i,r,t} \end{pmatrix}, \tag{3}$$

where the indies, i, r and t, denote the incident, reflected and transmitted waves, respectively, and $\vec{n}_p$ and $\vec{n}_s$ are the orts of the p- and s- polarizations.

We present the problem solution in the form:

$$\begin{bmatrix} E^p_r \\ E^s_r \end{bmatrix} = \begin{bmatrix} R_{pp} & R_{ps} \\ R_{sp} & R_{ss} \end{bmatrix} \begin{bmatrix} E^p_i \\ E^s_i \end{bmatrix}, \begin{bmatrix} E^p_t \\ E^s_t \end{bmatrix} = \begin{bmatrix} T_{pp} & T_{ps} \\ T_{sp} & T_{ss} \end{bmatrix} \begin{bmatrix} E^p_i \\ E^s_i \end{bmatrix}, \tag{4}$$

where $\hat{R}$ and $\hat{T}$ are the 2x2 reflection and transmission matrices of the system.

We carry our calculations as follows. First, we calculate the reflection and transmission matrices for the medium sublayer of the thickness equal to the wavelength of the ultrasound wave. For this purpose we divide the thickness, d, into a great number of narrower sub-sublayers with thicknesses: d_1, d_2, d_3,,,.....,d_N. If their maximal thickness is small enough, then it is possible to consider the parameters of each sub-sublayer constant. Then, according to [16, 17], the problem of determination of $\hat{R}$ and $\hat{T}$ for the sublayer with the thickness, $d=\Lambda$, is reduced to the solution of the following difference matrix equations:

$$\hat{R}_j = \hat{r}_j + \tilde{\hat{t}}_j \hat{R}_{j-1}\left(\hat{I} - \tilde{\hat{r}}_j \hat{R}_{j-1}\right)^{-1}\hat{t}_j, \quad \hat{T}_j = \hat{T}_{j-1}\left(\hat{I} - \tilde{\hat{r}}_j \hat{R}_{j-1}\right)^{-1}\hat{t}_j, \tag{5}$$

with $\hat{R}_0 = \hat{0}$, $\hat{T}_0 = \hat{I}$. Here $\hat{R}_j$, $\hat{T}_j$, $\hat{R}_{j-1}$, $\hat{T}_{j-1}$ are the reflection and transmission matrices for the media with j- and (j–1)-, sub-sublayers, respectively; $\hat{r}_j$, $\hat{t}_j$ are the reflection and transmission matrices of the j-th sub-sublayer; $\hat{0}$ is the zero matrix; $\hat{I}$ is the unit matrix. The tilde denotes the corresponding reflection and transmission matrices of the reversely traveling light. Then, to calculate the reflection or transmission of the whole system, we again apply the difference matrix equations (5), but now $\hat{R}_j$, $\hat{T}_j$, $\hat{R}_{j-1}$, $\hat{T}_{j-1}$ are the reflection and transmission matrices for the system, consisting of j- and (j–1)- sublayers, respectively; and $\hat{r}_j$, $\hat{t}_j$ are the reflection and transmission matrices for the j-th sublayer (with the thickness, d=Λ).

Thus, the problem is reduced to the calculation of the reflection and transmission matrices for a homogeneous gyrotropic layer. The analytic solution of this problem is known [18]. We proceed from the following equations for a homogeneous isotropic gyrotropic crystal:

$$\vec{D} = \varepsilon\vec{E} - \frac{\bar{\gamma}}{c}\frac{\partial\vec{H}}{\partial t}, \qquad \vec{B} = \mu\vec{H} + \frac{\bar{\gamma}}{c}\frac{\partial\vec{E}}{\partial t}. \qquad (6)$$

Now we pass to the eigen polarizations (EPs) and eigen values (EVs) of the amplitude. It is known, EPs are the two polarizations of the incident wave that do not change when passing through the system; and the eigen values are the reflection and transmission amplitude coefficients for the incident light, corresponding the respective EPs [16-17 and 19].

The EPs and EVs deliver much information about the interaction of light with the system; therefore, their calculation is important for every optical system. It follows from the definition of EPs that they must be connected with the polarizations of the internal waves (the eigen modes, EMs) aroused in the medium (they mainly coincide with the polarizations of the EMs).

Denoting the ratio of the field complex amplitude components at the entrance of the system by χ_i ($\chi_i = E_i^s / E_i^p$), and those at the exit of the system by χ_t ($\chi_t = E_t^s / E_t^p$) and taking into account that $\begin{bmatrix} E_t^p \\ E_t^s \end{bmatrix} = \begin{bmatrix} T_{11} & T_{12} \\ T_{21} & T_{22} \end{bmatrix}\begin{bmatrix} E_i^p \\ E_i^s \end{bmatrix}$, we get their connection:

$$\chi_t = (T_{22}\chi_i + T_{21})/(T_{12}\chi_i + T_{11}). \qquad (7)$$

The function, $\chi_t = f(\chi_i)$, is called *polarization transfer function* [19] and it carries information about the transformation of the polarization ellipse, when the light passes through the system. Every optical system has two EPs obtained from the definition of the EPs: $\chi_i = \chi_t$. Hence, according to (4), we have for the χ_1 and χ_2:

$$\chi_{1,2} = \frac{T_{22} - T_{11} \pm \sqrt{(T_{22} - T_{11})^2 + 4T_{12}T_{21}}}{2T_{12}}. \qquad (8)$$

3. The results and discussion

The optical properties of CPM, in the presence of one longitudinal ultrasound wave, are studied in detail in the works, [10, 13, and 14]. Therefore, we directly proceed to the study the effects caused by the presence of the two waves. The calculations are carried out for the CPM with the parameters: ε=2.5, μ=1.0, $\bar{\gamma}$=0.1; the layer thickness is: d=20 μm; $\alpha = 30^0$.

For a clear conception of the effects when dealing with the presence of the two ultrasound waves, we consider the case of the minimum influence of the dielectric borders, that is, we assume that $n_s = \sqrt{\varepsilon_s}$, where n_s is the refraction index of the medium, bordering the considered layer on both sides. Then, in the first step, we consider that ε, μ and $\bar{\gamma}$ do not depend on the frequency, and that their imaginary parts are very small and they do not depend on the frequency either; i.e. we do not consider the effects of optical dispersion and absorption.

In figure 1, it is presented the reflection spectra in the presence of only one longitudinal wave. As it is seen in this picture, for the given parameters of the problem, in the first order of reflection, three PBGs are formed; one gap is not selective regarding to the polarization of the incident light, and the others are selective (the shortwave gap is selective with regard to the polarization and is expressed weakly).

It is to be noted that the PBGs of the two types (the selective with respect to the polarization of the incident light and the non-selective one) are observed in periodical chiral PCs of very different types [3, 9, 20-22].

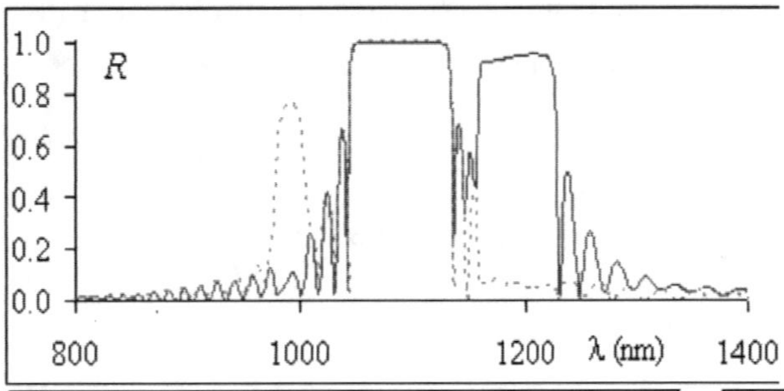

Figure 1. The dependence of the reflection coefficient, *R,* on the wavelength in the presence of only one ultrasound wave. The incident light has the left (the solid curve) and the right (the dashed curve) circular polarization. The problem parameters are: $\Delta\varepsilon=0.5$, $\Delta\mu=0.0$, $\Delta\bar{\gamma}=0.2$ and $\Lambda=400$ nm.

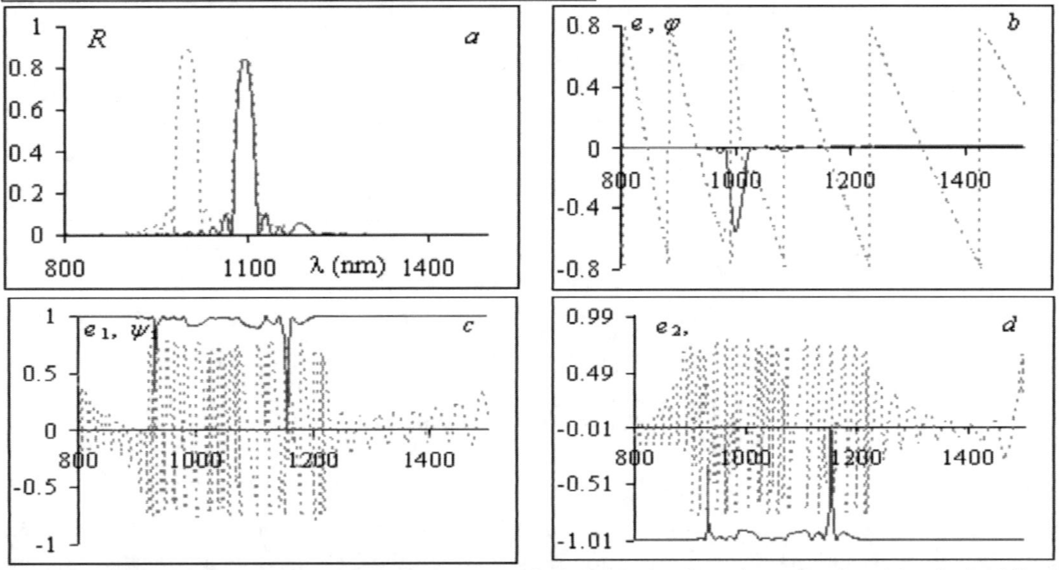

Figure 2. The reflection spectra (*a*); the spectra of: the polarization plane rotation (the dashed line) and the polarization ellipticity (the solid line) (*b*) and the spectra of the azimuth (the dashed line) and the ellipticity (the solid line) of the first and second EPs (*c* and *d*, respectively). The problem parameters are the following: $\Lambda_1=\Lambda_2=0.4$ μm; $\Delta\varepsilon_1=0.5$, $\Delta\varepsilon_2=0.4$; $\Delta\mu_1=\Delta\mu_2=0.0$; $\Delta\bar{\gamma}_1=0.2$, $\Delta\bar{\gamma}_2=0.15$. (*b*) $\Lambda_1=0.4$ μm, $\Lambda_2=0.5$ μm; $\Delta\varepsilon_1=\Delta\varepsilon_2=0.5$; $\Delta\mu_1=\Delta\mu_2=0.0$; $\Delta\bar{\gamma}_1=\Delta\bar{\gamma}_2=0.2$. The other parameters are the same as in figure 1.

In figure 2: the reflection spectra (*a*); the polarization plane rotation and the polarization ellipticity spectra (*b*); the spectra of the azimuth and the ellipticity of the first and second EPs (*c* and *d*, respectively) for the oblique light incidence are presented in the presence of the two counter-propagating longitudinal ultrasound waves.

In figure 2 *a,* the incident light has left (the solid line) and right (the dashed line) circular polarizations; and in figure 2 *b*, it has linear polarization. As it is seen in figure 2 *b*, the polarization ellipticity practically equals to zero everywhere, except the PBG frequency ranges, which are selective with respect to the polarization of the incident light. Then, as it is seen from figure 2 *c, d*, the system EPs are non-orthogonal and quasi-circular. Here it is supposed that the wavelength of the ultrasound waves coincide and they differ only by their modulation depth.

In figure 3 there are the same spectra as in figure 2, but are presented for the presence of the two counter-propagating longitudinal ultrasound waves, whose modulation depths coincide and the wavelengths differ. In figure 3 *a*, the incident light has left (the solid line) and right (the dashed line) circular polarizations, too.

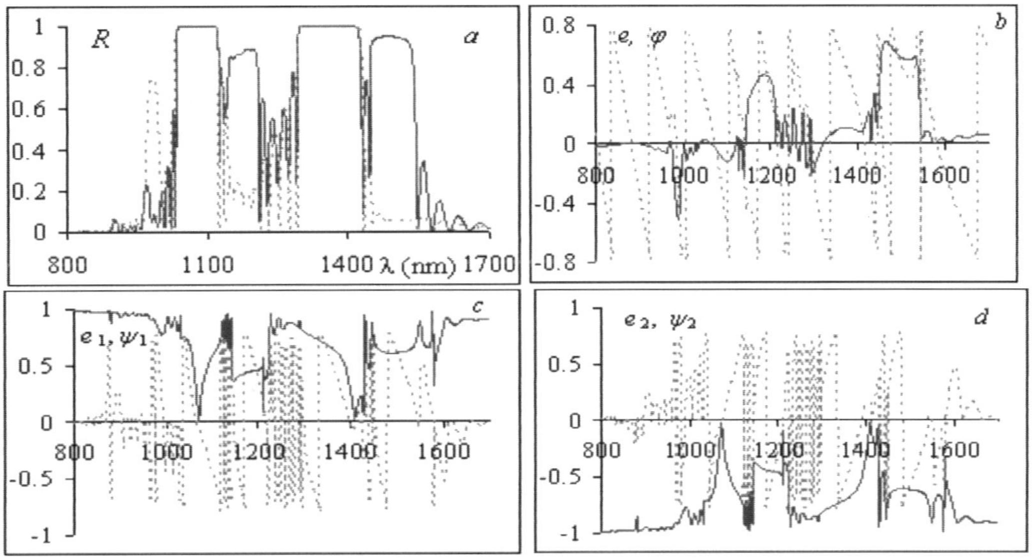

Figure 3. The same spectra as in figure 2, in the presence of two counter-propagating longitudinal ultrasound waves, whose modulation depths coincide and the wavelengths differ. The problem parameters are the following: Λ_1=0.4 μm, Λ_2=0.5 μm; $\Delta\varepsilon_1=\Delta\varepsilon_2$=0.5, $\Delta\mu_1=\Delta\mu_2$=0.0; $\Delta\bar{\gamma}_1=\Delta\bar{\gamma}_2$=0.2. The other parameters are the same as in figure 1.

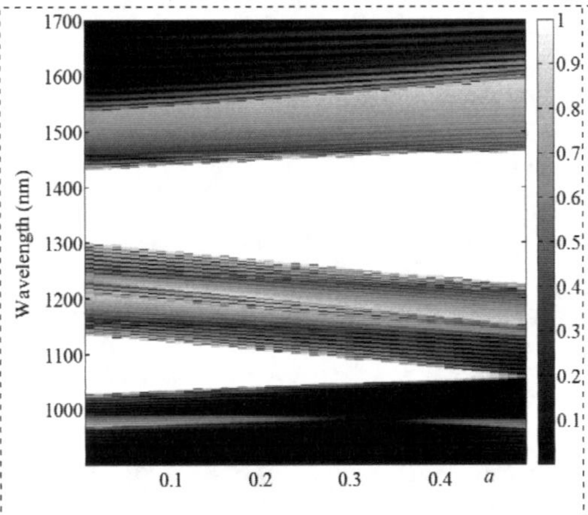

Figure 4. The evolution of the reflection spectra, when the parameter, *a*, is changing. The problem parameters are as follows: $\Delta\mu$=0.0; Λ_1=500 nm, Λ_2=400 nm; *d*=50000 nm. The other parameters are the same as in figure 1.

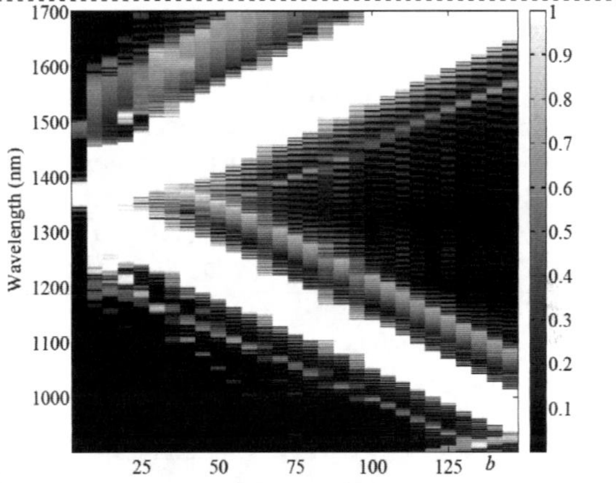

Figure 5. The evolution of the reflection spectra, when the parameter, *b*, changes. The problem parameters are as follows: $\Delta\mu$=0.0; $\Delta\varepsilon_1$=0.55, $\Delta\varepsilon_2$=0.4; $\Delta\bar{\gamma}_1$=0.15, $\Delta\bar{\gamma}_2$=0.1; *d*=50000 nm. The other parameters are the same as in figure 1.

And now, presenting the parameters: $\Delta\varepsilon_1$, $\Delta\varepsilon_2$, $\Delta\bar{\gamma}_1$ and $\Delta\bar{\gamma}_2$, in the form: $\Delta\varepsilon_{1,2} = 0.5 \pm a$ and $\Delta\bar{\gamma}_{1,2} = 0.1 \pm a/10$, we study the evolution of the reflection spectra with the alterations of the parameter, *a*, i.e. with the alteration of the modulation depths of the counter-propagating ultrasound waves. In figure 4, the evolution of the reflection spectra with alterations of the parameter, *a*, is presented. The incident light has linear polarization. The bright regions present the stronger reflection. As it is seen from the figure, the frequency locations and frequency widths of the PBGs are functions of the parameter, *a*.

Now we consider the situation when the wavelengths of the counter-propagating ultrasound waves are changed. We assume that these changes occur by the law: $\Lambda_{1,2} = \Lambda_0 \pm b$, with Λ_0=500 nm. In figure 5, the evolution of the reflection spectra if the parameter, b, changes is presented. The incident light has linear polarization. As it is seen from the figure, the frequency locations, frequency widths and the number of the PBGs are functions of the parameter, b, too.

These figures show that by changing the parameters of the ultrasound waves one can tune the system reflection, that is, one can change the PBGs (the diffraction efficiency in each order; the frequency width and its location in these regions; the polarization peculiarities of these ranges).

4. Conclusion

In this work we studied the peculiarities of the light diffraction in a gyrotropic layer in the presence of two counter-propagating longitudinal ultrasound waves. In this case, the diffraction pattern is much richer than in the presence of one ultrasound wave. New regions of diffraction reflection emerge within each diffraction order. The width of the regions of the diffraction reflection, their number, frequency location and frequency distance are determined by the parameters of the medium and those of the ultrasound waves; all they can be tuned.

Consequently, such systems can be used as: tunable polarization filters and mirrors; polarization mode converters; mode discriminators; multiplexers for circular polarized waves. They can also be used as sources of circular (or elliptic) polarization.

The present work was carried out with partial financial support of the Armenian National Science and Education Fund (ANSEF Grant No. Opt- 3517).

References

[1] Joannopoulos J, Meade R and Winn J 1995 *Photonic Crystals* (Prinseton: Prinseton Univ. Press)
[2] Sakoda K 2001 *Optical Properties of Photonic Crystals* (Berlin: Springer)
[3] Bassiri S, Papas C H and Engheta N 1988 *J. Opt. Soc. Am.* A **5** 1450
[4] Silverman M P and Badoz J 1994 *J. Opt. Soc. Am.* A **11** 1894
[5] Lakhtakia A, Varadan V V and Varadan V K 1989 *Int. J. Eng. Sci*, **27** 1267
[6] Tuz V R and Kazanskiy V B 2008 *J. Opt. Soc. Am.* A. **25** 2704
[7] Ivanov O V and Sementsov D I 2000 *Crystallography Report* **45** 487
[8] Konstantinova A F, Nabatov B V, et al. 2002 *Crystallography Report* **47** 815
[9] Flood K M and Jaggard D L 1996 *J. Opt. Soc. Am.* A **13** 1395
[10] Gevorgyan A H 2007 *Tech. Phys.* **52** 466
[11] Khoo I-Ch. 2007 *Liquid Crystals* (New Jersey: John Wiley and Sons)
[12] Balakshii V I, Parigin V N and Chirkov L E 1985 *Physical Backgrounds of Acousooptics* (Moscow: Radio and Sviaz)
[13] Harytyunyan E M and Gevorgyan A H 2006 *J. Contemp. Phys. (Arm. Ac. Sci.)* **41** 30
[14] Harutyunyan E M, Harutyunyan S H and Gevorgyan A H 2007 *J. Contemp. Phys. (Arm. Ac. Sci.)* **42** 16
[15] Ostrovskii L A and Stepanova B N 1971 *Izv. Vyssh.Uchebn. Zaved., Radiofiz.* **14** 484
[16] Gevorgyan A H 2011 *Phys. Rev. E,* **83** 011702
[17] Gevorgyan A H 2012 *Phys. Rev. E.,* **85** 021704
[18] Gevorgyan A H 2001 *Opt. Spectrosc.* **91** 762
[19] Azzam R M A and Bashara N M. 1977 *Ellipsometry and polarized light.*(New York: North-Holland)
[20] Ivanov O V and Sementsov D I 1997 *Pure Appl. Opt.* **6** 455
[21] Kim K, Yoo H, Lee D –H and Lim H. 2006 *Waves Rand. Comp. Media* **16** 75
[22] Tuz V R, Vidil M Y and Prosvirin S L 2010 *J. Opt.* **12** 095102

Transverse Dynamics of the Azimuthally Inhomogeneous Electron Bunch in a Multilayer Dielectric Cylindrical Waveguide

A M Altmark[1] and A D Kanareykin[1,2],
[1]St. Petersburg Electrotechnical University, St. Petersburg, 5 Prof. Popov St., 197376, St. Petersburg, Russia
[2]Euclid Techlabs LLC 400 Professional Dr. Ste. 220 Gaithersburg, MD 20879, USA

E-mail: aaltmark@mail.ru

Abstract. In reference [1], a complete analytical solution for Cherenkov wakefields generated by an azimuthally asymmetric annular beam propagating in a coaxial two-channel dielectric structure was presented. A drive bunch generates Cherenkov radiation (wakefield) inside the dielectric loaded waveguide and a second (witness) bunch passing through the structure at an appropriate delay with respect to the drive bunch is accelerated by the wakefield. Use of a ring beam in a multi-layer waveguide can significantly increase the transformer ratio by providing different paths for the ring driver and the accelerated bunch to pass through the structure. The main challenge of this scheme originates in the transverse dynamics of the drive bunch because of its high charge and relatively low energy. To hold the inner dielectric tube inside the waveguide metal (titanium) threads are used. The threads are located inside the drive beam section of the waveguide that leads to the segmentation of the drive beam. In this paper, we study the transverse dynamics of the annular beam with various types of azimuthally asymmetries that depend on the specifics of the beam generation and multilayer waveguide parameters. The different types of beam asymmetry and hybrid mode dependencies are presented using the original BBU-3000 [7] beam dynamics code.

1. Introduction

The efficiency of wakefield acceleration is related to the transformer ratio, defined as the peak energy transfer from the "drive" bunch to "witness" bunch [1, 2]. For traditional accelerating structure [2–4] with collinear drive and witness trajectories and a longitudinally symmetrical charge distribution the transformer ratio does not exceed a value of 2 [2]. In this paper we consider a partial non-collinear scheme of wakefield acceleration in a multilayer cylindrical waveguide [5, 6, 8, 9] (see in figure1a)).
The coaxial configuration considers the drive bunch is annular and is centred on the axis of a second accelerating channel that carries a witness bunch, figure 1a [1, 8, 9]. This technique allows to separate the driver and witness bunch channels that has been implemented in CLIC design at CERN [10] and previously demonstrated at the Argonne AWA facility [11]. A coaxial two-channel dielectric wakefield structure was examined for use as a high gradient accelerator in [8]. A THz design with mm range aperture potentially can provide GeV/m-level acceleration gradient and high transformer ratio [9]. An experiment is proposed for conduct at FACET (SLAC) to test a mm-scale THz coaxial

structure using its 30 μm short drive bunch. Meanwhile, a cm-scale GHz coaxial structure is currently under study at the Argonne Wakefield Accelerator facility, ANL [9].

Consider a coaxial high transformer ratio design. The driver is a uniform ring of charge that passes through vacuum layer (2), and the witness moves at a distance behind the drive beam in channel (0). The witness delay is chosen to correspond to an accelerating phase of the wakefield. The outer shell (3) improves the dielectric breakdown threshold, which can occur on the metal wall (d). The accelerating field E_z inside channel (0) increases due to the growth of the field in dielectric tube (1). This fact causes the transformer ratio coefficient to exceed 2 .The use of a drive bunch with an annular shape is associated with a strong transverse instability, which has been studied in detail in previous publication [1]. The transverse dynamics of a ring beam is much more complex in comparison with the instability of Gaussian bunches [7] because of the azimuthally as well as radial dynamics. We investigated in detail the transverse dynamics of a bunch with an annular shape consisting of several segments. This model of the bunch is chosen from technical considerations of the multilayer dielectric waveguide. The inner dielectric tube (1) is attached with special titanium threads, figure 1b). This segmented bunch passes through the waveguide without interacting with the titanium support elements.

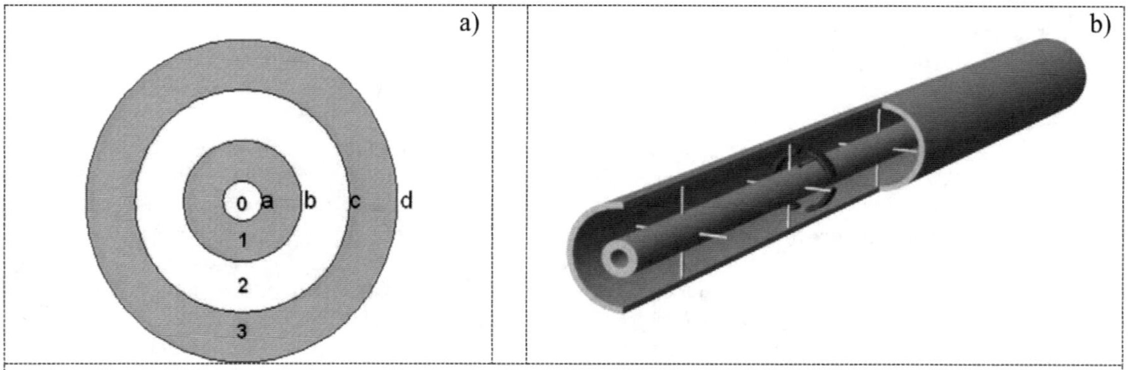

Figure 1. Cylindrical dielectric waveguide: a) cross-section. Radial layers are indicated by numbers (0,2 vacuum and 1,3 dielectric) and boundaries between layers with lower case letters. b) azimuthally segmented bunch passing through the cylindrical multilayer dielectric waveguide with threads

Figure 2 shows the transverse structure of the TM_{01} mode. It can be seen that the field E_z in dielectric tube 1 increases and the field in the vacuum channel 0 has a maximum value that provides efficient transfer of energy to witness bunch. Figure 3 shows the wakefield E_z (superposition of *TM* modes) behind the Gaussian bunch in channel 0 and vacuum layer 2. One can see the increase of the transformer ratio up to 3.5 value.

It should be noted that the field enhancement occurs only for the *TM* modes. The wakefield generated by a Gaussian bunch in layer 2 contains, in addition to *TM* modes, unwanted parasitic axially asymmetric hybrid electromagnetic (*HEM*) modes. This fact leads to the absence of the effect of increasing field. Therefore, transformer ratio is possible only for bunch with a ring charge distribution because the wakefield is mostly determined by *TM*- modes.

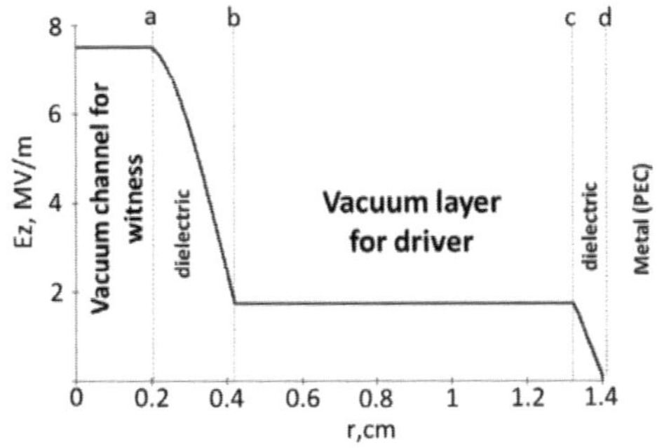

Figure 2. Radial structure of the TM_{01} mode

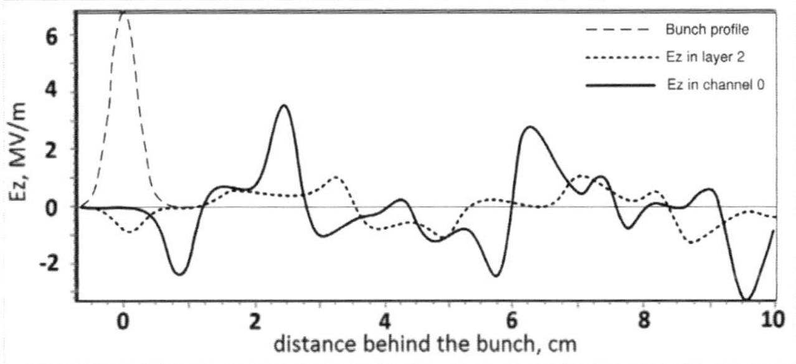

Figure 3. Wakefield in vacuum channel 0 and vacuum layer 2 for bunch #3 in Table 2 in the waveguide from Table 1.

2. HEM-mode structure

We showed in the previously published work [1] that the vacuum layer 2, through which the drive bunch is passing, can be divided into two areas: the area of the internal dielectric tube and the area of the outer dielectric layer. The radial displacement of the particles depends on the area of their passage through vacuum layer 2. This structure of the radial force is caused by the superposition of asymmetric *HEM* modes which can be divided into three types: positive, negative and alternating. The radial structure of each type of event is displayed in Figures 4-5. The sign of the radial force F_r of the alternating *HEM* mode depends on the particle position in vacuum layer 2 (figure 4). The position of a particle does not influence the sign of the positive and negative *HEM* modes (figure 5). In other words, the positive and negative modes are associated with deflection towards the outer and inner dielectric layers respectively.

a) b)

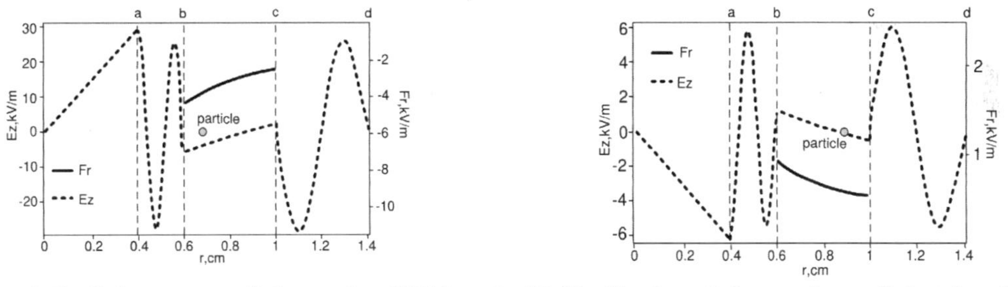

Figure 4. Radial structure of alternative *HEM*-mode (Table 1): a) particle near inner dielectric tube causes a negative sign; b) particle near outer dielectric layer causes a positive sign.

a) b)

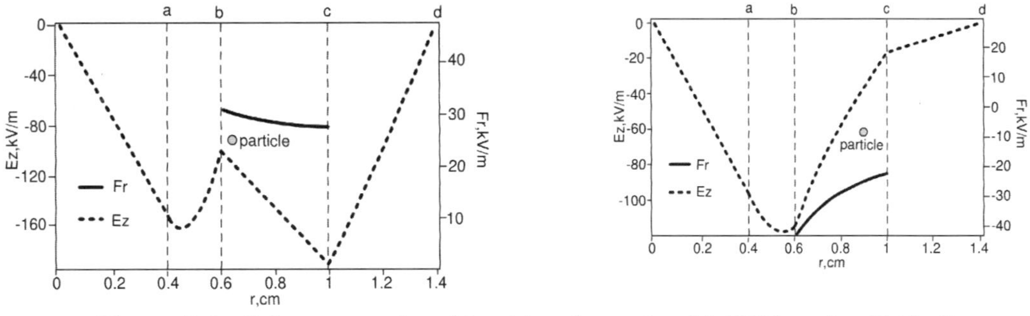

Figure 5. Radial structure of positive (a) and negative (b) *HEM*-modes (Table 1).

3. Beam dynamics modelling

With the in- house code BBU-3000 [7] we studied the dynamics of bunches, which consist of three and four azimuthally segments (Table 1). This code is developped for beam dynamics calculation in dielectric waveguides. The transverse dynamics of these bunches evolves quite rapidly despite the zero offset (transverse position of bunch center). The main contribution to the deflection forces comes from *HEM*-modes, the order of which is equal to coefficient of azimuthal filling N (the number of segments in the bunch). Fig.6 shows the quadruple *HEM* modes (ν=4) and Fig.7 corresponds to *HEM* modes with the order ν=3. We can see that the particles in tail of the bunch are displaced along angular coordinate to the centers of the segments and increase radial dynamics. The radial displacement is explained by location of the majority of particles in vacuum layer 2. The present results show that radial positions of particles close to area of inner dielectric tube.

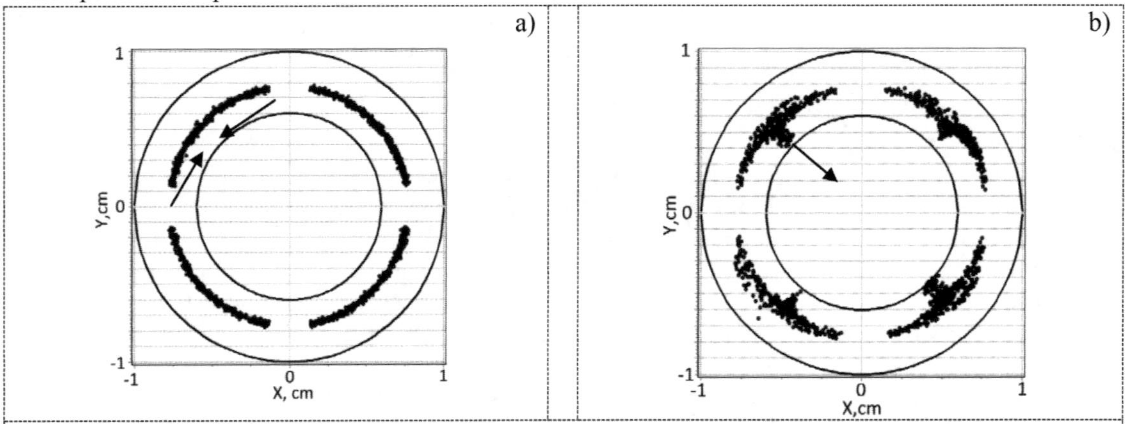

Figure 6. Results of dynamics calculation of azimuthally inhomogeneous (N=4) bunch #1 from Table 2 and waveguide from Table 1: a) initial particle positions; b) after propagating for 36 cm.

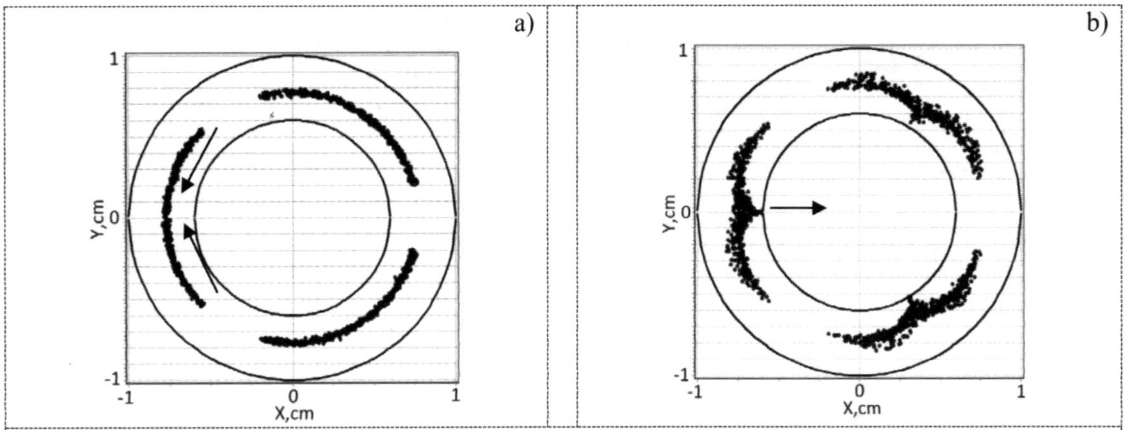

Figure 7. Results of dynamics calculation of azimuthally inhomogeneous (N=3) bunch #2 from Table 2 and waveguide from Table 1: a) initial particle distribution; b) after propagating for 34 cm.

Table 1. Parameters of the coaxial waveguide used in the transverse beam dynamics analysis.

#	a, cm	b, cm	c, cm	d, cm	ε_1	ε_2	f, GHz (TM_{02})
1	0.2	0.60	1.00	1.40	4.72	4.72	13.84

Table 2. Parameters of the annular drive beams used in the transverse beam dynamics analysis

#	beam type	*N*	Q, nC	σ_z, cm	σ_r, cm	offset, cm	radius, cm	W, MeV
1	asymmetrical annular beam	4	100	0.2	0.01	0	0.8	150
2	asymmetrical annular beam	3	100	0.2	0.01	0	0.8	150
3	Transverse point bunch with Gaussian longitudinal profile	-	10	0.2	0	0.8	-	15

4. Conclusions

A multilayer concentric annular dielectric wakefield device can be used to develop a transformer ratio greater than 2. We have modified the BBU-3000 code to investigate beam breakup effects from the annular electron drive beam in this device. The effects of azimuthal beam inhomogeneity originating from support structures for the inner dielectric tube were specifically studied.

- BBU effects from azimuthally inhomogeneous beams are a critical issue for dielectric based accelerator development.
- The number of segments in the annular bunch defines the highest order of the *HEM*-modes contributing to the deflecting force.
- Azimuthal displacement of particles results also in radial forces.
- There are three types of *HEM*- modes that are defined by the bunch position.

References

[1] Altmark A and Kanareykin A 2012 J. Phys.: Conf. Ser. **357** 012001
[2] Kanareykin A, Gai W *et al* 2003 Proc. PAC'03 1894
[3] Kanareykin A 2010 J. Phys.: Conf. Ser. **236** 012032
[4] Power J, Gai W, Kanareykin A, Sun X. 2001 Proc. PAC'01 114.
[5] Liu W and Gai W 2009 Phys. Rev. ST Accel. Beams **12** 051301
[6] Sotnikov G, Marshall T, and Hirshfield J 2009 Phys. Rev. ST-Accel. Beams **12** 061302
[7] Schoessow P, Jing C, Kanareykin A *et al* 2011 Proc PAC'11 1689
[8] Sotnikov G and Marshall C 2011 Phys. Rev. ST-Accel. Beams **14** 031302
[9] Shchelkunov S, Sotnikov G, Marshall T, and Hirshfield J 2012 AIP Conf. Proc. **1507** 945
[10] http://clic-study.org/accelerator/CLIC-ConceptDesignRep.php
[11] Conde M, Gai W, Konecny R, Power J *et al* 1998 AIP Conf. Proc. **472** 626

Device for X-ray spectral absorption analysis with use of acoustic monochromator

Yu Cherepennikov[1], **A Gogolev**

Tomsk Polytechnic University, Tomsk, Russia

E-mail: che@scalpnet.ru

Abstract. The method for X-ray elemental analysis and the device for its realization, based on the X-ray absorption spectroscopy technique, are proposed. Estimations of the method sensitivity have been carried out by means of numerical simulation. The simulation results indicate that the sensitivity of the proposed method for elemental analysis is not less than 10^{-4}.

Nowadays X-ray fluorescence analysis (XRF) [1] is the most common method of the X-ray elemental analysis. This technique is well known, simple and widely used for variety of applications. There are some problems, though, which are still impossible to solve by XRF method. One of them is ultrapure elemental analysis, during which radiated fluorescence is almost completely absorbed by investigated object. Variety of special methods of sampling, such as grinding or evaporation, is used to address this issue. Nevertheless these ways have some significant disadvantages: they are quite difficult to implement, time-consuming and the most important one - lead to destruction of an object.

Another solution, which has no aforementioned drawbacks, can be obtained by analyzing X-ray absorption degree at different wavelengths instead of XRF analysis for determination of elemental composition. Methods of absorption analysis are widely used in the X-ray absorption spectroscopy area and primarily in such kind of investigation as EXAFS (extended X-ray absorption fine structure) and XANES (X-ray absorption near edge structure) [2-3], but they can be used for elemental analysis. In contrast to XRF, where the information is carried by secondary radiation, in absorption analysis the information carrier is primary radiation, passing through the object and partly absorbed depending on radiation energy and elemental composition of the object. The presence of any admixture is found by the presence of intensity leap in the measured spectrum corresponding to the absorption edge of admixture element. The quantitative evaluation of the content of admixture can be obtained from the value of the leap when passing through the absorption edge. Obviously, the leap can be distinguished in a spectrum if its value is more than value of the statistical error. Thus the main limiting factor for the sensitivity of this method is intensity of the primary radiation beam and correspondingly the main task is producing the X-ray beams with satisfactory intensity.

There are two main methods of the X-ray energy spectra measurements in absorption analysis: energy dispersive and wave dispersive. In the first approach X-ray detectors with spectral resolution are used for energy dependence registration. In the second one, registration of energy spectral information is based on the use of monochromatic X-ray. Monochromatic line is isolated from continuous spectrum by means of using special devices (monochromators) before or after passing through the object and after that only this line is detected. Both of these approaches have their advantages and disadvantages. For example, energy dispersive method is easier and allows to register information across the whole energy range at the same time, without changing of the experimental setup parameters. However, sensitivity of this method to the scattered X-ray impact and significant limit of the speed according to relative low count rate of the modern energy dispersive detectors represent the main limitations. Wave dispersive method does not have these drawbacks and allow to achieve better resolution (equal to a few eV). The main disadvantage of this approach is significant loss in X-ray energy line intensity during monochromatization due to only partial reflection of radiation energy line to the direction of the diffraction and additionally absorption in the monochromator material. The last problem can be solved, though, using "pumping" and "transparency" effects [4-5], which are observed in crystal monochromators with super lattice initialized by external influences.

The "acoustic monochromator" device based on "pumping" effect was proposed by the authors in previous works [6] to produce more intensive monochromatic X-ray beams. Acoustic monochromator is a device consisting of crystal quartz plate with excited acoustic wave field. Production of monochromatic X-ray beam with certain characteristics which can be controlled by the changing of acoustic wave field parameters has a potential to reduce dramatically the beam loss. In this work the authors propose the device for X-ray spectral absorption analysis based on the use of X-ray tube and acoustic monochromator.

Functional scheme of the device is shown in figure 1. The device for X-ray spectral absorption analysis includes: (1) X-ray source based on X-ray tubes, (2) double-slit collimator, (3) acoustic monochromator, placed in the first holder of a goniometer (4), and absorber (5). The source, collimator, monochromator and absorber are placed at the one optical axis. An object under investigation (6) and an X-ray detector (7) are positioned at the second goniometer holder behind acoustic monochromator in the diffraction direction. Acoustic monochromator is connected with electrical signal amplifier (9) which enhances amplitude of the signal from electromagnetic waves generator (8). The X-ray source, goniometer, electromagnetic waves generator and detector are connected with a computer by means of control drivers. The goniometer holders with the set acoustic monochromator, object and detector can be turned and energy of monochromatic X-ray passing through the object can be changed in accordingly. At the same time detector registers spectrometric information, observation of which is the main purpose of the experiment.

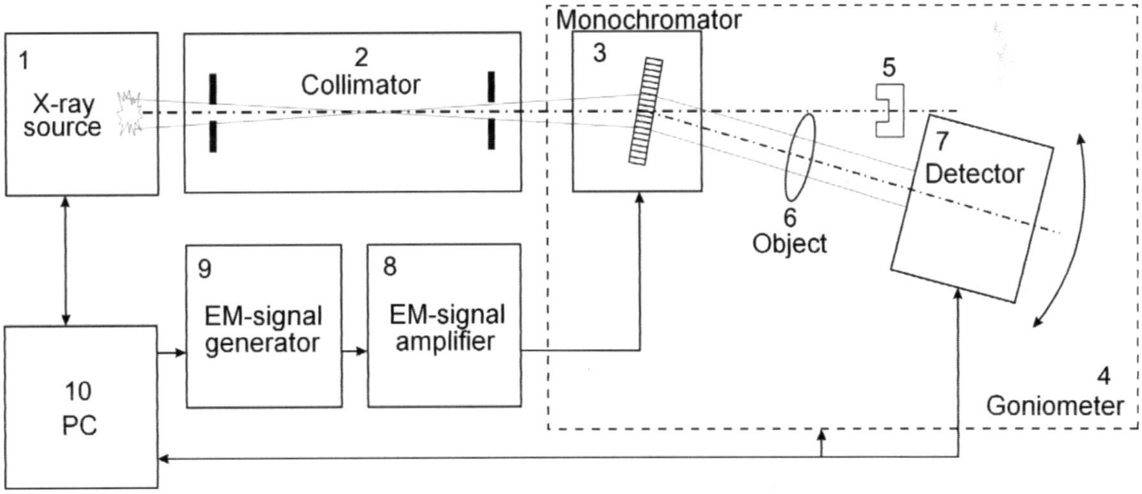

Figure1. Functional scheme of the device for X-ray spectral absorption analysis.

The aim of this work is to describe the sensitivity of the proposed method and evaluate value of admixture which can be distinguished in an experiment.

The estimations were carried out by means of the developed numerical model of the interaction between X-ray and investigated object material. Model was developed by using GEANT4 toolkit [7], which based on application of Monte Carlo method. Samples of SiO_2 matrixes with different values of Cr admixture concentration were chosen as the objects for simulated experiment. The samples were irradiated by X-ray beam with the energy close to Cr absorption K-edge. A detector with the transverse dimensions equal to 2×2 mm was placed at the distance equal to 20 mm at the beam axis. The detector registered all passed photons with energy resolution equal to 10 eV. Simulation of the X-ray spectra passed through the samples was carried out for samples with different admixture concentrations. These spectra were compared with spectrum after pure SiO_2 sample to find the leap of intensity.

X-ray spectra passed through the sample for admixture with mass fractions equal to $5\cdot10^{-2}$, 10^{-2}, 10^{-3}, 10^{-4}, 10^{-5} and 10^{-6} are shown in the figure 2. Simulation was carried out for 1 keV bandwidth from 5.5 keV to 6.5 keV according to the Cr absorption K-edge energy value, which equals to 5.97 eV. Number of initial photons histories was equal to 10^9 for all energy ranges or 10^7 per 10 eV channel. Figure 2a includes spectra for admixture mass fractions not less than 10^{-4} for all considered energy range while figure 2b - for shorter one for mass fractions not more than 10^{-3}. The minimal admixture concentration, which could be distinguished for that number of histories, equalled to 10^{-3} as can be seen from simulated spectra. Spectra for lower concentrations were almost equal to the X-ray spectrum passed through the pure SiO_2 sample.

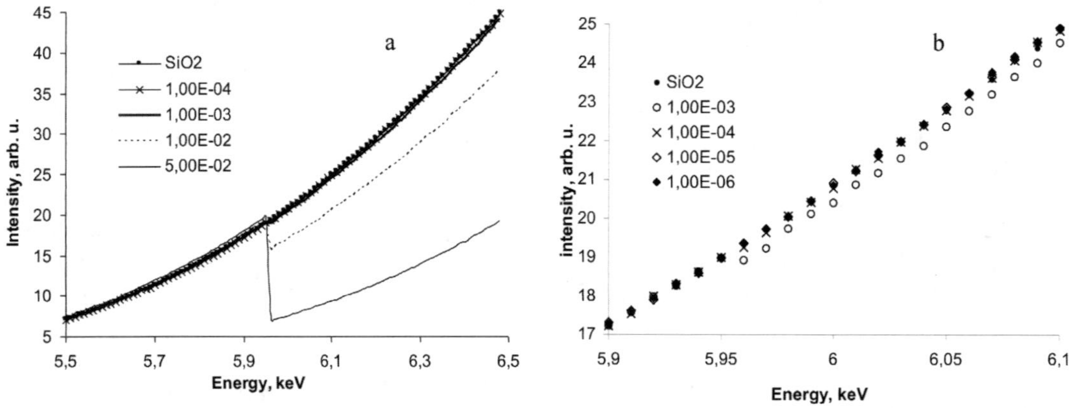

Figure 2. X-ray intensity distributions in the experiment, after passing through the samples with different admixture concentration.

It should be noted that "differential" spectra are more suitable for this kind of experiments. These spectra represent differences between intensities of X-rays passed through the pure sample and admixture contained one. Such spectra for mass fraction equal to 10^{-3} and 10^{-4} are shown in figure 3a. It is clearly seen from these spectra, that concentration 10^{-3} is distinguishable (a specific "leap" is present in the spectrum at the absorption edge), while concentration 10^{-4} is not seen (a spectrum only contains "noises", caused by a statistical error). The last proposition, though, is right for announced histories number. Rise of the initial X-ray intensity which passes through the samples can allow to distinguish lower concentrations.

The parameters of the model were slightly modified to distinguish 10^{-4} concentration in the experiment. At first, total number of particle histories was increased up to 10^{10}. Besides, simulation was carried out in the immediate vicinity of the absorption edge and width of simulated region was equal to 90 eV. All of that allows to increase particle histories per channel number up to 10^9 keeping the acceptable amount of time spending for simulation. Differential spectrum for admixture mass fraction equals to 10^{-4} in this simulation is shown in figure 3b. Well distinguished step at the absorption edge region is clearly visible and consequently for this value of the initial X-ray intensity the concentration equal to 10^{-4} can be successfully found in the sample. Thus, the proposed method of elemental analysis provides sensitivity not less than 10^{-4}. In the same time, it should be noted that further rising of the initial X-ray beam intensity might allow to find much lower concentration.

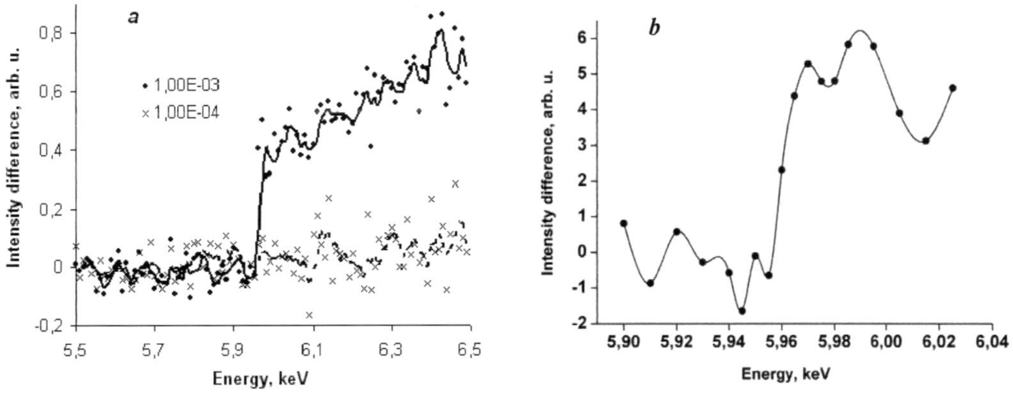

Figure 3. Differential spectra for SiO_2 sample contained Cr admixture. Number of the initial particles histories is equal to 10^7 (a) and 10^9 (b) per channel. Mass fraction is 10^{-3} and 10^{-4} (a), 10^{-4} (b).

With regard to the potential application of the presented device besides laboratory investigations environmental control can be an attractive area. For example, this device can be used for the analysis of water composition. The issue of the presence of harmful admixtures in sewage water or drinking water is very important now. To check the applicability of the proposed device for the control of water compositions simulations of radiation spectra passing through the sample consisting of a water layer containing different admixtures and through a layer of pure water were carried out. Elements from different areas of the periodic table which are the most common admixtures for natural water were chosen as admixtures. These elements were Ca for the lights, Fe for the mediums and Pb for the heavies. All of these elements had mass fractions equal to 10^{-4}. The region near to absorption K-edge was chosen for simulating for Ca and Fe admixtures while L-edge region was observed for Pb. Differential spectra are shown in figure 4 for all these cases. Presence of these admixtures with 10^{-4} mass fraction is well distinguished by proposed method as it can be seen from the figures.

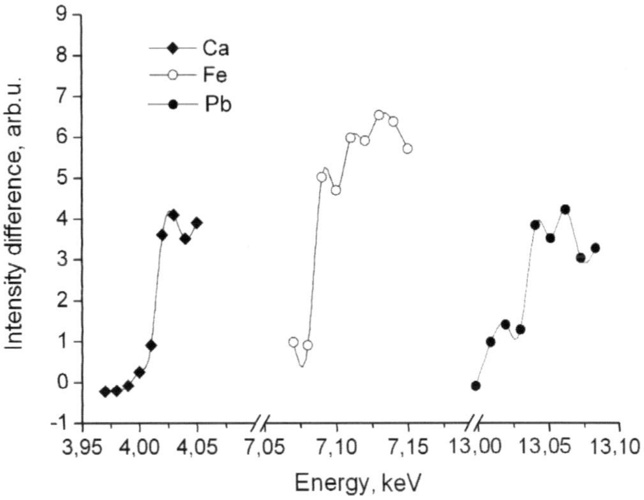

Figure 4. Differential spectra for water sample with Ca, Fe and Pb admixtures. Initial particles number is equal to 10^9 per channel.

Thus, the method for elemental analysis by different wavelength X-ray absorption in the object is proposed as well as device for X-ray spectral absorption analysis, which allows to realize this method. Developed numerical model of X-ray interaction with the object allows to estimate the proposed method sensitivities. The sensitivity is not less than 10^{-4} by Cr admixture in SiO_2 sample according to simulation results. Use of the device allows to analyze pure and ultrapure samples.

The estimations, carried out with the developed model for water samples consisting of a water layer, containing different admixtures, show retention of the method sensitivity at the same level for the wide range of admixtures – from Ca to Pb. This fact allows to conclude, that the proposed method can be applied in the analysis of air composition in ecological monitoring.

We would like to acknowledge Demyanenko Natalia for her useful help during the manuscript preparation.

This work was partially supported by Government of the Russian Federation "On the Federal Target Program" GK № 14.515.11.0102.

References

[1] Beckhoff B et al 2006 *Handbook of Practical X-Ray Fluorescence Analysis* (Berlin: Springer)
[2] Bunker G 2010 *Introduction to XAFS: A Practical Guide to X-ray Absorption Fine Structure Spectroscopy* (New York: Cambridge University Press)
[3] Stöhr J 2003 *NEXAFS Spectroscopy* (Berlin: Springer)
[4] Gabrielyan R G et al 1985 *Phys. Stat. Sol* .V (a) **92** 361-8
[5] Mkrtchyan A R, Navasardian M A, Gabrielyan R G et al 1986 *Solid State Communications* **59** 147-9
[6] Gogolev A S, Cherepennikov Yu M, Stuchebrov S G, Wagner A R and Potylitsyn A P 2011 X-ray scattering in deformed crystals *Journal of Physics: Conference Series* **357** 012031
[7] Information on http://geant4.web.cern.ch/geant4/

Channeling of protons in radially compressed carbon nanotubes

A Karabarbounis[1], S Sarros[1] and Ch Trikalinos[2]

[1] Department of Physics, Section of Nuclear and Particle Physics, University of Athens, Panepistimioupolis, Ilissia, 15771 Athens, Greece
[2] Department of Philosophy and History of Science, University of Athens, Panepistimioupolis, Ilissia, 15771 Athens, Greece

E-mail: stsarros@phys.uoa.gr

Abstract. Channeling of 10 MeV protons in radially compressed chiral carbon nanotubes is considered. Monte Carlo simulation program is used for the calculation of the trajectories, energy losses and angular distributions of protons in nanotubes of various lengths, where the potential in Doyle-Turner approximation is used to describe the interaction between a proton and a nanotube. Calculations were carried out for different incident angles between proton beam and the nanotube axis. The results show that a decreased angular distribution of the beam is observed, compared with propagation through a straight nanotube, in case when it enters from the compressed end of the nanotube. The energy and spatial distribution of channeled protons in compressed nanotubes is examined.

1. Introduction

Carbon nanotubes were discovered more than twenty years ago, multi-wall carbon nanotubes (MWCNT or MWNT) in 1991 [1] and single-wall carbon nanotubes (SWCNT or SWNT) in 1993 [2,3]. Since then, an advanced progress on the investigation of their properties and production methods has been achieved. Due to their unique physical properties, carbon nanotubes (CNTs) have attracted growing interest on various fields [4,5]. Carbon nanotubes are produced with various radii (from 0.2 nm up to several nm) and in a wide range of lengths (up to several μm and even mm). Although CNTs exhibit extremely high mechanical strength, recent studies have shown that they can undergo axial and radial compression [6]. This way it is feasible to obtain carbon nanotubes that are radially compressed at one end, that will resemble truncated nanocones.

Guidance of charged particle beams using CNTs has been investigated theoretically in recent years [7-11]. In those studies the possibility itself of channeling in single straight CNTs and nanotube bundles, as well the spatial and angular distribution of channeled particles have been examined. Also, in a recent study [11] the energy losses of channeled protons in straight chiral CNTs, presenting unusual results for their energy distribution, were examined.

Until now there have not been studies of propagation and channeling of charged particles in compressed CNTs. In particular, radially compressed carbon nanotubes at one end provide a suitable geometrical structure for particle focusing. Therefore, it is interesting to investigate the propagation and channeling of charged particles in such radially compressed CNTs and compare the results with the corresponding ones in uncompressed single CNTs.

2. Theory

Each nanotube can be described by two indices as (n,m) and it may be considered as a collection of atomic rows parallel to the axis of the nanotube and arranged in a specific way along the perimeter of the cylindrical surface. In this way, from simple geometrical consideration, each pair of these indices define the nanotube radius R and helicity or chiral angle θ (in practice, the angle under which the most closely packed rows of carbon atoms are wound on the cylindrical surface of the tube):

$$R = \left(l\sqrt{3}/2\pi\right)\sqrt{n^2+nm+m^2} \quad (1)$$

$$\theta = \arctan\left[\sqrt{3}m/(m+2n)\right] \quad (2)$$

where $l = 0.142$ nm is the length of the bond between the carbon atoms [4, 5]. SWNTs with $m = 0$ ($\theta = 0°$) are called zigzag, those with $m = n$ ($\theta = 30°$) are called armchair, and all the others ($0° < \theta < 30°$) are called chiral. In the present study we examine only the case of chiral nanotubes.

Let us first consider the case of motion of fast positively charged particles (e.g. protons) in a straight isolated carbon nanotube. If such a particle enters a single crystal at small enough angle θ_0 with respect to an atomic row, then it can be assumed to be governed mainly by the continuum potential, i.e. the actual periodic potential of the rows averaged over the direction parallel to the rows [12]. Applying the concept of continuum potential to chiral nanotubes, it can be considered as the continuum potential of the (rolled-up) graphite plane, and thus we may average the actual potential of a nanotube over the circumference, i.e. the azimuthal angles. In the Doyle-Turner approximation to the atomic form-factor, the axially symmetrical continuum potential of a chiral nanotube (without taking into account the thermal vibrations of the atoms) can be described by the following expression [8]:

$$U(r,\varphi) = 3^{-3/2}32\pi Ze^2l^{-2}R\sum_{j=1}^{4}\alpha_j b_j^2 \exp\left[-b_j^2(r^2+R^2)\right]I_0(2b_j^2Rr) \quad (3)$$

where $Z = 6$ is the atomic number of the carbon atom, r is the distance from nanotube axis and α_j, b_j are dimensional parameters in the Doyle-Turner approximation:
$\{\alpha_j\} = \{3.222,\ 5.270,\ 2.012,\ 0.5499\} \times 10^{-4}\ \text{nm}^2$, $\{b_j\} = \{10.330,\ 18.694,\ 37.456,\ 106.88\}\ \text{nm}^{-1}$.

In case of a nanotube that is radially compressed at one end, its radius at each transverse plane at a point z of the axis is defined as:

$$R = R_0 - z \cdot \tan\varphi \quad (4)$$

where R_0 is the radius of the uncompressed nanotube and φ is the angle of the slope of the walls of the nanotube after compression (see figure 1). This way, by introducing equation (4) into (3) instead of R, we can obtain the potential in this compressed nanotube (see figure 2). The above approach is valid, since we consider small radial compression of the nanotube, so that the geometry of the nanotube doesn't change significantly and therefore we can use the same expressions for the nanotube potential.

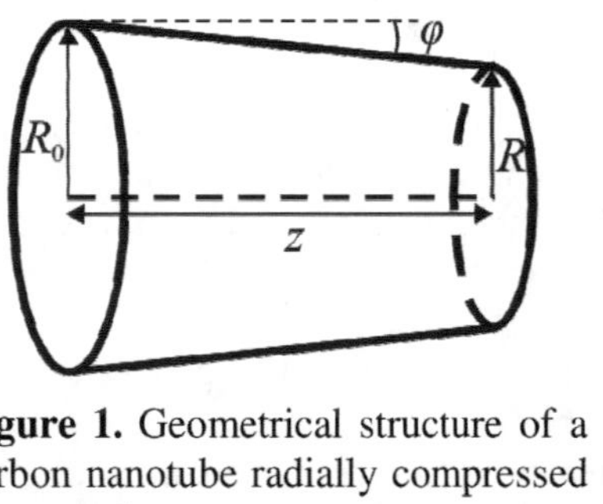

Figure 1. Geometrical structure of a carbon nanotube radially compressed at one end.

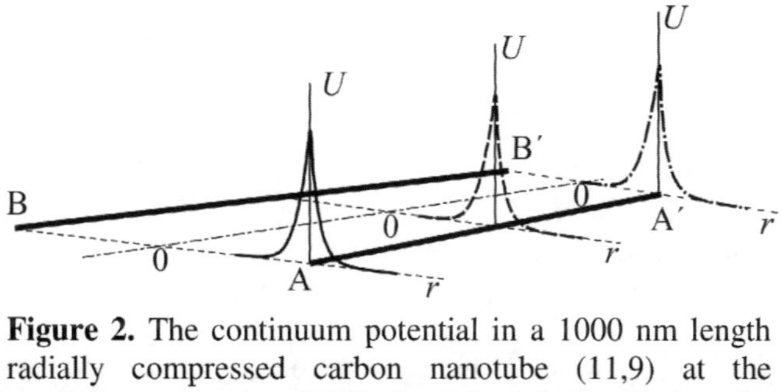

Figure 2. The continuum potential in a 1000 nm length radially compressed carbon nanotube (11,9) at the uncompressed end (solid line), at the middle of nanotube (dashed line) and at the compressed end (dot-dashed line), where AA′ and BB′ are the nanotube walls.

Because the de Broglie wavelength of a 10 MeV proton under consideration is very small compared to atomic dimensions, classical treatment of the problem can be adopted. We use the right-handed Cartesian coordinate system for the geometry of our problem, where z-axis coincides with the axis of the nanotube, and x, y axes lie on the transverse plane.

In order to investigate this problem we use a Monte Carlo simulation program for individual proton trajectories. The individual proton trajectories were calculated by solving the classical equations of ion motion and taking into account energy loss and multiple scattering at each integration step.

We take into account energy losses calculated by phenomenological expression for the local stopping power given by Lindhard [13]:

$$\frac{\Delta E}{\Delta z} = S(E) = \frac{4\pi Z_1^2 e^4 Z_{val}}{m\upsilon^2}\left[(1-\alpha)+\alpha n_e(r)\right]\ln\left(\frac{2m\upsilon^2}{I}\right) \tag{5}$$

where $Z_1 e$ and υ are the ion charge and velocity, respectively (for proton $Z_1 = 1$), Z_{val} is the number of valence electrons per atom (for carbon atom $Z_{val} = 4$), $n_e(r)$ is the normalized local valence electron density at a point in the space, through which the particle moves, and m is the electron mass. The coefficient α describes that part of close collisions which, according to Lindhard's model, is proportional to the electron density (in our case, is 0.5, according to Bohr's equipartition rule). The quantity I is the average excitation potential, $I = I_0 Z$, where $I_0 \cong 13.5$ eV [14] and Z is the atomic number of target atoms. The normalized electron density $n_e(r)$ is calculated by using the Fourier component of the electron density in carbon atom [8]:

$$f^{(e)}(\mathbf{k}) = Z\sum_{j=1}^{5}\alpha_j^{(e)}\exp\left[-k^2/(2b_j^{(e)})^2\right] \tag{6}$$

with the five pairs of fitting parameters:

$\{\alpha_j^{(e)}\} = \{0.3499,\ 0.3014,\ 0.2103,\ 0.0946,\ 0.0438\}$

$\{b_j^{(e)}\} = \{17.300,\ 11.400,\ 75.501,\ 155.24,\ 7.596\}\ \mathrm{nm}^{-1}$

Thus, by using equation (6) it is easy to show that for a chiral nanotube the normalized electron density averaged over the azimuthal angles is determined from the following expression:

$$n_e(r) = \frac{2NZ_{val}}{\pi d_R}\sum_{j=1}^{5}\alpha_j^{(e)}b_j^{(e)2}\exp\left[-b_j^{(e)2}(R^2+r^2)\right]I_0(2b_j^{(e)}Rr) \tag{7}$$

where N is the number of atomic rows in each sequence, and d_R is the distance between neighboring atoms within each row [8].

We also take into account the electronic multiple scattering after each integration step, by calculating a normal distribution of the scattering angle with standard deviation:

$$\theta_{ms}^2 = \frac{m\Delta E}{2m_1 E} \tag{8}$$

where E and ΔE are the energy and the energy loss at each integration step, respectively.

We assume that protons are dechanneled if during their propagation inside a nanotube they approach the nanotube wall less than the critical distance $r_c = \sqrt{2}u_\perp$, where $u_\perp$ is the thermal vibration amplitude in the transverse plane [15]. In our case, we can use the value $8.5\cdot 10^{-3}$ nm, which corresponds to that of graphite (at room temperature).

Our calculations were carried out for different angles between incident proton beam and the nanotube axis, and for different nanotube length. Beam protons are assumed to enter nanotube's cross-sectional area homogeneously. In all simulations we take into account energy losses and multiple scattering.

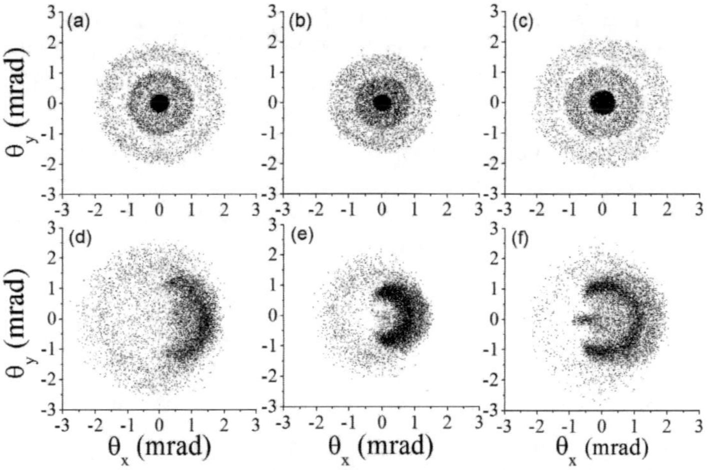

Figure 3. Angular distribution of 10 MeV protons channeled in a compressed (6,4) CNT for zero incident angle ((a), (b), (c)) and $0.5\cdot\psi_{cr.}$ ((d), (e), (f)) in cases: wide end entrance ((a) and (d)), narrow end entrance ((b) and (e)), compared with straight nanotube ((c) and (f)).

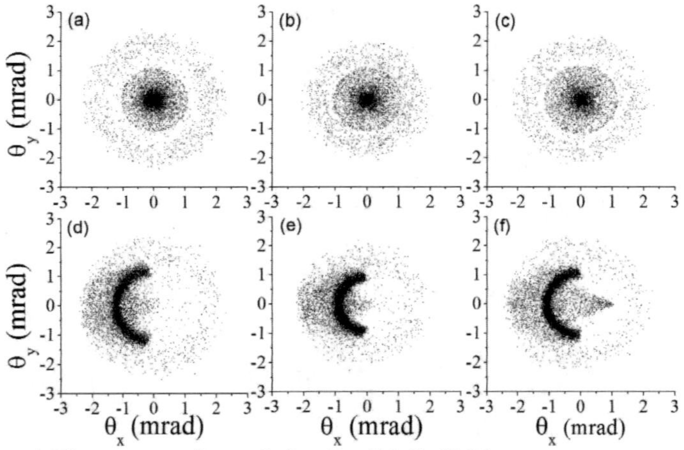

Figure 4. The same as figure 3, but for (11,9) CNT.

3. Results and discussion

We examined 10 MeV monoenergetic proton beam channeled in radially compressed at one end (6,4) and (11,9) SWCNTs of various length. The radii of these nanotubes at the uncompressed end are 0.341 and 0.679 nm, respectively, calculated with the use of more accurate value of length of the bond between the carbon atoms mentioned above. We investigated two cases of incident angle, zero and half of the critical angle. The cases of nanotube types examined are the following: straight, radially compressed with the angle of slope of the walls of 0.005 degrees with entrance at both wide and narrow ends. For proton energies of 10 MeV the critical angles $\psi_{cr.}$, which were calculated as $\psi_{cr.} = \left(U(r_c)/E\right)^{1/2}$, for (6,4) and (11,9) nanotubes are 2.181 and 2.169 mrad, respectively. The initial proton beam was assumed to be well collimated ($\Delta\theta = 0$).

Figures 3 and 4 show the angular distributions for three types of (6,4) and (11,9) nanotubes of 1000 nm length, respectively, and for different incident angles (0 and $0.5\psi_{cr.}$). We observe that in the case of the narrow end entrance, the beam exits more collimated than in the two other cases. This effect can be explained by examining the trajectory of the particle inside this nanotube in the x-z plane, where it moves in an increasing channel so that its motion resembles damping oscillation (see figure 5).

Journal of Physics: Conference Series **517** (2014) 012038 doi:10.1088/1742-6596/517/1/012038

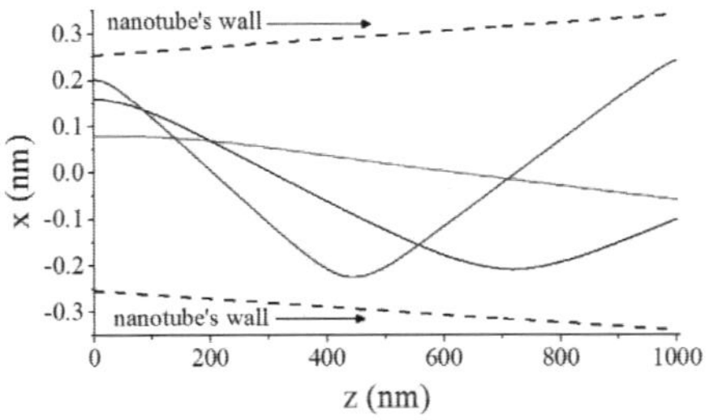

Figure 5. Trajectory in x-z plane of 10 MeV protons channeled in a compressed (6,4) CNT for various points of incidence and for zero incident angle.

Also, it is clear that a ring-shaped pattern of angular distribution is being formed when the incidence angle is $0.5\psi_{cr.}$ (doughnut scattering effect) with radius close to the corresponding incident angle of the beam. Another interesting feature, which was analyzed in our previous work [11], is the formation of "zones" in the angular distribution at zero incident angle of the beam, where each zone corresponds to different initial incident region. In particular, channeled protons incident in a region closer to nanotube's wall exit nanotube with larger angle. It is noteworthy, that these "zones" appear in the case of both types of radially compressed nanotubes as well.

The energy distributions of channeled protons for three types of (6,4) nanotubes are presented in figure 6. The incident angle of the beam considered was zero. In these figures the length evolution of energy distribution of protons is demonstrated. The appearance of several small peaks is observed in all three nanotube types, the number of which increases with larger nanotube length. This effect was observed for the first time in straight nanotubes [11]. An explanation of these results can be achieved by examining the particle trajectory inside a nanotube in x-z plane (see figure 5 in case of (6,4) CNT with particles entering the compressed end). As indicated in [11], the trajectory period strongly depends on the distance of incidence point from the nanotube wall. The closer this point is to the wall, the less is the period of proton motion between nanotube walls. This, in turn, forces the proton to spend more time in regions close to the wall, where the energy losses are significant, compared to the region around nanotube axis, where they are close to zero. Thus, protons that where incident in a zone close to the nanotube walls, correspond to the small peak with the minimum energy (maximum energy loss), while other protons incident in zones farther from the wall correspond to the rest small peaks. In the same figure it becomes clear that the exit angle of the protons is larger, when it is incident closer to the wall, which in turn corresponds to the zone with larger exit angles in the angular distribution (see figure 3). Therefore, we can correlate the zones in angular distributions (figure 3) with the corresponding energy peaks (figure 6). Regarding the main peak of maximum energy, it comes from the protons incident in the region around nanotube axis, where energy losses are negligible.

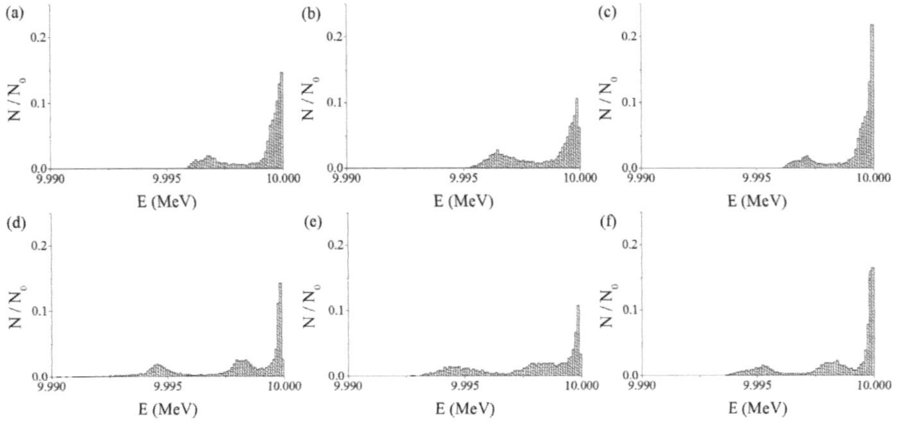

Figure 6. Energy distribution of 10 MeV protons channeled in a (6,4) CNT for zero incident angle, for CNT length 500 nm in cases of proton incident at: (a) uncompressed end, (b) compressed end, (c) straight nanotube, and for CNT length 1000 nm in cases of proton incident at: (d) uncompressed end, (e) compressed end, (f) straight nanotube.

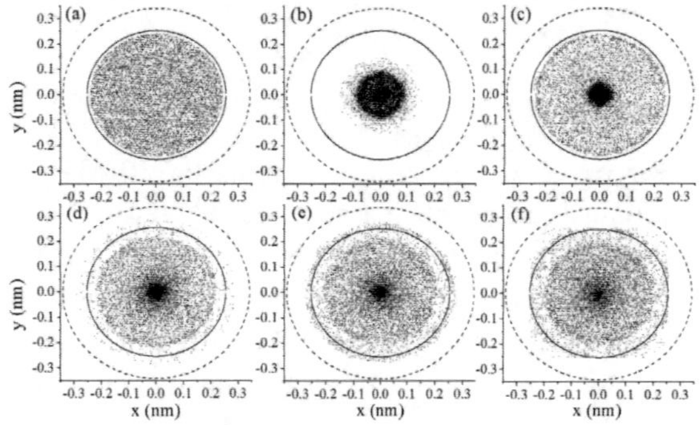

Figure 7. Spatial distribution of 10 MeV protons channeled in a (6,4) CNT, in case of incidence at compressed end and for zero incident angle, for various lengths: (a) 0, (b) 200 nm, (c) 400 nm, (d) 600 nm, (e) 800 nm, (f) 1000 nm. Solid line circle represents the circumference of the compressed end of CNT and dashed line circle represents the circumference of the uncompressed end of CNT.

Figure 7 shows the evolution of spatial distribution of protons propagating in a (6,4) radially compressed CNT, in case when particles enter the compressed end. Depending on the length of the compressed nanotube, focusing of particles in smaller area is different. This feature can be useful, in order to choose the proper length of CNTs to obtain the desirable focusing.

4. Conclusions

We studied proton channeling in radially compressed carbon nanotubes. The presented results of the angular distribution of channeled protons in compressed nanotubes are similar to the corresponding ones for straight nanotubes. The main difference is observed in the case when particles enter the compressed end, where the angular distribution is more focused than in other cases.

Moreover, we have studied energy losses of channeled protons in compressed nanotubes. The energy distribution of channeled protons incident with zero angle inside a compressed nanotube, in both cases of beam entering compressed and uncompressed end, appears to have many small peaks, depending on the length of the nanotube. Similar pattern was observed in straight nanotubes, and there is an indication that these peaks are correlated to the zones in the corresponding angular distribution plots.

Finally, we presented results of the evolution of spatial distribution of channeled protons incident at the compressed end of a (6,4) radially compressed CNT. The observed fluctuations in the focusing of the beam require farther investigation.

References

[1] Iijima S 1991 *Nature* **354** 56-8
[2] Iijima S and Ichihashi T 1993 *Nature* **363** 603-5
[3] Bethune D S, Kiang C H, de Vries M S, Gorman G, Savoy R, Vazquez J and Beyers R 1993 *Nature* **363** 605-7
[4] Jorio A, Dresselhaus G and Dresselhaus M S 2008 *Carbon nanotubes: Advanced Topics in the Synthesis, Structure, Properties and Applications* (Berlin: Springer)
[5] Wong H-S P and Akinwande D 2011 *Carbon Nanotube and Graphene Device Physics*, (Cambridge: Cambridge University Press)
[6] Barboza A P M, Chacham H and Neves B R A 2009 *Phys. Rev. Lett.* **102** 025501
[7] Zhevago N K and Glebov V I 1998 *Phys. Lett.* A **250** 360-68
[8] Artru X, Fomin S P, Shul'ga N F, Ispirian K A and Zhevago N K 2005 *Phys. Rep.* **412** 89-189
[9] Matyukhin S I 2008 *Kinetics of Ion Channeling in Carbon Nanotubes* (Orel: OrelGTU)
[10] Borka D, Lukic V, Timko V J and Borka Jovanovic V 2012 *Nucl. Instrum. Meth. Phys. Res.* B **279** 198-201
[11] Karabarbounis A, Sarros S and Trikalinos Ch 2013 *Nucl. Instr. Meth. Phys. Res.* B **316** 160-170
[12] Gemmell D S 1974 *Rev. Mod. Phys.* **46** 129-227
[13] Lindhard J 1965 *Kong. Danske Vid. Selsk., Mat.-Fys. Medd.* **34** 14
[14] Sigmund P 2006 *Particle Penetration and Radiation Effects* (Berlin: Springer)
[15] Lenkeit K, Trikalinos Ch, Balashova L L, Kabachnik N M and Shulga V I 1990 *Phys. Status Solidi* B **161** 513-524

Simulation of positron energy spectra generated by channeling radiation of GeV electrons in a tungsten single crystal

B Azadegan[1], S A Mahdipour[1] and W Wagner[2]
[1]Hakim Sabzevari University, Sabzevar, P.O. Box 397, Iran
[2]HZDR Dresden, P.O. Box 510119, 01314 Dresden, Germany

E-mail: azadegan@hsu.ac.ir

Abstract. Positron production based on the generation of channeling radiation by relativistic electrons channeled along the (110) crystallographic plane of a W crystal and the subsequent conversion of radiation into e+e−-pairs in an amorphous tungsten target is described. Electron dechanneling is considered by solving of the Fokker-Planck equation. The continuous potential of the channeling plane is calculated using the Doyle–Turner approximation to the atomic scattering factor taking into account thermal vibrations of the crystal atoms. The trajectories, velocities and accelerations of planar channeled electrons are obtained by solving the classical equation of motion. In the framework of classical electrodynamics, the spectral-energy distribution of radiation is obtained from the Fourier transforms of realistic electron trajectories, velocities and accelerations within the W crystal. The calculations of channeling radiation and dechanneling are carried out by means of our *Mathematica* codes. The conversion of radiation into e+e−-pairs and the energy distributions of produced positrons are simulated using the *GEANT4* package.

1. Introduction

Channeling radiation (CR) is generated by relativistic charged particles which traverse a single crystal parallel to a major crystallographic plane or axis [1]. Intense γ-radiation directed into a narrow cone can be obtained by channeling of high-energy (GeV) electrons and may be applied for the creation of positron beams. In conventional positron sources, e−e+-pairs are produced by conversion of ordinary bremsstrahlung (BS) generated in relatively thick amorphous targets which simultaneously serve as radiation converters. It has, however, been shown that aligned crystal targets are more effective radiators than amorphous ones. The enhancement of the positron yield due to CR has been confirmed experimentally at CERN [2-4] and KEK [5, 6]. The shower process of pair production in a single W crystal as well as combined target assemblies consisting of a crystalline radiator (W, Si, Ge, diamond) and an amorphous converter have been studied in [7-9].

For optimization of the type and thickness of the crystal radiator of a non-conventional positron source, consideration of dechanneling in thick crystals is essential. Applying the method explained in [10], this work aims at calculation of dechanneling of electrons planar channeled in a W single crystal by solving the Fokker-Planck equation. The dynamics of the particle distribution density is investigated in dependence on the electron energy and the initial scattering distribution. Radiation spectra of CR are obtained for thick W crystals using the theory already described in [11-13]. The conversion of the calculated radiation into e−e+-pairs in an amorphous W target has been simulated by means of the *GEANT4* package. Positron energy spectra resulting from the conversion of CR generated in the crystalline W radiator are presented.

2. Basics of channeling

The shape of the continuous periodic transverse potential, $U(x)$, depends on the structure and lattice parameters of the used crystal. Our calculations are based on the Doyle–Turner approach [14] already applied in [11-13]. Figure 1 shows the continuous potential calculated for electrons channeled along the (110) plane of a W single crystal. For convenience, it was shifted with its minimum to zero. The equation of motion of a relativistic particle in the one-dimensional planar potential $U(x)$ reads

$$\gamma m\ddot{x}(t) = F = -\frac{\partial U(x)}{\partial x} \tag{1}$$

where m is the particle rest mass and γ is the Lorentz factor. The initial conditions for solving of equation (1) are (i) the point of incidence of the particle into the crystal, $x(0) = x_0$, and (ii) its transverse momentum, $p_x(0) = p\theta_0$, where θ_0 is the angle of incidence with respect to the (110) plane. The initial transverse energy is given by

$$E_x = \frac{p^2\theta_0^2}{2\gamma m} + V(x_0). \tag{2}$$

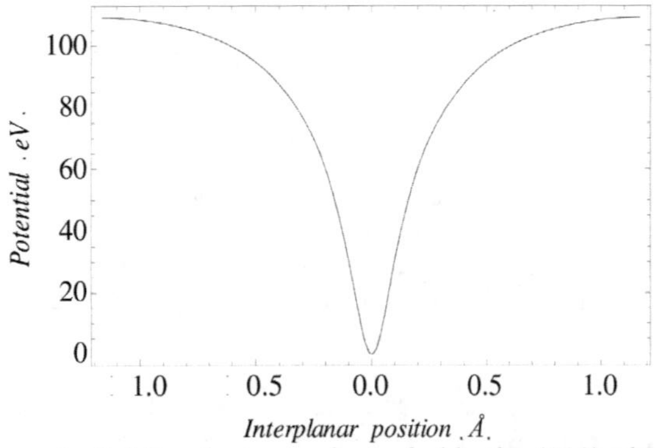

Figure1. Continuous potential for electrons channeled in the (110) plane of a W single crystal.

3. Solution of the Fokker-Planck equation

3.1. Basic background

The general expression of the Fokker-Planck equation which describes the evolution of the probability density $F(z, E_\perp)$ in time (or with the traversed distance z in the crystal, respectively) reads [10]

$$\frac{\partial F(z, E_\perp)}{\partial z} = \frac{\partial^2}{\partial E_\perp^2}[D^2(E_\perp)F(z, E_\perp)] - \frac{\partial}{\partial E_\perp}[D^1(E_\perp)F(z, E_\perp)]$$

$$\frac{\partial F(z,0)}{\partial z} = 0, \frac{\partial F(z, E_{\perp,c})}{\partial z} = 0 \tag{3}$$

where the drift coefficient $D^1(E_\perp) = \langle\frac{\Delta E_\perp}{\Delta z}\rangle_T$ describes the mean increase of the transverse energy and the diffusion is given by $D^2(E_\perp) = \frac{1}{2}\langle\frac{(\Delta E_\perp)^2}{\Delta z}\rangle_T$. These coefficients are averaged over one oscillation period of the electron within the continuous potential. Using the Kitagava-Ohtsuki approximation [15], drift and diffusion coefficients are calculated by means of the integrals

$$D^1(E_\perp) = \aleph \int_{-x_d}^{x_d} \frac{\exp(-x^2/2u_1^2)}{\sqrt{2(E_\perp - U(x))/E}} dx \tag{4}$$

$$D^2(E_\perp) = \aleph \int_{-x_d}^{x_d} \frac{2(E_\perp - U(x))\exp(-x^2/2u_1^2)}{\sqrt{2(E_\perp - U(x))/E}} dx \quad (5)$$

where $\aleph = \frac{E_s^2}{4EvX_0} \frac{4}{T(E_\perp)} \frac{d_p}{\sqrt{2\pi}u_1}$ is a function defined by projectile and crystal parameters with E_s=13.6 MeV, E is the total electron energy, v is the electron velocity, u_1 is the thermal vibration amplitude, X_0 is the characteristic radiation length of the crystal, d_p is the inter-planar distance, x_d is the point corresponding to the solution $E_\perp = U(x)$ for $E_\perp < U(\frac{d_p}{2})$ and $x_d=d_p/2$ for $E_\perp \geq U(\frac{d_p}{2})$. The time parameter is defined by

$$T(E_\perp) = 2\sqrt{\frac{E}{2c^2}} \int_{-x_d}^{x_d} \frac{dx}{\sqrt{(E_\perp - U(x))}}. \quad (6)$$

As initial conditions at $z = 0$, required for the numerical solution of equation (3), a uniform electron distribution across the transverse x-coordinate and a Gaussian scattering distribution tilted by an angle θ_0 and with a standard deviation σ_y for the angular divergence are assumed. The probability density is then given by the integral

$$F_0(E_\perp) = \frac{1}{4\sqrt{2\pi}(\sigma_y/\vartheta_c)U(\frac{d_p}{2})} \int_{-x_d}^{x_d} \frac{1}{\theta(E_\perp)} \left(\exp\left[-\frac{(\theta(E_\perp)-\theta_0)^2}{2(\sigma_y/\vartheta_c)^2}\right] + \exp\left[-\frac{(-\theta(E_\perp)-\theta_0)^2}{2(\sigma_y/\vartheta_c)^2}\right] \right) \quad (7)$$

with $\theta(E_\perp) = \sqrt{\frac{2(E_\perp - U(x))}{E}}$ and the critical channeling angle $\vartheta_c = \sqrt{2U(d_p/2)/E}$.

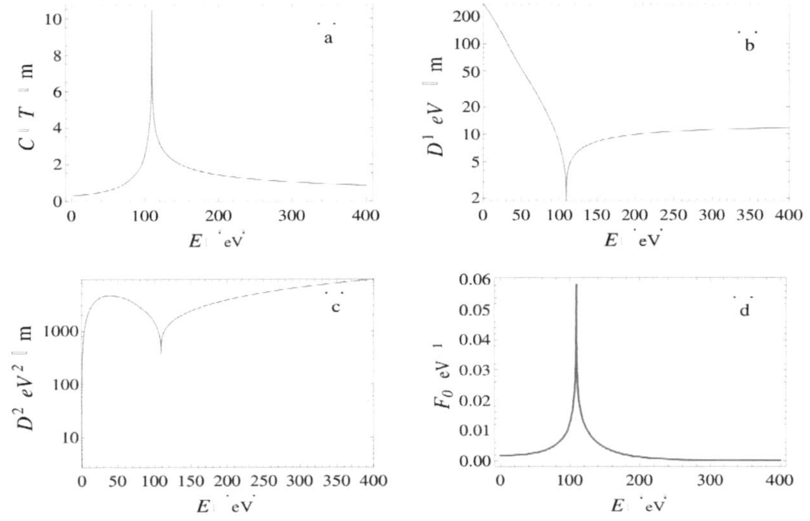

Figure 2. (a) Time parameter cT, (b) drift coefficient D^1, (c) diffusion coefficient D^2, (d) initial probability density distribution F_0 at $z=0$, calculated with standard deviation $\sigma_y=0.5\vartheta_c$ and $\theta_0=0$, all as function of transverse energy$E_\perp$for 2 GeV electrons channeled along the (110) plane of a W single crystal.

3.2. Numerical results

The time parameter cT, drift coefficient $D^1(E_\perp)$, diffusion coefficient $D^2(E_\perp)$ and the initial probability density distribution $F_0(E_\perp)$ are shown in figures 2(a)-(d) as functions of transverse energy $E_\perp$ for electrons of energy2 GeV channeled in the (110) plane of a W single crystal. Equation (3) has been solved numerically within the bounds $0 \leq z \leq 200\mu m$ and $0 \leq E_\perp \leq 10U(\frac{d_p}{2})$ using *Mathematica 7*. The solution is depicted in figure 3.

The fraction $f_{ch}(z) = \int_0^{U(\frac{d_p}{2})} F(z, E_\perp) dE_\perp$of channeled electrons with transverse energies $0 \leq E_\perp \leq U(dp/2)$ is illustrated in figure 4. The dechanneling length is obtained from

$$J(z,E_\perp) = -\frac{\partial}{\partial E_\perp}[D^2(E_\perp)F(z,E_\perp)] + D^1(E_\perp)F(z,E_\perp) = J_{diff} + J_{drift} \tag{8}$$

$$L_d(z) = \frac{f_{ch}(z)}{J_{drift}\left(z, E_\perp = U(\frac{d_p}{2})\right)} \tag{9}$$

where z follows from the condition $f_{ch}(z) = \frac{1}{e} f_{ch}(0)$ [10]. For 2 GeV electrons channeled in the (110) plane of the W crystal, a dechanneling length of 5.56 μm has been obtained for z=15 μm. The dechanneling length as function of the electron energy is shown in figure 5. It scales with the beam energy such as $L_d = 2.78\ \mu m \times E/GeV$.

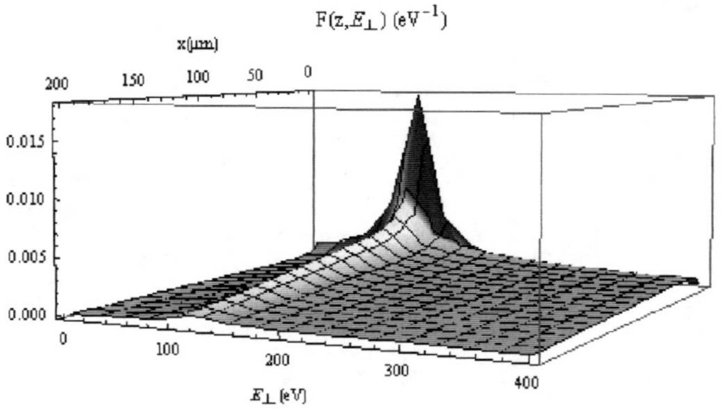

Figure 3. Probability density for 2 GeV electrons channeled along the (110) plane of a W single crystal for a standard deviation of the scattering angle σ_y=0.5ϑ_c and the angle of incidence θ_0=0.

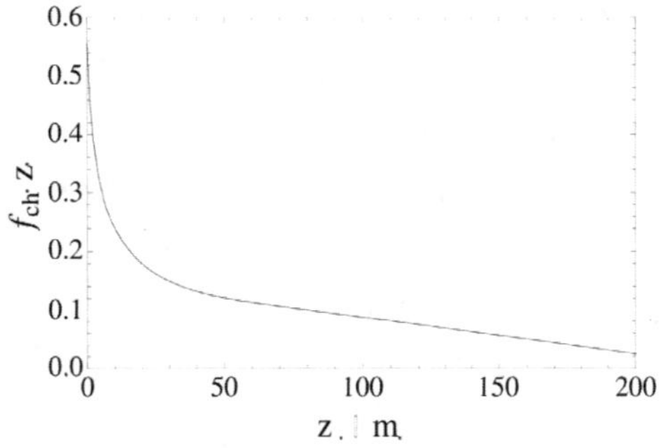

Figure 4. Electron dechanneling function $f_{ch}(z)$ for (110) planar channeling of 2 GeV electrons in a W single crystal.

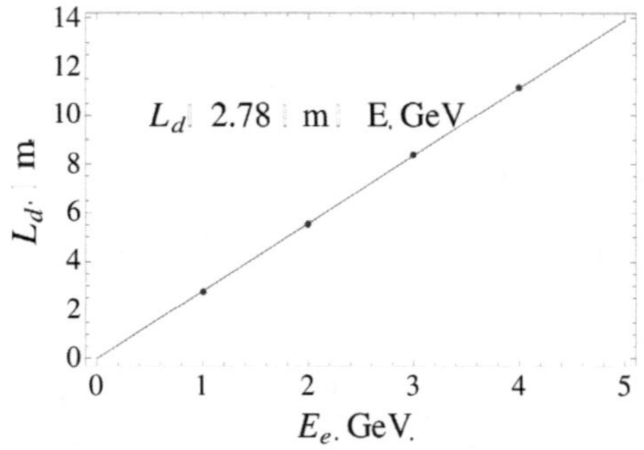

Figure 5. Dechanneling lengths as function of the electron energy for (110) planar channeling of electrons in a W single crystal.

4. Channeling radiation generated in a thick W single crystal

For planar channeling, the total energy radiated into the solid angle 1/γ per unit path length in the crystal reads

$$\frac{dE}{d\hbar\omega\Delta z} = \frac{e^2}{4\pi^2\hbar c^4}\sum_{n=1}^{\infty}\Theta[1-\eta_n]\omega\,\tilde{\omega}^2(\eta_n^2 - \eta_n + \frac{1}{2})\cdot|\dot{x}_{\tilde{\omega}}|^2 . \tag{10}$$

$$\eta_n = \frac{T\omega}{4\pi\gamma^2 n};\ \tilde{\omega} = \frac{2\pi n}{T};\ \dot{x}_{\tilde{\omega}} = \int_0^T \dot{x}e^{i\tilde{\omega}t}dt$$

where $\Theta[1-\eta_n]$ is the theta-function, $\dot{x}_{\tilde{\omega}}$ is the Fourier component of the transverse velocity, and T is the period of electron oscillations. The CR intensity for thick crystals is given by

$$I_{thick}(\hbar\omega, E_\perp) = \frac{1}{L}\int_0^L \frac{dE}{d\hbar\omega\Delta z} F(z, E_\perp)\, dz. \tag{11}$$

These equations are valid for a single particle. Averaging over all possible electron trajectories (or over the complete set of transverse energies $E_\perp$) reflects a uniform distribution overall points of incidence into the crystal. Figure 6 shows the dependence of the CR spectrum on the crystal thickness for 2 GeV electrons channeled in the (110) plane of a W single crystal.

5. Simulation of positron spectra

The scheme of the considered positron source (figure 7) consists of a 2 GeV electron beam, a W crystal radiator being aligned with its (110) plane parallel to the beam and an 8 mm thick amorphous W converter. Taking the calculated CR spectra as input data, positron energy distributions obtained at the converter exit have been simulated by means of the *GEANT4* Monte Carlo package and are shown in figure8.

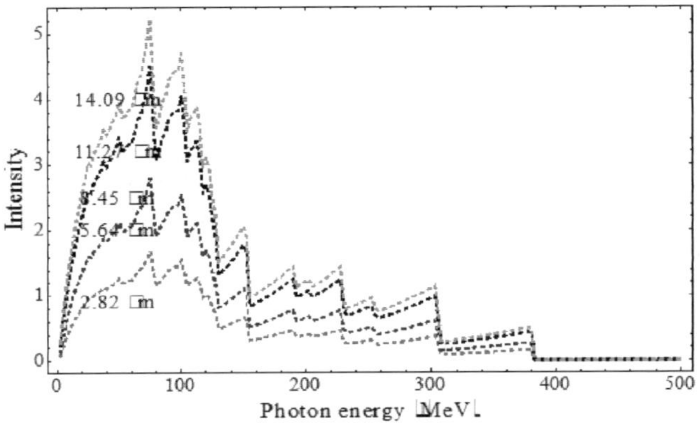

Figure 6. Relative thickness dependence of integrated CR spectra of 2 GeV electrons channeled along the (110) plane of a W single crystal for a standard deviation of the scattering angle $\sigma_y = 100$ μrad and the angle of incidence $\theta_0 = 0$.

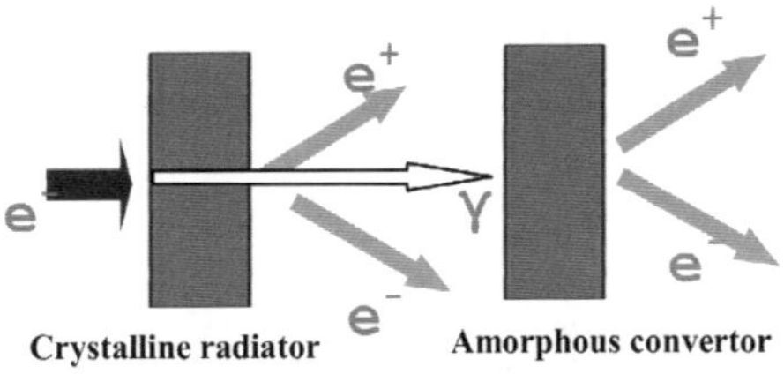

Figure 7. Scheme of a non-conventional positron source consisting of a crystalline radiator for generation of CR combined with an amorphous conversion target for pair production.

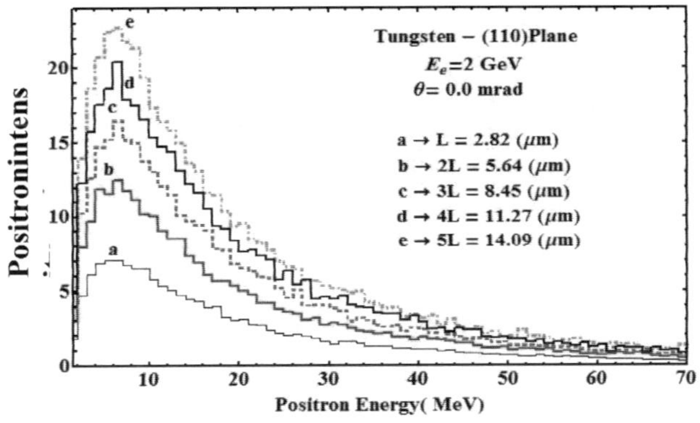

Figure 8. Positron energy distributions simulated for conversion of CR generated by 2 GeV electrons channeled in the (110) plane of a W crystal radiator of different thickness. L is the calculated dechanneling length.

Here it is important to note that we considered only that part of positrons which is produced by the amount of CR generated in the crystalline W radiator ($\theta_0 = 0$). Therefore, the results shown in figure 8 are for now qualitative because for quantitative statements and comparison with experimental data one has to account yet for realistic beam parameters (emittance), precise geometry of the source set up and additional (not small) contributions to the positron spectra originating from ordinary and coherent BS. Nevertheless, from figure 6 one can already conclude that CR leads to a weakening of the positron spectrum compared with BS. Furthermore, it can be seen that due to dechanneling the obtained CR yield does not increase linearly with increasing radiator thickness. Note also that the structures in the CR spectra of figure 6 occurring due to integration over the total solid angle $1/\gamma$ of radiation emission are smeared out in the respective positron spectra due to the partitioning of the photon energy between electron and positron of created e−e+-pairs.

A series of positron-production experiments with axially-oriented single-crystal W targets has been performed at 4 and 8 GeV electron energy at the KEKB facility [5, 6].The thickness of the used W crystals amounted to 2.2 ÷ 14.2 mm. Our calculations showed that typical dechanneling lengths for W crystals are much smaller. Moreover, the dechanneling length decreases with increasing atomic number of crystal material. Therefore, main part of positrons obtained in the KEK experiments, was produced by BS. Theoretically, a diamond radiator would probably be a good alternative.

6. Conclusion

We presented simulations of positron energy spectra obtained by CR induced e−e+-pair production in an amorphous W converter. Planar CR is generated by ultra-relativistic electrons in a separate crystalline W radiator aligned with its crystallographic (110) plane to the direction of the electron beam. We for the first time calculated dechanneling of high-energy electrons planar channeled in a W crystal and deduced the beam-energy scaling of the dechanneling length. Since dechanneling has remarkable consequences on the CR yield, it should be taken into account for optimization of the radiator thickness of a hybrid positron source. For comparison with experimental data, our formalism will be adapted for simulation of realistic non-conventional positron sources where, in a next step, the effect of BS accompanying CR on the produced positron spectra will be investigated.

7. References

[1] Kumakhov M A 1976 *Phys. Lett. A* **57** 17
[2] Chehab R *et al 2002 Phys. Lett. B* **525** 41
[3] Artru X *et al* 2003 *Nucl. Instrum. Methods B* **201** 243
[4] Artru X *et al* 2005 *Nucl. Instrum. Methods B* **240** 762
[5] Suwada T *et al* 2003 *Phys. Rev. E* **67** 16502
[6] Satoh M *et al* 2005 *Nucl. Instrum. Methods B* **22** 73
[7] Baier V N *et al* 1995 *Nucl. Instrum. Methods B* **103** 147
[8] Azadegan B, Mahdipour A, Dabagov S B, Wagner W 2013 *Nucl. Instrum. Methods B* **309** 56
[9] Azadegan B and Mahdipour A 2013 *Nucl. Instrum. Methods B* **316** 105
[10] Backe H, Kunz P, Lauth W and Rueda A 2008 *Nucl. Instrum. Methods B* **266** 3835
[11] Azadegan B, Wagner W and Pawelke J 2006 *Phys. Rev. B* **74** 045209
[12] Azadegan B and Dabagov S B 2011 *Eur. Phys. J. Plus* **126** 58
[13] Azadegan B, Dabagov S B and Wagner W 2011 *Nucl. Instrum. Methods B* **269** 2098
[14] Doyle P A and Turner P S1968 *Acta Crystallogr. A* **24** 390
[15] Kitagava M and Ohtsuki Y H 1973 *Phys. Rev. B* **83** 117

Coherent X-ray Cherenkov radiation produced by microbunched beams

M A Aginian[1], K A Ispirian[1], M K Ispiryan[1], M I Ivanyan[2]
[1]Alikhanian National Scientific Laboratory, Yerevan Physics Institute, Br. Alikhanians 2, Yerevan, 0036, Armenia[2] CANDLE Synchrotron Research Institute, Acharyan 31, 0040 Yerevan, Armenia

E-mail: karo@mail.yerphi.am

Abstract. Analytical and numerical results on the coherent X-ray Cherenkov radiation (CXCR) produced by microbunched beams in the region near the K-, L-edges of materials are obtained. The results show that CXCR can serve as a suitable mechanism for production intense beams of photons in the "water window" region as well as for studying the important microbunching process at FLASH TESLA, LCLS and other FELs.

1. Introduction

At the end of the long undulators of free electron lasers (FELs) [1-4] the electron beams undergo longitudinal periodical density modulation or microbunching (MB) with period equal to the wavelength of the FEL SASE radiation, $\lambda_r = 2\pi c/\omega_r \approx L_{und}/2\gamma^2(1+K^2/2)$, where L_{und} and K are the period and parameter of the undulator, $\gamma = E/mc^2 = [1-(V/c)^2]^{1/2}$ is the electron relatyivistic factor. Using the unique parameters of the SASE beams many important experiments can be carried out, among which it is worthwhile to mention the experiments requiring water window photons with λ = (2.3–4.4) nm having relatively large and short absorption lengths for water and other elements, respectively. Recently an experiment devoted to the first observation of atomic inner-shell laser photon production at 1.46 nm has been carried out at LCLS [5]. The beams of [5] are more stable, reproducible with narrower spectral and angular distributions than the SASE beams, and these properties are necessary for some experiments.

At present the FELs' microbunched beams, the MB of which is spoiled after some distance, are sent to dumps without any applications. Only rarely these microbunched electron beams (MBEB) with energies less than 1 GeV have been used (see [1]) for production of soft coherent transition radiation (CTR) for the measurement of the parameters of MB according to the proposal of [6]. To our knowledge the MB of MBEB with energies higher than 1 GeV has not been studied, though a method based on coherent X-ray transition radiation (CXTR) has been proposed [7] and even some preparatory works [8, 9] have been carried out. Recently, it has been theoretically studied other types of coherent X-ray radiation of MBEB: coherent bremsstrahlung and resonance transition radiation in [10]; coherent X-ray diffraction radiation in [11]; coherent X-ray crystalline undulator radiation in [12]; coherent X-ray channeling radiation in [13]; coherent X-ray parametric radiation in [14, 15]. A short review of theory of these processes is given in [16].

In [17] it has been shown that in narrow regions close to K-, L- edges for some materials the refraction index $n(\omega)$ or the dielectric constant $\varepsilon(\omega)=n^2(\omega)$ are greater than 1, and therefore X-ray Cherenkov radiation (XCR) can be produced by charged particles. XCR has been experimentally studied in [18 – 21]. Reviews on XCR are given in [22, 23].

Taking into account the growing demand for intense beams of water window photons for biology and medicine, and that at present there is a possibility to study coherent X-ray Cherenkov radiation (CXCR) produced by MBEB at DESY FLASH TESLA [2,3] and SLAC LCLS [4] FELs this work is devoted to the theoretical study of CXCR.

2. The Calculation of CXCR

It is usual to present the complex values of $\varepsilon(\omega)$ and $n(\omega)$ in the forms: $\varepsilon(\omega)=n^2(\omega)=\varepsilon'(\omega)+i\varepsilon''(\omega)=1+\chi'(\omega)+i\chi''(\omega)$ and $n(\omega)=1+\delta(\omega)+i\Delta(\omega)$ where χ' and χ'' are the real and imaginary parts of the susceptibility, and absorption length $L_{abs}=1/\mu(\omega)=c/(\omega\chi''(\omega))$ where μ is the linear absorption coefficient. In order to calculate CXCR first it has been calculated χ' and $\chi''$by the methods described in [22]. Then the theory of CXCR produced by MBEB is developed by the method described in [6, 16] using the well-known formula

$$\frac{d^2N_{CXCR}}{d\omega d\theta}\cong N_b^2F(\omega,\theta)\frac{d^2N_{XCR}}{d\omega d\theta}, \tag{1}$$

where N_b is the number of electrons in a macrobunch, the form factor $F(\omega,\theta)$ of the MBEB are given in the works [6, 16]. By inserting the values into the XTR formula (1.59) of [24] one obtains the following formula for the spectral-angular distribution of CXCR per bunch produced at the exit interface of a plate (CXCR produced at the entrance is totally absorbed):

$$\begin{aligned}\frac{d^2N_{CXCR}}{d\hbar\omega d\theta}\cong N_b^2b_1^2\frac{2\theta^3}{137\pi\hbar\omega}\frac{(\chi'^2+\chi''^2)}{(\theta^2+\gamma^{-2})^2[(\theta^2+\gamma^{-2}-\chi')^2+\chi''^2]}\exp\left(-(k\sigma_r\theta)^2\right)\times\\ \times\exp\left(-\left(\frac{\omega}{V}-k_r\right)^2\sigma_z^2\right),\end{aligned} \tag{2}$$

where b_1 is the modulation depth of MB [6, 16], $k=\omega/c$, $k_r=\omega_r/c$, $\sigma_{r,z}$ are the radial and longitudinal standard deviations of the Gaussian macrobunch. It is necessary to choose the parameters of the undulator in such a way that provides the equality of $\hbar\omega_r$ to the energy of the K- or L-edge of the radiator material. Formula (2) is our basic one, and as it has been mentioned above it is necessary to carry out integrations of (2) over $\hbar\omega$ and θ.

The angular distribution of CXCR photon number is obtained integrating (2) over ω in an narrow interval around ω_r following the works [6, 16]:

$$\frac{dN_{CXCR}}{d\theta}\cong N_b^2b_1^2\frac{2\theta^3V}{137\sqrt{\pi}\sigma_z\omega_r}\frac{(\chi'^2+\chi''^2)|_{\omega=\omega_r}}{(\theta^2+\gamma^{-2})^2[(\theta^2+\gamma^{-2}-\chi'|_{\omega=\omega_r})^2+\chi''^2|_{\omega=\omega_r}]}\exp\left(-(k_r\sigma_r\theta)^2\right). \tag{3}$$

The spectral distribution of CXCR obtained by integration of (2) over angles is given by the expression

$$\frac{dN_{CXCR}}{d\omega}\cong N_b^2b_1^2\frac{2(\chi'^2+\chi''^2)}{137\pi\omega}I(\omega,\theta_0)\exp\left(-\left(\frac{\omega}{V}-k_r\right)^2\sigma_z^2\right), \tag{4}$$

where

$$I(\omega,\vartheta_0)=\int_0^{\vartheta_0}\frac{\theta^3}{(\theta^2+\gamma^{-2})^2[(\theta^2+\gamma^{-2}-\chi')^2+\chi''^2]}\exp\left(-(k\sigma_r\theta)^2\right)d\theta \tag{5}$$

and $\theta_{diff} << \theta_0 << 1$, the diffraction angle $\theta_{dif} = c/\omega\sigma_r$.

The calculation of the integral $I(\omega,\theta_0)$ is done by means of its representation in the form of the sum of the four integrals to be calculated analytically:

$$I(\omega,\theta_0)=\left(aI_1+bI_2+cI_3+dI_4\right)/2 \tag{6}$$

where after the notation $\tilde{\theta}=\theta^2$:

$$I_1=\int_0^{\theta_0^2}\frac{e^{-\tilde{\theta}(k\sigma_r)^2}d\tilde{\theta}}{\tilde{\theta}+\gamma^{-2}}=e^{(k\sigma_r/\gamma)^2}\left\{\Gamma\left(0,k^2\sigma_r^2/\gamma^2\right)-\Gamma\left(0,k^2\sigma_r^2\left(\gamma^{-2}+\theta_0^2\right)\right)\right\}, \tag{7}$$

$$I_2=\int_0^{\theta_0^2}\frac{e^{-\tilde{\theta}(k\sigma_r)^2}d\tilde{\theta}}{\left(\tilde{\theta}+\gamma^{-2}\right)^2}=\gamma^2-\frac{e^{(k\sigma_r\theta_0)^2}}{\gamma^{-2}+\theta_0^2}-(k\sigma_r)^2I_1, \tag{8}$$

$$I_{3,4}=\int_0^{\theta_0^2}\frac{e^{-\tilde{\theta}(k\sigma_r)2}d\tilde{\theta}}{\tilde{\theta}+\gamma^{-2}-\chi'\pm j\chi''}= \\ =e^{\left(\gamma^{-2}-\chi'\pm j\chi''\right)(k\sigma_r)^2}\left\{\Gamma\left(0,-(k\sigma_r)^2\left(\gamma^{-2}-\chi'\pm j\chi''\right)\right)-\Gamma\left(0,-(k\sigma_r)^2\left(\gamma^{-2}-\chi'\pm j\chi''+\theta_0^2\right)\right)\right\}, \tag{9}$$

$$a=\frac{\chi'^2+\chi''^2-2\gamma^{-2}\chi'}{\left(\chi'^2+\chi''^2\right)^2},\ b=-\frac{1}{\gamma^2}\frac{1}{\chi'^2+\chi''^2}, \tag{10}$$

$$c,d=\frac{\chi''\mp j\left(\gamma^{-2}-\chi'\right)}{2\left(\chi'\mp j\chi''\right)^2\chi''}, \tag{11}$$

Here $\Gamma(a,z)=\int_z^{\infty}t^{a-1}e^{-t}dt$ is incomplete Gamma function.

For $\theta_0 <<1$ one obtains:

$$I(\omega,\theta_0)=\frac{(\theta_0\gamma)^4}{4q}-\frac{\alpha(\theta_0\gamma)^6}{6q^2}, \tag{12}$$

where

$$\alpha=\gamma^{-4}\left(4+\left(\frac{k\sigma_r}{\gamma}\right)^2\right)-2\gamma^{-2}\chi'\left(3+\left(\frac{k\sigma_r}{\gamma}\right)^2\right)+\left(\chi'^2+\chi''^2\right)\left(2+\left(\frac{k\sigma_r}{\gamma}\right)^2\right), \tag{13}$$

$$q=\left(\gamma^{-2}-\chi'\right)^2+\chi''^2.$$

The total number of CXCR photons obtained by integration of (3) over θ, or of (4) over ω is equal to

$$N_{CXCR}=N_b^2b_1^2\frac{2V(\chi'^2+\chi''^2)|_{\omega=\omega_r}}{137\sigma_z\sqrt{\pi}\omega_r}I(\omega_r,\theta_0). \tag{14}$$

The properties of the angular and spectral distributions as well as of the total number of the CXCR photons will be discussed in the next section after numerical calculations.

3. Numerical Results

For numerical calculations it has been taken for carbon, C, ρ_C =2.2 g/cm^3 and $\hbar\omega_r = 284.2$ eV and for titanium, Ti, ρ_{Ti}= 4.54 g/cm^3 and $\hbar\omega_r = 453.8$ eV with dependences $\chi'(\hbar\omega)$ and $\chi''(\hbar\omega)$ close to [21] and to [20c] for C and Ti, respectively.

The below given numerical results obtained with the help of (3) – (14) are for the FLASH2 with electron energy E = 1.2 GeV [2, 3, 25] and at XFELs SACLA [26] and EuroFEL [27] with E = 8 and 10 GeV, respectively. In all case the parameters of the FELs are tuned to provide the necessary $\hbar\omega_r$. Unfortunately, according to [25] $\hbar\omega_r$= 453.8 eV is out of the achievable microbunching region. In all the cases it is taken $\sigma_z = 15\,\mu$m and $\sigma_r = 13\,\mu$m.

3a. The Angular Distributions of CXCR

Figure 1a and b show the angular distributions of CXCR for C and Ti. It is necessary to note that in figure 1, as well as in figures 2 and 3 the the values on the ordinate axis are divided on γ^4. As it is seen in figure 1

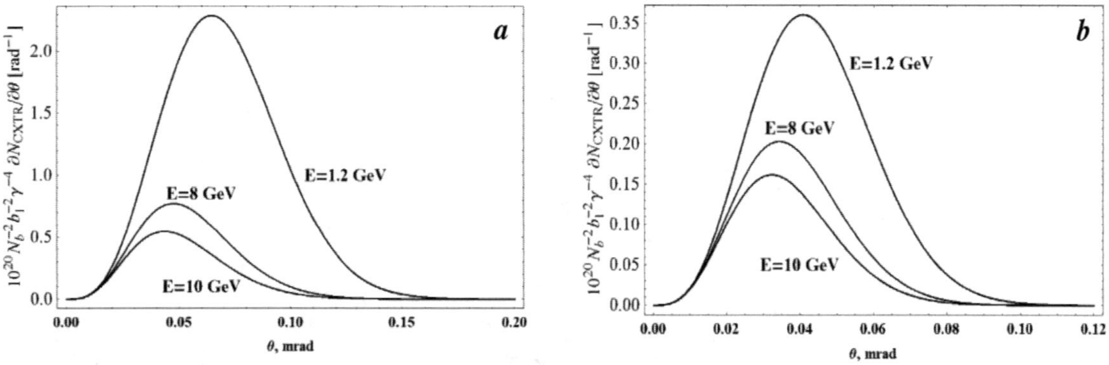

Figure 1. Angular distributions of CXCR photon number for C ***a***) and Ti ***b***).

just as the other types of radiations of microbunched beams [19] the CXCR photons are emitted mainly under angles much smaller than $\theta = 1/\gamma$, the angles of radiation of relativistic particles. The dependence of the angle of maximal intensity on energy is very weak, while the maximal intensity of CXCR increases strongly with the increase of the particle energy and approximately is proportional to $\sim \gamma^4$.

3b. The Spectral Distributions of CXCR

Figures 2a and b show the spectral distributions of CXCR for C and Ti.

As expected the spectral distributions of CXCR are very narrow as the well known spectral distributions of PXR and are concentrated around $\hbar\omega_r$. From figure 2a and b it is seen that the absolute width of CXCR for Ti is broader than that for C. However, all the curves have similar relative widths approximately equal to $\Delta\omega/\omega < 0.00012 = 0.012\%$ (FWHM). The numerical integration of the corresponding curves of figure 1 and figure 2 gives total number of CXCR photons equal to those given in figure 3.

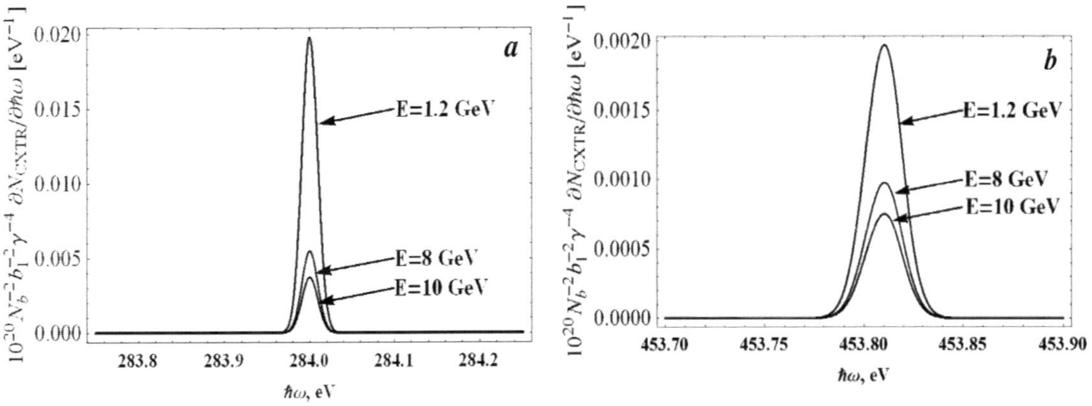

Figure 2. The spectral distributions of CXCR for C ***a***) and for Ti ***b***).

As expected the spectral distributions of CXCR are very narrow as the well known spectral distributions of PXR and are concentrated around $\hbar\omega_r$. From figure 2a and b it is seen that the absolute width of CXCR for Ti is broader than that for C. However, all the curves have similar relative widths approximately equal to $\Delta\omega/\omega < 0.00012 = 0.012\%$ (FWHM). The numerical integration of the corresponding curves of figure 1 and figure 2 gives total number of CXCR photons equal to those given in figure 3.

3c. The Energy Dependence of CXCR

Figures 3a and b show the dependence of the total number of CXCR photons upon electron energy for C and Ti, respectively.

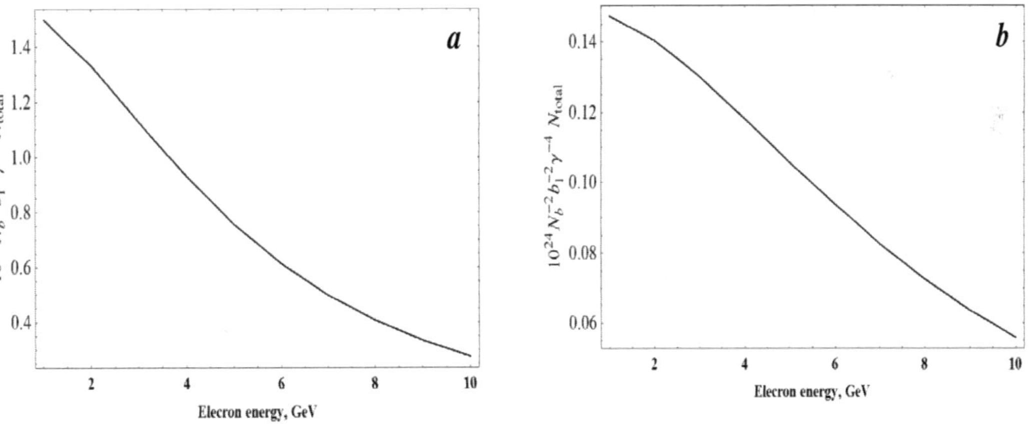

Figure 3. Total number of photons vs electron energies for C ***a)*** and for Ti ***b***).

From figure 3 it is seen that the intensity of CXCR increases with the increase of particle energy slightly weakly than $\sim\gamma^4$. As it follows from figures 2 and 3 theoretically the CXCR yields from carbon radiators is higher than that from titanium radiator. However, it is necessary to take into account the experimental difficulties connected with the C radiators (see [19, 21])

4. Discussion and Conclusions

It is of interest to compare the main characteristics (the total number and width) of CXCR with those of incoherent XCR produced by not microbunched electrons and of SASE produced by the same charge q = 1 nC or $N_b = 6.24x10^9$ electrons per pulse at E= 1.2 GeV. According to [3, 25] for SASE photon beam of FLASH2 it is expected to have $N_{SASE}=10^{11} - 10^{13}$ with $\Delta\omega/\omega_r \approx (0.5-2)\%$ (FWHM). Let us take the mean expected values: $N_{SASE}= 10^{12}$, $\Delta\omega/\omega_r \approx 1\%$ and $\Delta\omega \approx 2.8$ eV. On the other hand, for a carbon, C, radiator according to the Table 2.1 of [21] the expected total number of the incoherent XCR photons is equal to $N_{XCR} = N_b x K_{XCR} = 3x10^6$ with $\Delta\omega/\omega \approx 0.0035 = 0.35\%$ and $\Delta\omega \approx 1$ eV (here K= $4.8x10^{-4}$ is the number of XCR photons per electron). According to the above obtained results for CXCR one expects N_{CXCR} (b_1= 1) = $1.66x10^9$, $\Delta\omega/\omega_r = 0.00012 = 0.012\%$ and $\Delta\omega \approx 0.033$ eV. From these values it follows that assuming b_1= 1, the CXCR beam will have ~ 600 times less intensity and 80 times narrower width than the SASE beam as well as 550 times higher intensity, and 30 times narrower width than the XCR. For real value of b_1<1 the CXCR intensity will be reduced proportional to b_1^2, however, just as in the case of [5] the CXCR width will remain by the same times narrower than SASE and XCR, which is very important for many experiments.

Taking into account these advantages and disadvantages it is worthy to use the microbunched beam after the FLASH2 undulator for production CXCR beam. The situation becomes much better for higher energy electrons and Ti radiator, since the CXCR intensity is approximately proportional to γ^4. On the other hand, in our knowledge, the real value of b_1 has not been measured, and with the help of CXCR it can be easily measured, if by another simple method, say measuring the current, one knows the value of N_b. Besides this application as it has been shown above the second application of CXCR is the production of additional monochromatic photon beams with parameters comparable with those of PXR. Let us also note that that just as in the case of [5] CXCR photon beams have stable $\hbar\omega_r$.

References

[1] Tremaine A *et al* 1998 *Phys. Rev. Lett.* **81** 5816
[2] Ayvazyan V *et al* 2006 *Eur. Phys. J.* D **37** 217
[3] Honkavaara K *et al* 2012 *Proc. of IPAC2012*, **TUPP052**
[4] Emma P *et al* 2009 *Proc. of PAC09*, 397*;* 2010 *Nature Photonics* **4** 641
[5] Rohringer N *et al* 2012 *Nature* **481** 488
[6] Rosenzweig J, Travish G, Tremaine A 1995 *Nucl. Instr. and Meth.* A **365** 255
[7] Gazazian E D, Ispirian K A, Ispirian R K, Ivanian M I 1999 *Pisma Zh. v Eksp. Teor. Fiz.* **70** 664; 2001 *Nucl. Instr. and Meth.* B **173** 160
[8] Lumpkin A H, Fawley W M, Rule D W 2004 *Proc. FEL* 515
[9] Lumpkin A H *et al* 2006 *SLAC-PUB-12451*
[10] Ispirian K A 2011 *Proc. of the Intern. Conf. on Electron, Positron, Neutron and X-Ray Scattering under External Influences* (*Yerevan-Megri, Armenia*) p. 197
[11] Aginian M A, Ispirian K A, Ispiryan M K 2012 *Proc. of NAS of Armenia, Physics* **47** 83
[12] L.A. Gevorgian L A, Ispirian K A, Shamamian A H 2013 *Nucl. Instr. and Meth.* B **309** 63
[13] Aginian M A, Gevorgian L A, Ispirian K A 2012 *Proc. of the Intern. Conf. Channeling* (Alghero, Italy)
[14] Artru X, Ispirian K A 2002 Preprint ArXiv ph/0208018
[15] Aginian M A, Gevorgian L A, Ispirian K A 2012 *Proc. of the Intern. Conf. Channeling* (Alghero, Italy) p.194
[16] Ispirian K A 2013 *Nucl. Instr. and Meth.* B **309** 4
[17] Bazilev V A *et al* 1976 *Pisma c Zh. Eksp. Teor. Fiz.* **24** 406
[18] Bazilev V A *et al* 1981 *Pisma c Zh. Eksp. Teor. Fiz.* **34** 103
[19] Bazilev V A *et al* 1981 *Zh. Eksp. Teor. Fiz.* **81** 1664
[20] Knulst W *et al* a) 2001 *Appl. Phys. Lett.* **79** 2999; b) 2003 *Appl. Phys. Lett.* **83** 4050; c) 2004 *Proc. of SPIE* **5196** 393
[21] Knults W 2004 *Dissertation* Eindhoven University of Technology
[22] Bazilev V A, Zhevago N K 1987 *Izluchenie Bistrikh Chastits v Veshchestve i vo Vneshnikh Polyakh* (Moscow, Nauka)

[23] Ispirian K A 2005 *Proc. of NATO ARW Adv.photon Sources and Apps, NorHamberd, Armenia, 29 August-3 Sep, 2004, NATO Sc. Ser. II, Math. Phys Chem.* **199** 217
[24] Garibian G M, Yan Shi 1983 *Rentgenovskoe Perekhodnoe Izluchenie* Publishing House of Academy of Sciences of Armenia, Yerevan, Armenia
[25] Vogt M *et al* 2013 *Proc. IPAC* TUPEAO04
[26] Ishikawa T *et al* 2012 *Nature Photonics* **6** 540
[27] *TESLA Technical Design Report* 2001 DESY 011

Undulator-Based and Crystal-Based Gamma Radiation Sources for Positron Generation

A Potylitsyn[a,b]
[a] National Research Tomsk Polytechnic University, 634050, Tomsk, Russia
[b] National Research Nuclear University "MEPhI", 115409, Moscow, Russia

E-mail: potylitsyn@tpu.ru

Abstract. Oriented Si or diamond crystals with thicknesses about 10-20 mm can be considered as an effective "solid-state undulator" if electron energy will be higher than 10 GeV. The scheme where such a crystal (photon radiator) placed upstream of the amorphous positron convertor with thickness less than 1 radiation length (hybrid scheme) may provide the intensity of produced and accelerated positrons comparable with an intensity of the initial electron beam. In this case the deposited energy in both parts of the hybrid scheme positron source may be much less than the damage threshold.

1. Positron sources for future colliders, such as, for instance, the International Linear Collider (ILC) [1], have to provide a high-energy positron beam intensity which is at least, equal to an intensity of the counter propagated electron beam. Currently there are two approaches to construct a positron source with the required parameters. The first one is the conventional source based on an electromagnetic cascade-shower process whereby a heavy-metal target is irradiated by high energy electrons from the linear accelerator [2]. The positron yield is naturally proportional to the incident electron intensity, but when the intensity exceeds a certain limit, the target may be destroyed due to the excessive heat load. The second approach is rather new and based on a long undulator (~100 m) where an electron beam with an energy of $E_0 > 100$ GeV is generating an undulator radiation (UR) beam with mean photon energy $\bar{\omega} \sim 10$ MeV. For such a long undulator the photon flux is enough to create a few positrons per each initial electron in a thin amorphous target ($\sim 0.4 X_0$, X_0 is radiation length) [3, 4]. In this case the heat power deposited in the target and, consequently, the temperature stress remains at an acceptable level.

An oriented crystal may be considered as some kind of a solid-state undulator with extremely low period. In principle, the appropriate crystal may be used as a "photon radiator" instead of an undulator with huge length [5]. The oriented crystal target (photon radiator) can be separated from an amorphous positron convertor (so-called hybrid scheme). Experiments with electron energy $E_0 \sim 1$ GeV [6, 7] demonstrated the possibility to obtain intense radiation in oriented crystal targets with thickness $t \sim (0.1 \div 0.2) X_0 >> L_d$ (L_d stands for the dechanneling length [8]) to produce photon beam with spectrum enriched by "soft" photons.

In the work [7] authors demonstrated an applicability of a hybrid scheme using 4 mm tungstem crystal and 4mm amorphous converts for electrons with $E_0 \leq 10$ GeV. Authors of the papers [9] showed that, electrons with an energy $E_0 \sim 6$ to 10 GeV interacting with oriented thick tungsten crystals may produce positrons with characteristics appropriated for acceptance in acceleration.

2. Let's compare the main characteristics of emitted spectra for ordinary bremsstrahlung (BS), coherent bremsstrahlung (CBS), and channeling radiation (ChR). Radiation losses for BS may be estimated as

$$\Delta E_{BS} \approx \frac{t}{X_0} E_0, \; t << X_0. \tag{1}$$

The mean photon energy in the BS spectrum is:

$$\bar{\omega}_{BS} \approx \frac{\Delta E_{BS}}{\bar{N}_{BS}}, \tag{2}$$

where $\bar{N}_{BS}$ is the mean number of photons emitted by an electron [10]:

$$\bar{N}_{BS} \approx \frac{t}{X_0} \int_{\gamma\omega_p}^{E_0} \frac{d\omega}{\omega} = \frac{t}{X_0} \ln \frac{mc^2}{\omega_p} \sim 10 \frac{t}{X_0}. \tag{3}$$

Here γ is the Lorentz-factor, ω_p is the plasmon energy of target material ($\omega_p \approx 30$ eV for Al and Si). From (1), (2), (3) one may obtain

$$\bar{\omega}_{BS} \approx 0.1 E_0 . \tag{4}$$

The mean energy for ChR process may be estimated from the following expression (see [8]):

$$u = \frac{\bar{\omega}_{ChR}}{E_0 - \bar{\omega}_{ChR}} \approx 2\gamma^{1/2} \sqrt{\frac{V_0}{mc^2} \frac{\lambda_e}{a_s}} \Big/ 1 + \frac{1}{2}\gamma \frac{V_0}{mc^2}, \tag{5}$$

where V_0 is the axis or plane potential, a_s is a screening radius, $mc^2 = 0.511$ MeV, $\lambda_e = 3.86 \cdot 10^{13}$ m. For the <111> axis of Si target and $E_0 \sim 1$ GeV $\bar{\omega}_{ChR} \sim 15 \text{ MeV} < \bar{\omega}_{BS}$. For the coherent bremsstrahlung the same value is:

$$u = \frac{\bar{\omega}_{CBS}}{E_0 - \bar{\omega}_{CBS}} \sim \frac{\pi\gamma\theta}{a}, \tag{6}$$

where θ is the orientation angle, a is the interplanar distance. For the axial orientation of a thick monocrystalline target a rough estimation of the angle θ can be obtained as:

$$\theta \sim \bar{\theta}_{ms} = \frac{21}{E_0} \sqrt{\frac{1}{2} \frac{t}{X_0}}, \quad E_0 \text{ in MeV}.$$

For instance, for Si, $t = 10$ mm and $E_0 \sim 1$ GeV $\bar{\omega}_{CBS} \sim 0.5 E_0 \sim 50$ MeV, and

$$\bar{\omega}_{ChR} < \bar{\omega}_{CBS} < \bar{\omega}_{BS} . \tag{7}$$

The radiation losses for an axial orientation of the crystal can be considered as consisting of two parts

$$\Delta E = \Delta E_{BS} + \Delta E_{cr},$$

where the first term is connected with a radiation from an amorphous target with the same thickness, but the second one is determined by coherent radiation processes (channeling radiation together with coherent bremsstrahlung). Ratio between these parts was measured for electron energy $E_0 = 900$ MeV [11]. Ratio $R = \Delta E_{CR} / \Delta E_{BS}$ was measured for electron energy $E_0 = 900$ MeV for a thick diamond target ($t = 10$ mm, <100>, $R_{Di} = 2.5$) and for a thick silicon target ($t = 10$ mm, <111>, $R = 1.8$) [11]. The thicknesses chosen for these crystal targets are close enough to the optimal thickness as estimated and proposed in the work [12]. Having in mind the relation (7) one may obtain:

$$N_{cr} \approx \frac{\Delta E_{cr}}{\bar{\omega}_{cr}} >> N_{BS} \tag{8}$$

because of $\Delta E_{cr} > \Delta E_{BS}$, $\bar{\omega}_{cr} < \bar{\omega}_{BS}$.

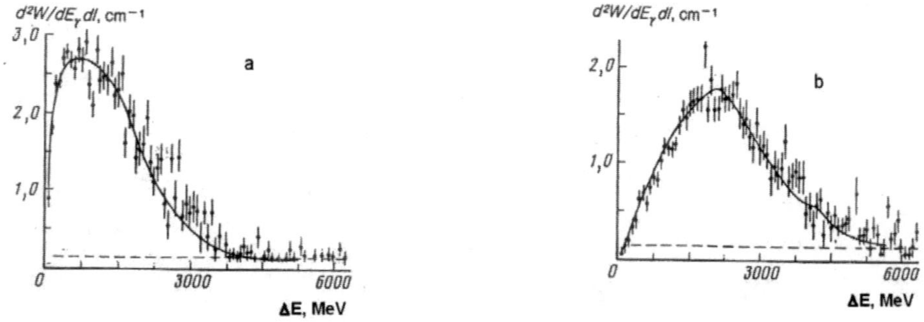

Figure 1. Radiation losses spectra for channeled electrons with $E_0 = 10$ GeV in Si crystal with thickness 0.8 mm (**a**) and 3.0 mm (**b**). Solid lines are the simulation results [13].

Authors of the experiment [13] that was carried out with electron energy $E_0 = 10$ GeV had measured the energy losses for <111> axial orientation of Si crystals with thickness $t = 0.8$ and 3.0 mm (figure 1). From Monte-Carlo simulation they have obtained the mean photon number (photon multiplicity)

$$\bar{N} = \bar{N}_{BS} + \bar{N}_{cr}$$

for each case:

$$\bar{N}(0.8\text{mm}) = 1.8\text{ph}/e^-, \quad \bar{N}(3.0\text{mm}) = 5.4\text{ph}/e^-.$$

The measurements and simulations were carried out for the threshold photon energy $\omega_t = 20$ MeV. The contribution to the multiplicity from bremsstrahlung is negligible (see eq. (4)):

$$N_{BS}(0.8) \approx 0.9 \cdot 10^{-3}\text{ ph}/e^-, \quad N_{BS}(3.0) \approx 3.2 \cdot 10^{-3}\text{ ph}/e^-.$$

In the paper [14] an approach was proposed to estimate the photon multiplicity from the measured distribution of the energy losses (energy straggling). Neglecting the bremsstrahlung losses, the photon multiplicity and the mean photon energy may be estimated in the following manner:

$$\bar{N}_{cr} \approx \frac{\bar{Q}^2}{\sigma^2}, \bar{\omega}_{cr} \sim \frac{\Delta E_{cr}}{\bar{N}_{cr}}, \tag{9}$$

where $\bar{Q}$ is the mean value of the radiation losses calculated from energy losses distribution, σ is the distribution variance, ΔE_{cr} is the total energy losses for axial orientation. Fitting the experimental results [13] by a smooth curve one may obtain for $t = 3$ mm (see figure 1).

$\bar{Q} \approx 2.3$ GeV, $\sigma \approx 1.1$ GeV and, consequently:

$\bar{N}_{cr} \approx 4\text{ph}/e^-$ and $\bar{\omega}_{cr} = 0.2\text{GeV} < \bar{\omega}_{BS} = 1$ GeV.

The multiplicity estimated from this model agrees reasonably with the Monte-Carlo simulation. The model of photon multiplicity developed in [8] gives for the considered case ($E_0 = 10$ GeV, $t = 3$ mm, Si <111>) (see figure 17.3 there):

$$\bar{N}_{tot} \approx 24\,ph/e^-.$$

It should be noted that the estimation of photon multiplicity accordingly to the model [8] gives much higher a photon number emitted by each electron due to low threshold energy.

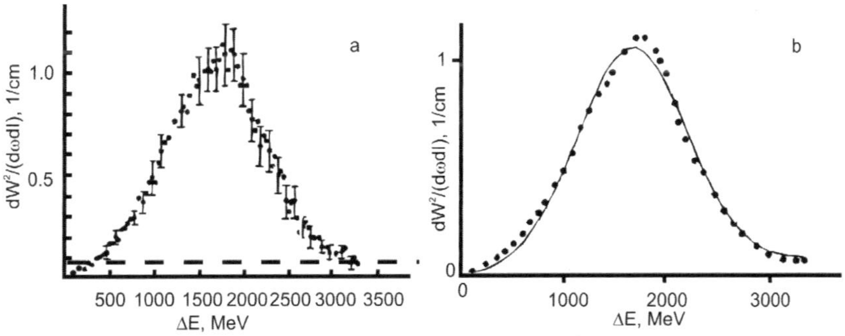

Figure 2. a) Radiation losses spectra for channeled electrons with $E_0 = 4.5$ GeV in diamond crystal with thickness 10 mm [15]; b) Fit by a gaussian with $\bar{Q} = 1660$ MeV, $\sigma = 536$ MeV, $\Delta E = 1300$ MeV.

Using the same approach let's estimate the photon multiplicity in the experiment [15]. This experiment was performed at the electron beam with energy $E_0 = 4.5$ GeV for axial <111> orientation of the natural diamond target with thickness $t = 10$ mm (see figure 2a). Figure 2b shows the fit by a Gaussian ($\bar{Q} = 1660$ MeV, $\sigma = 536$ MeV, $\Delta E_{cr} = 1300$ MeV), which gives the following result:

$$\bar{N}_{cr} \approx 10\ ph/e^-,\ \omega_{cr} \approx 130\,\text{MeV}. \tag{10}$$

It allows to think that the photon multiplicity may reach the value $\bar{N}_{cr} \sim 20$ if one uses an electron beam with an energy $E_0 \sim 10$ GeV and a diamond target with thickness $t \approx 20$ mm. Data from Dubna's experiment [16] show that the radiation losses of 10 GeV electron moving along <111> axis in the 10-mm silicon (~ 0.11 rad. length) reach 30 %. According to [8] losses from a diamond of the same radiation thickness (~ 13 mm) increases in $a_{Si}/a_{di} = 1.5$ times. Due to this reason the axis potential (and, subsequently, radiation intensity) is much higher for a diamond crystal. The usage of diamond with thickness $t \sim 20$ mm leads to radiation increasing also. As a result one may expect that each electron will emit about 50% of the initial energy ($\Delta E \approx 5$ GeV).

The well known dependence of the multiple scattering angle on the energy and on the target thickness makes possible to estimate the mean photon energy for the considered case using formula (6):

$$\bar{\omega}_{cr}(E_2, t_2) \approx \sqrt{\frac{t_2}{t_1}}\,\bar{\omega}_{cr}(E_1, t_1). \tag{11}$$

For the considered case $E_2 = 10$ GeV, $t_2 = 20$ mm, $E_1 = 4.5$ GeV, $t_1 = 10$ mm, $\bar{\omega}_{cr}(E_1, t_1) \approx 130$ MeV one can obtain $\bar{\omega}_{cr}(E_2, t_2) \approx 180$ MeV and $\bar{N}_{cr} \sim 14\ ph/e^-$. This photon flux may be used to produce positrons in a thin amorphous target in a complete analogy with an undulator radiation beam.

3. The positron yield for an initial electron with energy E_0 interacting with a rather thin amorphous target ($t \le 1$ rad. length) can be calculated using a simple analytical model [17]. In this model, an electron crossing the layer with thickness $t/3$ emits bremsstrahlung photon which creates e^+e^- pair at the next layer with thickness $t/3$ and, at last, the created positron moves through the remaining layer ($t/3$) changing its outgoing angle due to multiple scattering. The positron spectrum (neglecting ionization and radiation losses) can be written as

$$\frac{dN_+}{d\varepsilon_+}=0.07\left(\frac{t}{X_0}\right)^2\left[\ln\left(\frac{1}{\lambda}\right)-0.5\right]\left[\frac{1}{\varepsilon_+}-\frac{1}{E_0}\right],\left(\frac{1}{\text{MeV}}\right) \quad , \tag{12}$$

where ε_+ is the positron energy. This formula is valid for the case $E_0>\frac{2}{\lambda}mc^2, \varepsilon_+>\frac{2mc^2}{\lambda}$, $\lambda=\frac{z^{1/3}}{111}$ is the screening parameter.

Accuracy of the expression (12) can be estimated upon comparison with the Monte Carlo simulations [17] (see figure 3).

This model can be developed for an initial photon beam. For the sake of simplicity let us consider a "flat" photon spectrum with a photon multiplicity N_{ph} ("triangle" intensity spectrum closed to UR and CBS ones):

$$\frac{dN_{ph}}{d\omega}=\begin{cases}\dfrac{N_{ph}}{\omega_{max}}, \ \omega\le\omega_{max}, \\ 0, \ \omega>\omega_{max}.\end{cases} \tag{13}$$

Radiation losses may be calculated easily from (13):

$$\Delta E=\int_0^{\omega_{max}}\omega\frac{dN_{ph}}{d\omega}d\omega=N_{ph}\frac{\omega_{max}}{2}=N_{ph}\bar{\omega}, \quad \omega_{max}=2\bar{\omega}.$$

The positron spectrum produced by photons with the energy ω in the converter with thickness t may be approximated by the expression [17]:

$$\frac{dN_+(\omega)}{d\varepsilon_+}=0.14\frac{t}{L_{rad}}\frac{\ln\frac{1}{\lambda}-0.5}{\omega}, \quad \omega>\frac{2}{\lambda}mc^2. \tag{14}$$

After convolution of this spectrum with photon one (13) it is possible to obtain the analytical expression for the output positron spectrum:

$$\frac{dN_+}{d\varepsilon_+}=0.14\frac{t}{X_0}\frac{N_{ph}}{\omega_{max}}\ln\left(\frac{\omega_{max}}{\varepsilon_+}\right)\left[\frac{1}{2}\ln\left(\frac{\varepsilon_+\omega_{max}}{m^2c^4}\right)-1.2\right]. \tag{15}$$

Flottmann's calculations [3] for UR with $\omega_{max}=21$ MeV, $N_{ph}=2\text{ph}/e^-$ give the result $\frac{dN_+}{d\varepsilon_+}=0.020$ for $\varepsilon_+=10$ MeV and $t=1$ rad. length. Using (15) one obtain the value $\frac{dN_+}{d\varepsilon_+}=0.021$ for the same parameters.

Figure 4 shows spectra of positrons generated in the 1 rad. length amorphous converter by a photon flux with $N_{ph}=20$ created by an initial electron with energy $E_0=10$ GeV in a diamond target (upper curve) and in a silicon target (lower curve). The thicknesses of the targets was 20 mm for both cases. The positron yield for an energy interval 5 MeV $\le\varepsilon_+\le 25$ MeV reaches $\Delta N_+\sim 3.1$ per each initial electron integrating with diamond crystal. When taking into account the positron beam divergence one may expect that about 30 % of the yield can be accepted for acceleration.

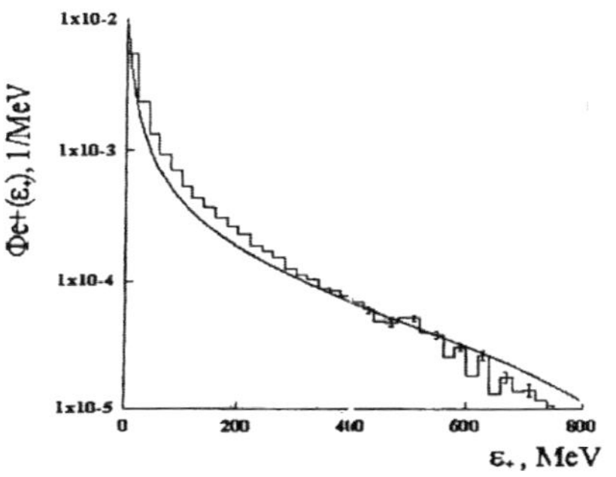

Figure 3. Comparison of positron spectra from Monte Carlo simulation (histogram) and analytical expression (13). Initial electron energy $E_0 = 1$ GeV, converter thickness $t = 0.5$ rad. length.

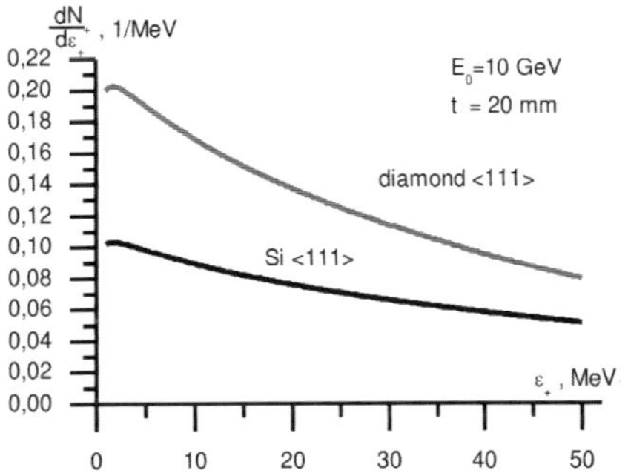

Figure 4. Positron spectra initiated by 10 GeV electrons in a hybrid target (20 mm thickness crystal and 1 rad. length amorphous target).

In summary, it may be noted the considered scheme may provide an efficiency for the positron source as high as 1 e^+/e^- at least. In the work [18] authors considered a conventional positron source for ILC based on a long undulator. Author of the work [19] proposed to design a positron source for the JLAB facility and showed that utility of the 2 mm (0.6 X_0) tungsten target can provide a continuous positron current up to 3μA generated by the 10 mA electron beam with an energy 126 MeV. The hybrid scheme looks as an attractive variant for such a positron source.

References

[1] The ILC: Technical Design Report, http://www.linear collider.org.
[2] Rinolfi L, Kamitani T *The positron production at CLIC* 2002 CLIC Note 465, CERN
[3] Flottmann K Investigations toward the development of polarized and unpolarized high intensity positron sources for linear colliders 1993 Preprint DESY 93-161
[4] Wenming Liu and Wei Gai Positron Source for the International Linear Collider 2011, in: Beam Dynamics Newsletter (54)
[5] Artru X, Baier V V, Chehab R et al. 1994 *Nucl. Instr. and Methods A* **344** 443
[6] Anami S, Andreyashkin M Yu, Enomoto A et al. 2001 *Nucl. Instr. and Methods B* **183** 459
[7] Artru X, Baier V, Beloborodov K et al. 2003 *Nucl. Instr. and Methods B* **201** 243
[8] Baier V N, Katkov V M, Strakhovenko V M 1998 *Electromagnetic Processes at High Energies in Oriented Single Crystals.* World Scientific
[9] Chehab R, Cizeron R, Sylvia C et al. 2002 *Phys. Letters B* **525** 41
[10] Potylitsyn A P Electromagnetic Radiation from Relativistic Particles 2010 Springer Tracts in Modern Physics **239**
[11] Amosov K Yu, Vnukov I E, Kalinin B N et al. 1991 *Sov. Phys. Journal* **34** 472
[12] Baier V N, Katkov V M, Strakhovenko V M 1986 *Phys. Stat. Sol. (b)* **133** 583
[13] Bavizhev M D, Nilsen Yu, Yur'ev B A 1989 *Sov. Phys. JETP* **68** 803
[14] Kolchuzhkin A, Potylitsyn A 2001 *Nucl. Instr. and Methods B* **173** 126
[15] Avakian R, Avetisyan A E, Asatryan R A et al. 1988 *Sov. Tech. Phys. Letters* **14** 395
[16] Tsyganov E I 1989 *Fizika elementarnyx chastits i atomnogo yadra* **20** 5 (in Russian)
[17] Potylitsyn A P 1997 *Nucl. Instrum. and Methods A* **398** 395
[18] Omori T, Takahashi T, Riemann S et al. arXiv: 1110.1450
[19] Golge S http://www.jlab.org/accel/inj_group/docs/2010/peppo/slides15.pdf

Crystal devices for beam steering in the IHEP accelerator

Yu A Chesnokov, A G Afonin, V T Baranov, G I Britvich, P N Chirkov, V A Maisheev, V I Terekhov, I A Yazynin

IHEP, 142281, Protvino, Moscow region, Russia

E-mail: chesnokov@ihep.ru

Abstract. Different crystal devices are described, which provide an extraction and splitting of beams for a long period of time at the U-70 accelerator of IHEP. The modes of channeling and volume reflections in the bent crystals are used for these tasks. In regular accelerator runs crystals produce the particle beams in a wide range of intensity, from 10^6 up to 10^{12} particles in a cycle. Novel crystal techniques suitable for charged particle beams deflection and focus as well as photon generation are presented also.

1. Introduction

Ideas of use the particle channeling in bent crystals for steer the beams have been checked up and advanced in many experiments (see [1, 2, 3] and references herein). This method has found the widest practical application in U-70 accelerator of IHEP, where crystals are used in regular runs for beam extraction and forming. The questions of crystal channeling physics were described in [4, 5]. Here we concentrate on aspects of different bent crystal devices peculiarities.

2. Various options of beam deflection by bent crystals

Today, about 10 locations on the IHEP 70-GeV main ring of the accelerator facility are equipped by crystal extraction systems, serving mostly for routine applications rather than for research and allowing a simultaneous run of several particle physics experiments, thus significantly enriching the IHEP physics program (see figure 1).

In a figure 2 few developed crystal devices for beam steering are presented:

1 – The first long bent crystal for beam splitting in U-70 beam lines (1987) [6].
2 – The working model of the bent crystal for beam extraction in beam lines N2 and N4 [7].
3 – A thin crystal of a strip-type for a highly-efficient crystal extraction of a beam in beam line N8 [8, 9].
4 – The device for beam collimation, based on multiple reflection on a chain of the bent crystals [10, 11].
5 – The device for focusing of particle trajectories in ultrahigh energy region [12].

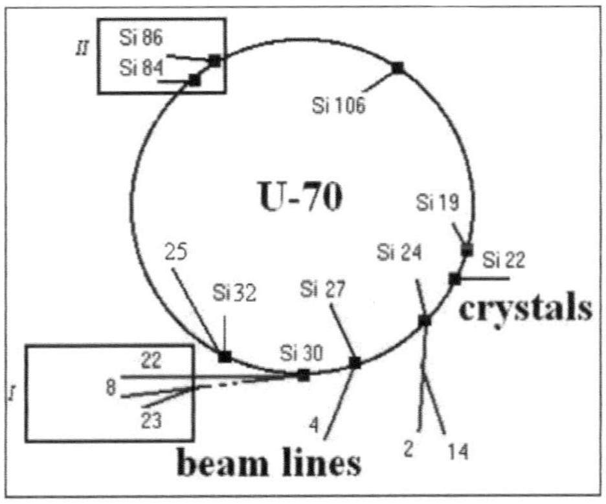

Figure 1. Crystal location at U-70 ring.

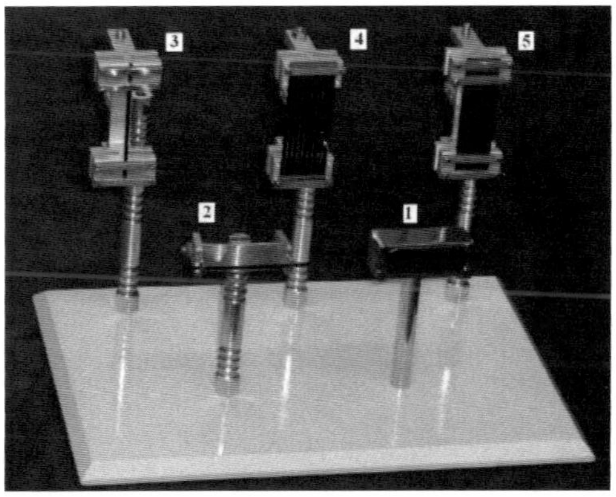

Figure 2. The arrangement of crystals at the stand.

On figures 3÷6 principles of work and the characteristic of these devices are briefly described.

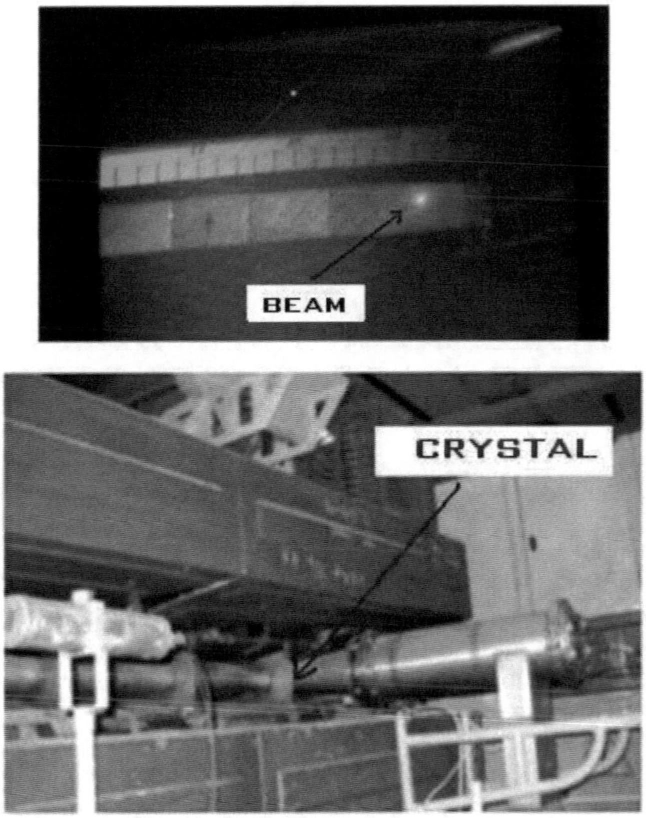

Figure 3. The extraction of a beam from U-70 ring by crystals 1 and 2 in a direction of beam lines N2 and N4.

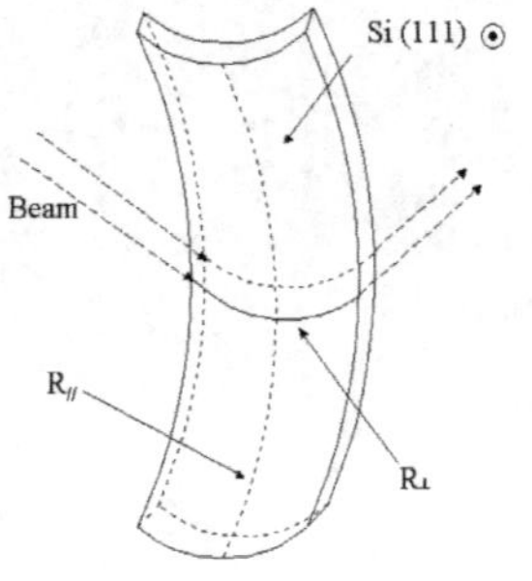

Figure 4. The Principle of a bend of a strip-type crystal (the device 3).

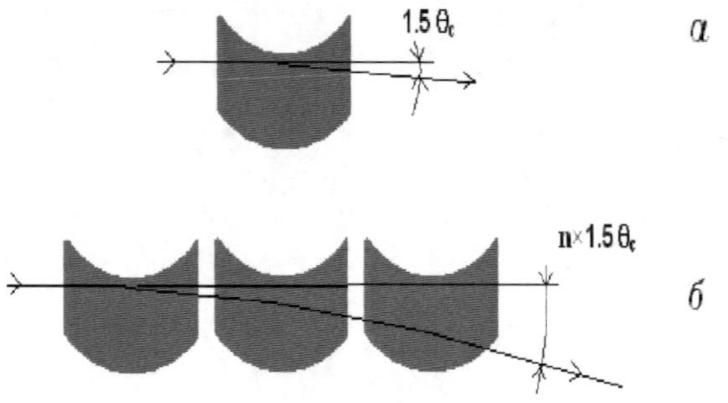

Figure 5. Amplification of deflection angle by multiple reflection of the particles on a chain of the bent crystals (the device 4).

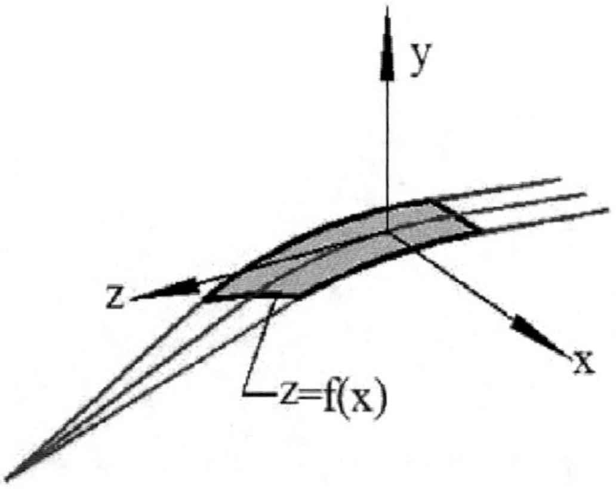

Figure 6. The principle of beam focusing by the device 5.

Employing the devices 1 and 2, the first extracted charged particle beams were obtained with a moderate intensity of up to 10^7 particles per pulse at extraction efficiencies of $10^{-4}\div10^{-3}$. In 1989, at the IHEP U-70 accelerator, a proton beam with energies of 50 and 70 GeV was extracted into one of the existing beam lines. To direct the extracted beam into this beam line, it was necessary to bent a (111) silicon crystal 65×15×0.6 mm in size to a large angle, about 90 mrad. As can see on figure 2, the crystals 1 and 2 were bend by metallic holders in longitudinal direction [7]. The moderate extraction efficiency in experiments with long crystals was caused by the fact that particles are captured into the channeling regime mostly during the first passage through the crystal.

It was predicted [13] that efficiency of crystal channeling extraction can be boosted to much higher values by multiple particle encounters with a shorter crystal (about 1mm in length) installed in a circulating beam (see figure 7). To clarify this mechanism a new experiment was started at IHEP at the end of 1997, with intention to test very short crystals and achieve very high efficiencies of extraction [8, 9]. Radical increase of extraction efficiency up to 85% was achieved with the help of the device 3, figure 2 known as "strip crystal". This method of bending is based on the anisotropic properties of crystal lattices. From the theory of elasticity, it is known that bending a crystal plate in a longitudinal direction gives rise to deformations that assume a saddle- or barrel-like shape, depending on the particular anisotropic properties and orientation of the crystal (e.g., see [4, p. 85]). In silicon crystals, the largest lateral deformations are produced for the (111) orientation and take on a saddle-like shape (see figure 4). In this case, bending the crystal in height through an angle of 100 *mrad* ensures beam deflection through an angle of ~1 mrad in a transverse direction, which is sufficient for the beam extraction. This principle of crystal bend appeared fruitful for creation of multi-crystals (device 4, figure 2) and focusing crystals (device 5, figure 2). This design was modified at further in precision experiments with short crystals in CERN.

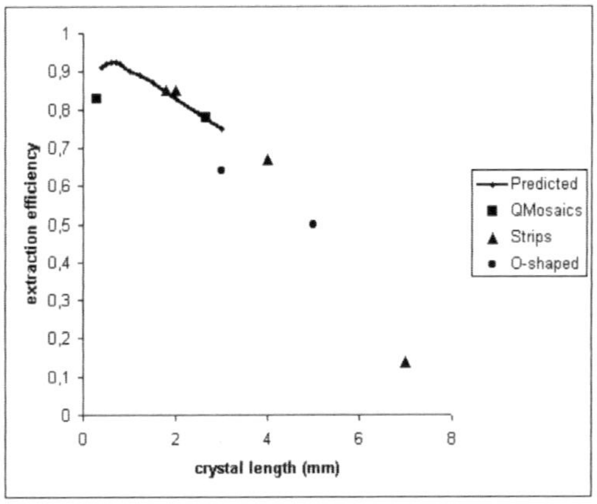

Figure 7. Extraction efficiency dependence versus crystal length.

After the success [9] research collaboration in CERN with participation of IHEP experts starts study of high energy particle beam bending by short crystals. The main goal of these researches will consist in study of crystal collimation of the LHC beam for increasing of its luminosity [14]. The big attention is given research of beam bending with the help of reflection in short crystals [15, 16]. Volume reflection is caused by interaction of incident particle with potential of the bent atomic lattice and occurs on small length in the vicinity of a tangent to the bent atomic plane, leading to deflection of a particle aside, opposite to a bend (figure 8). The probability of reflection is very high and comes nearer to unit at the energies bigger 100 GeV. The phenomenon of reflection occurs in wide area of angles and is more effective, than usual channeling. Therefore, there are real prospects to use of reflection for extraction and collimation of beams from U-70 accelerator [10, 11].

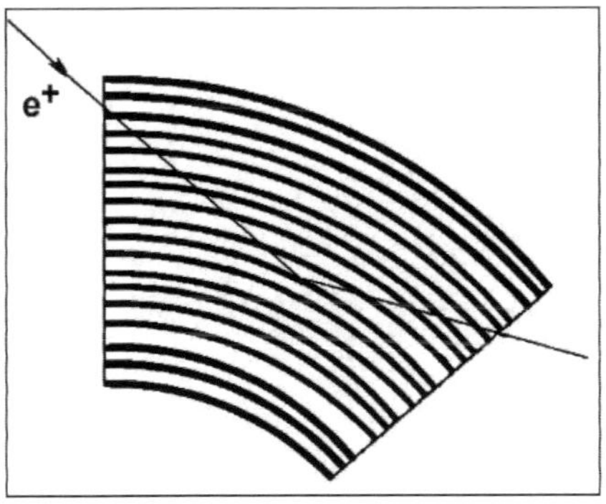

Figure 8. Scheme of volume reflection.

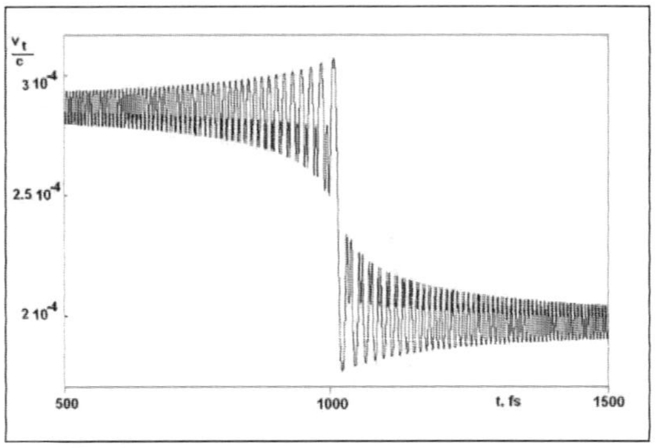

Figure 9. Behavior of velocity near reflection point.

3. New sources of radiation based on short bent crystals

In 2008 in [17] we have paid attention for the first time, that the short bent crystals are a new source of intensive radiation of photons at passage through them of charged particles in mode of reflection. At volume reflection particles cross a row of bent crystallographic planes (figure 8) and the behavior of its transversal velocity has aperiodic oscillation character (figure 9). On the basis of consideration of this motion in [17] the new type of intensive coherent radiation for electrons and positrons was predicted for these conditions. Due to high value of a Lorentz factor γ this radiation is more significant for light leptons than for other particles. The first experiment on detection of this radiation has been carried out in IHEP with 10 GeV positrons [18]. Positrons were directed on 0.65 mm silicon crystal bent at 0.5 mrad angle.

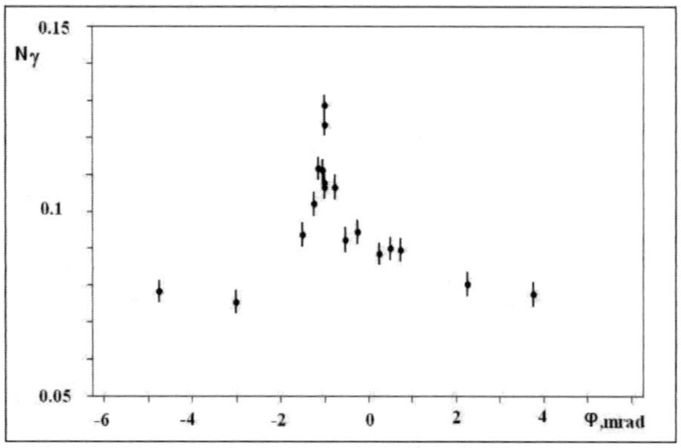

Figure 10. Orientation curve of photon yield.

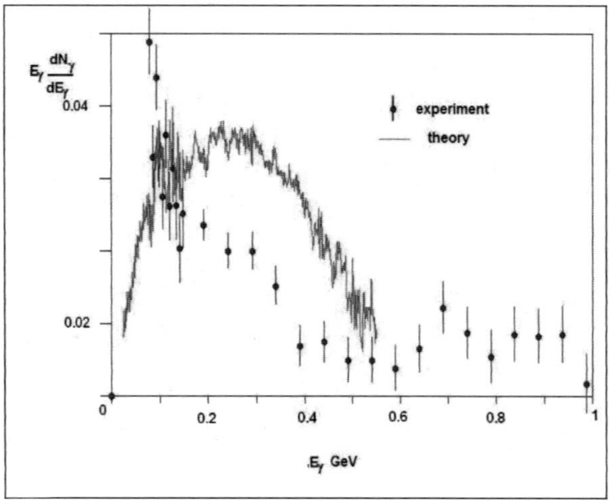

Figure 11. Differential energy losses.

Initially the orientation dependence of radiation intensity was found. Figure 10 illustrates the result of these measurements. One can see that the orientation curve has a clear maximum. Figure 11 illustrates the measured radiation energy losses $E_\gamma dN_\gamma / dE_\gamma$ of positrons in single crystal. For comparison, one can point that value of energy losses for amorphous silicon (or disoriented crystal) is 0.007 and it is practically independent of gamma-quantum energy. Thus presented here experimental results show on valuable radiation process of 10 GeV positron beam at volume reflection in bent single crystals. The more detailed researches of volume reflection radiation were carried out in CERN with 180 GeV positrons [19]. Even more impressing data are received recently with application of crystals of axial orientation in [20]. The main positive feature of volume reflection radiation is independence (within the limits of whole crystal bend) of radiation intensity from an incident angle of a particle. Such type of radiation can be applied at accelerators for production of intensive gamma-quantum beams and positron sources. In [21] the opportunity of bent crystal radiators application was proposed for beam cleaning systems in the ILC.

In the near future, we plan to show enhancement of radiation intensity from applications of chain of bent crystals (the device 4 on figure 2). It is expected that intensity of radiation will be proportional to amount the short bent crystals in a chain while in straight crystals intensity of radiation slows down the growth with length of a crystal [22, p.321].

Acknowledgments
The research is supported by the IHEP directorate and Russian Funding for Basic Research, grants 13-02-90434 and 13-02-00019.

References

[1] Elishev A F *et al* 1979 Phys. Lett. B **88**:387; 1979 JETP Lett. **30** 442
[2] Shiltsev V *et al* 2010 IPAC-2010-TUOAMH03, FERMILAB-CONF-10-127-APC
[3] Scandale W *et al* 2010 Phys. Lett. B **692** 78
[4] Biryukov V M, Chesnokov Yu A and Kotov V I 1997 *Crystal channeling and its application at high-energy accelerators* pp 219 (Berlin, Germany: Springer)
[5] Chesnokov Yu A, Afonin A G, Baranov V T *et al* 2011 Nuovo Cim. C **34** N4 407
[6] Bavizhev M D *et al* 1989 *IHEP preprint 1989-77* (Protvino)
[7] Arkhipenko A A, Afonin A G, Baranov V T *et al* 2008 JETP Lett. **88** 229-231
[8] Afonin A G *et al* 1998 JETP Lett. **68** 568-572
[9] Afonin A G *et al* 2001 Phys. Rev. Lett. **87** 094802
[10] Afonin A G *et al* 2009 Atomic Energy **106** 409
[11] Afonin A G *et al* 2011 JETP Lett. **93** 187
[12] Afonin A G *et al* 2012 JETP Lett. **96** 424
[13] Biryukov V 1991 Nucl. Instrum. Methods B **53** 202;
Taratin A M, Vorobiev S A, Bavizhev M D and Yazynin I A 1991 Nucl. Instrum. Methods B **58** 103
[14] Scandale W 2012 UA9 Status Report EuCARD-REP-2012-002
[15] Ivanov Y M, Petrunin A A, Skorobogatov V V *et al* 2006 Phys. Rev. Lett. **97** 144801
[16] Scandale W, Still D, Carnera A *et al* 2007 Phys. Rev. Lett. **98** 154801
[17] Chesnokov Yu A, Kotov V I, Maisheev V A and Yazynin I A 2008 JINST 3:P02005
[18] Afonin A G *et al* 2008 JETP Letters ***88*** (7) 486
[19] Scandale W *et al* 2009 Phys. Rev. A **79** 012903
[20] Mazzolari A et al this volume
[21] Seryi A 2010 Nucl. Instrum. and Meth. **A** 623 23-28
[22] Baier V N, Katkov V M and Strakhovenko V M 1998 *Electromagnetic Processes at High Energies in Oriented Single Crystals* (World Sci., Singapore)

Single and Multiple Volume Reflections of Ultra-Relativistic Electrons in a Bent Crystal as Tools for Intense Production of Electromagnetic Radiation

L Bandiera[1,2], **A Mazzolari**[1,2*], **E Bagli**[1,2], **A Berra**[4,5], **D Lietti**[4,5], **D De Salvador**[7,8], **V Guidi**[1,2], **M Prest**[4,5], **V Tikhomirov**[3], **E Vallazza**[6]

[1] Dipartimento di Fisica, Università di Ferrara, Via Saragat 1, 44122 Ferrara, Italy
[2] INFN Sezione di Ferrara, Via Saragat 1, 44122 Ferrara, Italy
[3] Research Institute for Nuclear Problems, Belarus State University, Bobruiskaya 11, 220030 Minsk, Belarus
[4] Università degli Studi dell'Insubria, Via Valleggio 11, 22100 Como, Italy
[5] INFN Sezione di Milano Bicocca, Piazza della Scienza 3, 20126 Milano, Italy
[6] INFN Sezione di Trieste, Via Valerio 2, 34127 Trieste, Italy
[7] Dipartimento di Fisica Università di Padova, Via VIII Febbraio 1848 2, 35121 Padova, Italy
[8] INFN Laboratori Nazionali di Legnaro, Viale Università, 35020 Legnaro (PD), Italy

E-mail: andrea.mazzolari@unife.it

Abstract. Electromagnetic radiation emitted by ultrarelativistic volume-reflected and multi-volume-reflected electrons in a bent crystal is very intense and takes place over a broad angular range of the incident beam. Such radiation results to be nearly independent from the particles trajectory and charge. This paper describes the possible applications these features allow, from the production of a gamma or a positron source to the crystal-assisted collimation of future linear e^+/e^- colliders and crystal-based electromagnetic calorimeters.

1. Introduction

In the middle 50s, the scientific researches of Ferretti, Ter-Mikaelian, Dyson and Uberall [1-3] on the influence of the crystalline lattice on the motion and on the radiation production by charged particles crossing a crystal led to the discovery of the so-called *coherent bremsstrahlung* (CB) by Diambrini-Palazzi et al. [4]. In 1964, Jens Lindhard introduced the concept of channeling [5] that fostered the interest to the possibility to generate intense radiation by channeled $e^{\pm}$ inside a straight crystal, thus called *channeling radiation* (CR). Channeling consists in the trapping of charged particles with energy E and with an angle of incidence with the crystal planes (axes) smaller than $\vartheta_{critical} = \sqrt{2U_0/E}$ in the planar (axial) potential well with depth U_0. Radiation generation by electrons/positrons in straight crystals has been investigated for decades [6], mainly in view of its possibility to be used as a powerful hard X-ray and gamma-radiation source. However, the generation of radiation in straight crystals takes place in a narrow angular region of orientation between crystal and beam direction [7], which becomes still narrower with the increasing of beam energy, thus limiting the possible applications. An opportunity to overcome such limitation is offered by the volume reflection (VR) effect in a bent crystal. In 1976, Tsyganov proposed the idea that in a bent crystal channeled particles could deviate from their initial direction just following the crystal channel [8]. Important applications

of bent-crystal-assisted manipulation of charged particle trajectories are beam extraction and halo cleaning [9]. VR consists in the reversal of transverse momentum of over-barrier particles by the interaction with the planar potential barrier at the tangency point of the trajectory with the curved crystalline planes [10]. As a result, volume-reflected particles are deflected by an angle of the order of the Lindhard angle (i.e., the critical angle for channeling) to a direction opposite to that of crystal bending. VR has a smaller deflecting power than channeling in a bent crystal, but it has the great advantage of larger deflection efficiency and wider angular acceptance, the latter being equal to the bending angle of the crystal. In order to increase the deflecting power of VR while maintaining its advantages, one can profit from the effect of multiple volume reflection in one crystal (MVROC) [11]. MVROC occurs in a bent crystal when the particles move at few channeling angles with respect to a crystal axis, when correlated string of strings (SOS) scattering assures the possibility of repeated VR from the planes sharing the same axis, leading to a multi-reflection process. VR and MVROC have been experimentally proven to work for either positively or negatively charged particles [12-16]. The e.m. radiation emitted by charged particles under VR and MVROC conditions has been also studied both theoretically and experimentally [17-26]. The large angular acceptance makes radiation accompanying VR and MVROC interesting for applications such as a gamma or a positron source or as an innovative scheme for crystal-assisted collimation in future electron-positron linear colliders or for crystal-based e.m. calorimeters.
In this paper we present a study of features and possible applications for radiation accompanying VR and MVROC.

2. Features of radiation accompanying VR

It is known that the type of e.m. radiation generated by charged particles in an external field depends on the peculiarities of particles motion in such a field. In general [6, 7, 27, 28] the main features of the radiation process is determined by the ratio of the deflection angle θ and the typical radiation angle $1/\gamma$, γ being the Lorentz factor. One limit is at $\theta \ll 1/\gamma$, when mainly one harmonic is emitted (dipole radiation) and the other limit appears when $\theta \gg 1/\gamma$, which corresponds to the case of synchrotron-like radiation. The same approximations are valid for quasi-oscillatory particles motion in straight crystals. Following the formalism of [6, 7], if the angle between the particles trajectory and the crystal planes (axes) is larger than U_0/m, the radiation has a dipole nature; in the opposite case the radiation is synchrotron-like. U_0/m is independent on the energy, while the critical angle decreases with energy as $1/\sqrt{E}$. As energy increases, CR becomes more and more synchrotron-like. Under the VR condition, the type of radiation changes during the particle motion inside the crystal. In fact, the average misalignment angle between the particle trajectory and the crystalline planes decreases as the particle approaches the reflection point [19]. Therefore, either the amplitude or the period of the particle motion between planes increases and the radiation regime turns from dipole- to synchrotron-like. As a result of multiple reflections on several planes, the planar radiation is more synchrotron-like for MVROC conditions than for a single VR. By a combination of several planar reflections occurring for MVROC with the contribution of atomic strings, the spectral intensity is very strong as compared to that of an individual VR. Furthermore, since the axial electric field is stronger than that for planes, the SOS contribution considerably increases the probability of hard-photon emission [29].

An example of the experimental results on VR and on MVROC for 120 GeV/c electrons is shown in figure 1 and 2. Figure 1 shows the experimental distribution of reflected electrons inside a bent Si crystal 2mm long, with a curvature radius R= (2.71 ± 0.07) m. The crystal is bent along the [111] direction and the bent (110) planes are used to steer particles. The VR deflection angle is quite small, being equal to θ_{vr} = (-11.4 ± 0.7) μrad, while the MVROC deflection angle is larger, being θ_{mvr} = (-43.1±2.2) μrad. On the contrary, the deflection efficiency is higher for single than for multiple VR, being ε_{vr} = (95.5 ± 0.8)% and ε_{mvr} = (85 ± 3)%, respectively. Nevertheless, both the deflection efficiency and the angular acceptance for VR/MVROC are larger than that for channeling. More details on the experiment can be found in [21, 26].

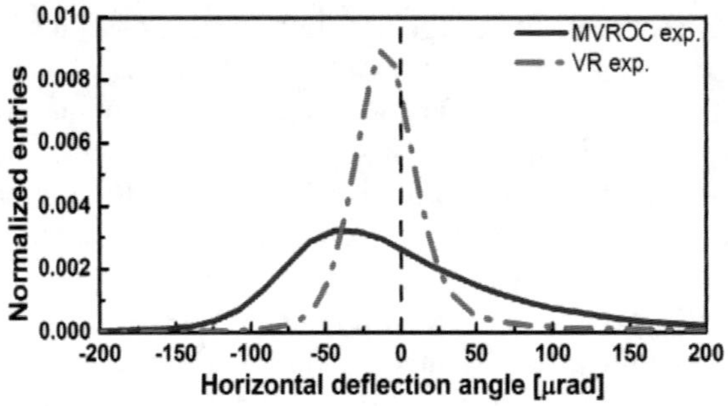

Figure 1 Experimental distribution of the horizontal deflection angle of particles interacting with the 2 mm bent Si crystal aligned on VR (dot-dashed line) [21] and MVROC (solid line) [26].

Figure 2 displays the experimental energy-loss distributions under VR (circles with bars) and MVROC (squares with bars) for the 120 GeV/c electrons in the 2 mm long crystal. The details about the procedure carried out to determine the distribution, (dn/dE)E vs. the energy lost by electrons, E, can be found in [20, 21]. The energy loss spectral intensity for either VR or MVROC is much larger than the Bethe-Heitler (BH) value (see figure 2 dashed line), which is the typical one for bremsstrahlung in amorphous materials [30]. The greater intensity for MVROC than for VR in the soft-medium region of the spectra is simultaneously determined by both the SOS scattering and the planar field contribution. Harder photons are emitted due to the SOS CB-like process in the field of axis [25, 26, 29], explaining the much stronger intensity of radiation accompanying MVROC in the hard region of the spectrum. The mean number of photons, with $E\gamma > 1$ GeV, emitted by each particles (multiplicity factor) is about 1.35 and 2.2 for VR and MVROC cases, respectively [25].

The high intensity of radiation accompanying VR and MVROC, combined with their large angular acceptance that exceeds those for CB and CR in a straight crystal [19–26], can be exploited for many applications. As described above, the radiation mechanism changes during the particle motion depending on the direction of the particle trajectory with respect to the crystal planes and axes. The main point is that the dynamics of particles under VR and MVROC conditions is similar to each other for different incidence angles within the total angular acceptance. In other words, particles trajectories under VR or MVROC conditions remain more or less in the same interval of directions of motion with respect to the crystal planes over the whole angular acceptance. Similar particle dynamics means similar radiation generation. Another advantage of such type of radiation is its weak dependence on particle charge [19, 25].

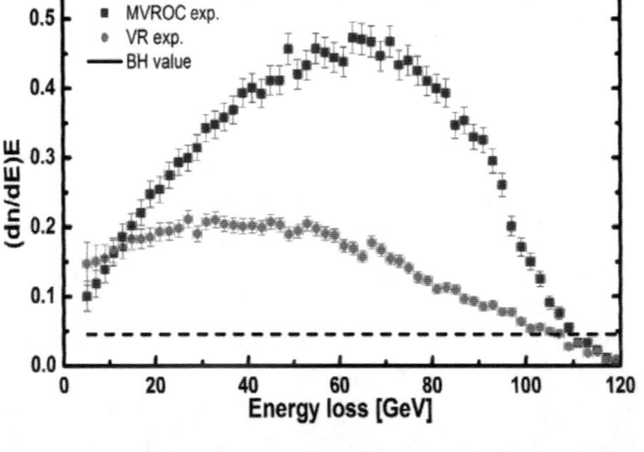

Figure 2. Energy loss spectra for 120 GeV/c electrons in the bent Si crystal [26]; VR (circles with bars); MVROC (squares with bars); amorphous estimation (dashed line).

3. Possible applications

3.1. Gamma and positron source

All the features presented in Sec. 2, such as the high-intensity and the wide angular acceptance, make the radiation accompanying VR and MVROC very attractive for several applications at once.

Additional advantage of radiation generation relying on VR is its robustness vs. crystalline imperfections as compared to channeling-based effects [31]. In contrast to channeling, the presence of crystal defects is not crucial in the case of VR and the deflection efficiency is practically the same as for a perfect bent crystal. Moreover, the local nature of the reflection effect will also assure the weak sensitivity to crystal mosaic structure.

Such possibility envisages also the usage of higher-Z materials are better than Si for e.m. generation because of their higher atomic potential [32], but they cannot be produced with the same perfection as Si. For instance, since the average electric field strength in W is about an order of magnitude stronger than in silicon, we expect that the advantages of the radiation accompanying VR and MVROC measured in Si at 120 GeV will be maintained in W at lower energies (about 10 GeV), available at many worldwide electron accelerators. In order to maintain the advantages of a broad angular acceptance at lower energies, one has simply to adjust the crystal curvature.

For all these reasons, e.m. radiation under VR and MVROC in a tungsten crystal could be exploited to convert high-emittance electron beams to high-intensity γ-beam.

In order to obtain an high-intensity gamma source, MVROC could be a preferable choice than VR due to the contribution of both crystal axis and multiple reflecting planes. A further opportunity for exploitation of such type of radiation is as a positron source starting from the idea of Ref. [33], where the authors proposed to use axial channeling to produce a high-intensity gamma beam to be then converted in positron-electron pairs through the BH mechanism.

Channeling application in positron sources is quite limited by several factors, like the W crystal mosaicity, the primary electron beam divergence and crystal-to-beam alignment precision. Moreover, the electron channeling role in positron sources is essentially limited by dechanneling in W crystals with thicknesses necessary exceeding the radiation length [33–35]. Because of these reasons, the main mechanism of γ emission in e+ sources is not really axial channeling radiation, but axial radiation emitted by e- moving at angles considerably exceeding the axial channeling angle. Due to the high-multiplicity contribution of multiple reflecting planes, the MVROC effect could be exploited for an additional amplification of γ-emission, in experimental conditions similar to those already used in positron sources based on axial channeling [33–35].

3.2. Crystal-assisted collimation

One of the most attractive applications that would take advantage of the features of VR and MVROC from both the points of view of radiation and deflecting power could be a crystal-assisted collimation device for future linear electron-positron colliders, e.g., the International Linear Collider (ILC) [36]. ILC is a proposed electron-positron collider with a planned collision energy of 500 GeV and a possible upgrade to 1000 GeV. The beam will be delivered in pulsed bunches; these bunches are not uniformly distributed and have 'halo' particles which need to be removed to avoid unacceptable background in the detectors. To remove the halo particles, the beam will be collimated in its delivery system. The baseline design requires two-part collimators made up of small aperture spoilers with 0.5-1 radiation length (X_0) thickness close to the beam, with downstream absorbers 30 X_0 thick [37]. Since the spoilers are placed very close to the beam, the beam-spoiler interactions result in wakefield perturbations [38]. The insertion of a short (1-2 mm ~ 0.02 X_0) Si crystal instead of a Ti spoiler of some cm of length would diminish the wakefield perturbations. A proposal in this sense has already been put forward by Seryi [39]. Such collimation scheme consists in replacing one or more spoilers with bent crystals oriented with the beam under VR. The choice of VR instead of channeling is connected to its larger angular acceptance and larger deflection efficiency with respect to channeling. Particles deflection by VR would increase halo-cleaning efficiency per unit of length as compared to the case of an amorphous spoiler. Furthermore, the larger energy loss distribution in VR than in amorphous media would improve the discrimination of halo particles, which will be deflected by forward magnets. The larger deflection and the greater energy loss in MVROC condition than the case of an individual VR, makes MVROC radiation even more suitable for crystal-assisted collimation of future linear colliders [26]. The choice between VR and MVROC will depend only on the required deflection efficiency and on the possibility of crystal-beam alignment in one or two directions. Fig. 3 shows a possible scheme for crystal assisted collimation through VR-MVROC and briefly summarizes the various steps described above. Finally, it should be reminded that the energy loss distribution under VR-MVROC conditions is nearly independent on the particle charge [25, 26] and so it is almost the same for either electrons or positrons, which makes these two coherent effects even more attractive for collimation in $e^{\pm}$ colliders.

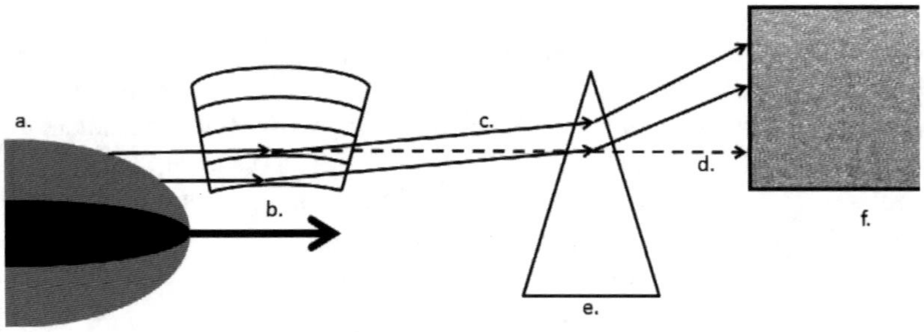

Figure 3. Concept of linear collider collimation system based on VR-MVROC radiation phenomenon [39]. In the figure **a.** is the beam, having a core and a halo; **b.** is the bent crystal oriented in VR-MVROC conditions; **c.** solid lines are the particles from the beam halo that have been deflected and have lost energy due to VR-MVROC; **d.** dashed-line represents the photons produced during VR-MVROC; **e.** is the bending magnet, separating halo particles that have lost a significant part of their energy during VR-MVROC and **f.** is the absorber of halo particles and photons.

3.3. Crystal-based electromagnetic calorimeter

In high energy physics and γ-astronomy, $e^{\pm}$ and γ energies are measured by electromagnetic calorimeters. Typically, they are composed of scintillating crystals, which collect the light emitted by electromagnetic showers inside the calorimeter material itself [40, 41]. Since the full length of the showers depth can exceed 20 X_0 for ultrahigh energies (>100 GeV), large-Z element scintillators are used to reduce the X_0 and, so, the calorimeter size. Usually the scintillator crystals are used just for their properties of scintillation and shower development, while the orientational effects are not taken into account. However, it has been demonstrated that an enhancement of the e.m. shower due to coherent interactions, namely *coherent radiation* and *coherent* $e^{\pm}$ *pair production,* with respect to random incidence occurs if the beam impinges on the crystal along a main axis direction [6, 7, 42]. This fact has as a consequence a much faster development of the e.m. shower as compared to what happens in amorphous material. Thereby, one could still reduce the calorimeter thickness through e.m. shower development acceleration. The limitation also in this case is the narrow angular region in which all coherent effects manifest themselves, which imply that one should align precisely each scintillator crystal axis with the incoming $e^{\pm}$ and γ direction from the interaction region and take into consideration the details of the orientation dependence of the radiation and pair production processes probabilities when data are processed.

The radiation and pair production under VR-MVROC conditions allow one to overcome these difficulties, permitting the acceleration of e.m. shower development in a greater angular range of orientation between the particle direction and the crystal. In order to achieve this goal, all the scintillator crystals that compose the calorimeter have to be bent by angles somewhat larger than the local divergences of $e^{\pm}$ and γ arriving from the interaction point and to align the crystals to ensure VR of all the incident particles. In fact, it could not be necessary to bent the crystal over its whole length, which is normally quite large (tens of cm), to obtain an enhancement of the e.m. shower, but just the first layer of some mm, which ensures VR condition to be valid. Alternatively, one could bend some mm thick crystals via the well-established technique of superficial grooving to obtain self-bent crystals [43]. Some of these crystals can be stuck together [44] and then stuck to the entrance surface of a larger straight scintillator crystal.

5. Conclusions

We have described the radiation accompanying volume reflection in a bent crystal and some possible applications of this type of radiation. The most peculiar characteristics of VR-MVROC radiation are the high intensity, the wide angular acceptance and the weak dependence on the particle charge and trajectory. The combination of these features makes these types of radiation suitable for the conversion of high-emittance electron beams into hard gamma quanta or for a positron source in high-Z crystals (like tungsten). By combining their high deflecting capability with their high radiative power, VR and MVROC can be envisaged as good candidates for crystal-assisted collimation in future $e^{\pm}$ linear collider. Finally, VR or MVROC could be exploited to drastically reduce the thickness of crystalline electromagnetic calorimeters in High-Energy-Physics, by reducing the longitudinal length of the electromagnetic shower.

References

[1] Ferretti B 1950 *Il Nuovo Cim.* **7** 118
[2] Ter-Mikaelian M L 1953 *JETP (USSR)* **25** 289
[3] Dyson F H and Uberall H 1955 *Phys. Rev.* **99** 604
[4] Diambrini-Palazzi G 1968 *Rev. Mod. Phys.* **40** 611 31
[5] Lindhard J 1964 *Phys. Phys. Lett.* **12** 126
[6] Baryshevskii V G, Tikhomirov V V 1989 *Usp. Fiz. Nauk* **159** 529
[7] Baier V N, Katkov V M and Strakhovenko V M 1998 *Electromagnetic Processes at High Energies in Oriented Single Crystals*World Scientific, Singapore
[8] Tsyganov E 1976 *Fermilab Preprint* **TM-684** http://lss.fnal.gov/archive/tm/TM-0684.pdf
[9] Scandale W *et al* 2010 *Phys. Lett.* B **692** 78-82
[10] Taratin A M and Vorobiev S A 1987 *Phys. Lett. A* **119** 425
[11] Tikhomirov V V 2007 *Phys. Lett. B* **655** 217
[12] Ivanov Y M *et al* 2006 *Phys. Rev. Lett.* **97** 144801
[13] Scandale W *et al* 2009 *Physics Letters* B **681** 233 6
[14] Scandale W *et al* 2009 *Phys. Lett.* B **682** 274
[15] Scandale W *et al* 2010 *Phys. Lett.* B **693** 545
[16] Scandale W *et al* 2011 *Europhys. Lett.* **93** 56002
[17] Afonin A G *et al* 2008 *JETP Letters* **88** 414 7
[18] Chesnokov U A, Kotov V I, Maisheev V A, Yazynin I A 2008 *JINST* **3** 02005
[19] Scandale W *et al* 2009 *Phys. Rev.* A **79** 012903
[20] Lietti D *et al* 2012 *Nucl. Instrum. Methods Phys. Res.* B **283** 84
[21] Bandiera L *et al* 2013 *Nucl. Instrum. and Meth.* B **309** 135140
[22] Maisheev V A 2011 *Il Nuovo Cimento* C **34** 73
[23] Bondarenco M V 2011 *Il Nuovo Cimento* C **34** 381
[24] Bellucci S, Maisheev V A 2012 *Phys. Rev.* A **86** 042902
[25] Guidi V, Bandiera L, Tikhomirov V 2012 *Phys. Rev.* A **86** 042903
[26] Bandiera L *et al* 2013 *Phys Rev. Lett.* **111** 255502
[27] Jackson J D 1975 *Classical Electrodynamics* Wiley & Sons
[28] Akhiezer A I and Shulga N F 1996 *High-Energy Electrodynamics in Matter* Gordon & Breach, (New York)
[29] Strakhovenko V M 2003 *Phys. Rev.* A **68** 042901
[30] Bethe H A, Heitler W 1934 *Proc. Phys. Soc. Lond.* **146** 83

[31] Bagli E *et al* 2013 *Phys Rev. Lett.* **110** 175502
[32] De Salvador D *et al* 2013 *J. Appl. Phys.* **114** 154902
[33] Artru X, Baier V N, Chehab R, Jejcicm A 1994 *Nucl. Instrum. and Meth.* A **344** 443
[34] Artru X *et al* 2005 *Nucl. Instrum. and Meth. in Phys. Res.* B **240** 762
[35] Suwada T *et al* 2007 *Phys. Rev. ST Accel. Beams* **10** 073501
[36] Nigel S Lockyer 2008 *International Linear Collider*. The Logic of Nature, Complexity and New Physics: pp. 399-418
[37] Ellwood G and Greenhalgh J *Beam impact studies on ILC collimators, Proceedings of PAC07* (Albuquerque, New Mexico, USA)
[38] Toader A M, Barlow R J, Jackson F, Angal-Kalinin D *Effect of collimator wakefield in the beam delivery system of the International Collider, Proceedings of EPAC08* (Genoa, Italy)
[39] Seryi A 2010 *Nucl. Instrum. and Meth.* A **623** 23
[40] The CMS Collaboration *et al* 2008 *JINST* **3** S08004
[41] Atwood W B *et al* 2009 *ApJ* **697** 1071
[42] Baurichter A *et al* 1996 *Nucl. Instr. and Meth.* B **119** 143
[43] Camattari R, Guidi V, Lanzoni L, Neri I 2013 *Meccanica* **48** 1875
[44] Neri I, Camattari R, Bellucci V, Guidi V and Bastie P 2013 *J. Appl. Cryst.* **46** 953

The study of the guiding process for 10 keV electrons by planar Plexiglass surfaces

K. A. Vokhmyanina[1], P. N. Zhukova[1], A. S. Kubankin[1], Le Thu Hoai[1], R. M. Nazhmudinov[1], A.N. Oleinik[1] and G. P. Pokhil[2]

[1] Laboratory of Radiation Physics, Belgorod State University, Belgorod, Russia
[2] Skobel'tsyn Research Institute of Nuclear Physics, Moscow State University, Moscow, Russia

E-mail: vokhmyanina @bsu.edu.ru

Abstract. Experimental study of electron beam reflection from a single planar surface of Plexiglas was made. The distinct guiding effect for the part of the beam was observed for 10 keV electrons within angles of incidence from 0 to +3 degrees. The experiments using Poly plates showed a number of features of the process such as the dependence of the reflection on the plate surface quality and material of the surfaces, the divisions of the beam into two parts with different behaviour depend on tilt angle and the beam current value, the effect of an elevation angle of the beam in compare with initial beam trace at negative and zero tilt angles of the plate.

1. Introduction

In recent years the possibility to guide electrons by insulating nanocapillaries attracts considerable attention because of its potential practical use in different areas of technology and science. Ion and electron transmission through the channels with large aspect ratio has been studied after the discovery of the guiding of 3 keV ions of Ne7+ transmitting through PET [1]. But the behavior of electrons passing through the capillaries is somewhat different from the behavior of ions indicating the inequality in the mechanism of the interaction of charged particles with insulating surface. Works in this area were carried out since 70 years of the last century, but the studies were focused mainly on high-current electron beams of high energy [2–4].

The set of experimental studies for the electrons with energy up to 1.4 keV was made previously [5–9]. Most of the experiments were made for the capillaries created in Mylar, SiO2, Al2O3 and for straight and tapered glass capillaries. All the results indicate the formation of the self-consistent charge distribution on the inner walls of that channels providing partially contactless transition of electrons even for non-zero angles of incidence.

Straight transmission of electrons with energies from 2 to 10 keV through glass polycapillaries, tubes, and tapered capillaries revealed that the energy of the electrons was almost unchanged. It follows from the measurement results that the electron transmission through capillaries is controlled by the potential at the inner channel surface induced by the charge injected during irradiation.

In the last our work [7] a similar experiments were carried out for a planar channel formed by two glass plates. The distinct guiding effect was observed for 10 keV electrons within angles of incidence from 0 to +3 degrees. To reveal the mechanism of the non-contact transmission of electrons through different insulating channels, an experimental study of beam reflection from a single planar surface was made. The experiments using plates showed a number of features of the process, such as the deviation of the reflection angle from the specular angle and the dependence of the reflection on the plate length, the division of the beam into two parts with different behaviour depend on tilt angle, and the dependence of the beam trace position on the beam current value.

Clear understanding of the mechanism will give the opportunity to select the most suitable materials and geometry of the capillaries to optimize the guiding process.

2. Experiment

In order to reveal the mechanism of electron-surface interaction the experiment with planar surface was made. The material of Plexiglas was used as a target. We used 10 keV energy for electrons. The diagram of experimental setup is shown in figure 1.

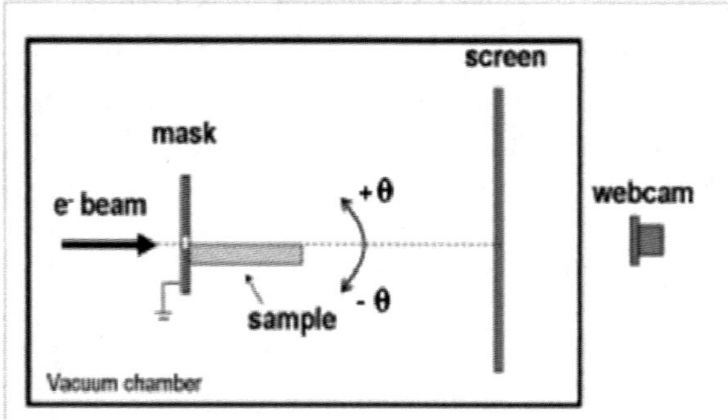

Figure 1. The diagram of experimental setup in vacuum chamber (θ – tilting angle).

The sample in the special holder can be tilt with respect to the beam axis while the entrance of the sample stands still. The holder can be shifted horizontally letting explore several samples and check the primary beam. In front of the sample there is a copper mask that allows regulate beam size (height and width of the beam cross section). The electrons passing though the slit of the mask move close to the insulating surface interacting with it and fall on the glass screen covered with thin layer of scintillation powder. The sample was sequentially tilted and a snapshot was taken at each angle.

Several types of Poly surfaces were used: rough (seed size of about 20um) and smooth plates with different lengths, ~6 mm of thickness. The diameter of the initial beam was 2 mm; the beam current was 170 nA. The monochromaticity of the beam is better than 0.1%.

The tilting of the plates was made in small angle interval typically from -2 degrees to +5 degrees. When the plate surface reaches the positive angles in most experiments with Plexiglas without mask the spot on the screen is split into two parts: central narrow static spot (which can be seen without mask) and the halo that can be in form of arc or "cloud" (figure 2 a, b). The halo can be guided by the plate (figure 2 b).

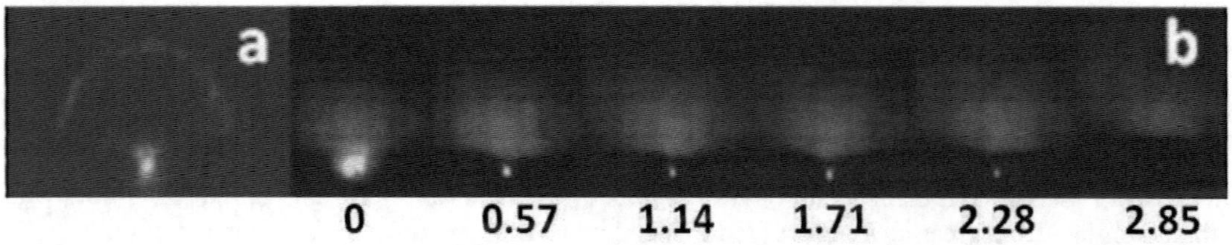

Figure 2. Types of the form of the halo: a- "arc", b- "cloud", that can be guided by the Plexiglass plate (On the photo the appropriate tilting angles (in degrees) of the plate are indicated).

The corresponding dependence is shown in figure 3 (The dependence was measured for the lower edge of the halo from the straight beam direction). The "cloud" follows the planar surface from 0 to 3 degrees without movement of the "central" part. The similar situation was observed with glass plates earlier [10].

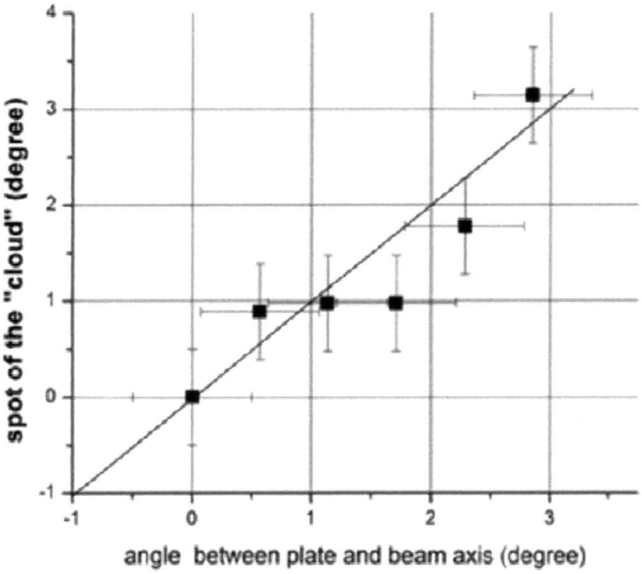

Figure 3. The guiding effect of the "cloud" by planar surface of Poly.

Studying guiding effect we discovered that some types of Poly surfaces were modified by the beam of 10 keV electrons (figure 4 a, b). That modification was different depend on current of the beam.

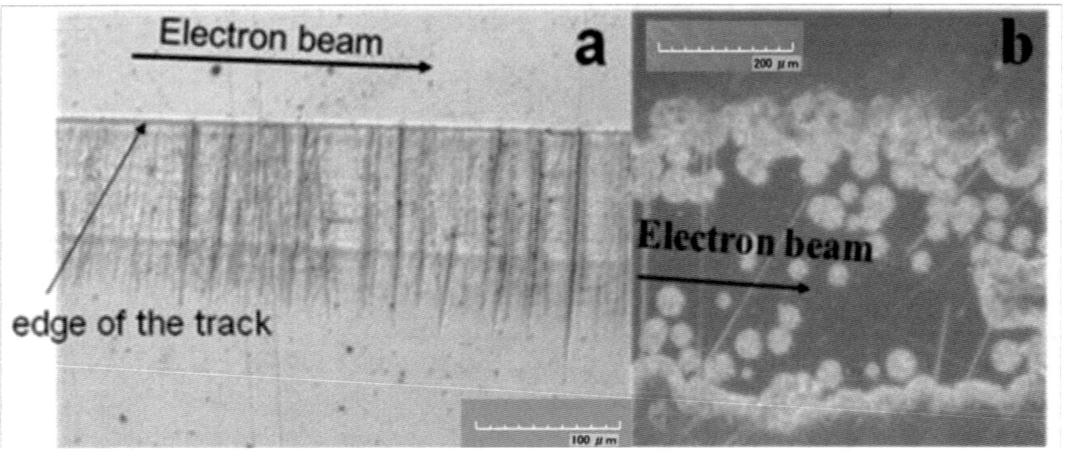

Figure 4. Modification of the Poly surface under the irradiation by 10 keV electrons a- beam current is 10 nA, b - beam current is 100 nA.

For sufficiently small currents (about 10 nA) on the edges of the track there were needle-like structure perpendicular to the direction of motion of the electrons, and with high current (~ 100 nA) we observed "drops" of dendrite structures on the track that remind insulator surface discharge.

3. Results

Experimental study of electron beam reflection from a single planar surface of Plexiglas was made. As for the ions the deviation of the reflection angle from the specular angle and moving of electrons along the surface was obtained. The experiments using Poly plates showed a number of features of the process such as the dependence of the reflection on the plate surface quality and material of the surfaces, the divisions of the beam into two parts with different behaviour depend on tilt angle and the beam current value.

Several types of Poly surfaces were used: rough (seed size of about 20um) and smooth plates with different lengths. The tilting of the plates was made in small angle interval typically from -2 degree to +5 degree. When the plate surface reaches the positive angles in most experiments with Plexiglas without mask the spot on the screen is split into two parts: central narrow static spot and the halo that can be guided by the plate. The same situation was observed with glass plates earlier. That behaviour cannot be explained by the bulk charge due to too little energy of the incident electrons (less than 100 эВ). Moreover, the shape and behaviour of the spot on the screen indicating some structure in the charge distribution on the surface that can be dynamically changed by tilting the plate.

The halo can be in form of arc or cloud. We assume it depends of initial beam current and quality of surface. To explain this effect it is necessary to simulate the formation of surface charge based on ionization, secondary electron emission and draining overcharge on the surface.

After the interaction of an electron beam with an energy of 10 keV with a Plexiglas surface at grazing incidence (-0.5 ÷ 0.5 deg.) in some experiments there was a noticeable modification of the surface under the action of electrons. And for sufficiently small currents (about 10 nA) on the edges of the track there were needle-like structure perpendicular to the direction of motion of the electrons, and with high current (~ 100 nA) we observed "drops" of dendrite structures on the track that remained surface discharge in dielectrics. It should be noted that despite of the substantial current density of the incident beam, the surface modification was observed only in some cases.

Formation of tracks on the surface of the Plexiglas for a long time irradiation does not affect the position of the pattern at the screen (the position was measured within 30 minutes and significant changes in the intensity or position of the spot have not been identified).

4. References

[1] Stolterfoht N, Bremer J-H, Hoffmann V, Hellhammer R, Fink D, Petrov A and Sulik B 2002 *Phys. Rev. Lett.* **88** 133201
[2] Zavadovskaya E, Annenkov Yu, Starodubtsev V, Vakhromeev V, Malofienko G 1975 *Russian Physics Journal* **17** (1) 144
[3] Kivenko E, Sergeev A, Yagushkin N 1991 *Russian Physics Journal* **33** 11 901
[4] Agafonov A, Bogachenkov V, Krastelev E 2004 Proceedings of 15th International Conference on High-Power Particle Beams 143
[5] Das S, Dassanayake B, Winkworth M, Baran J, Stolterfoht N and Tanis J 2007 *Phys. Rev.* A **76** 042716
[6] Das S, Dassanayake S,Tanis J, and Stolterfoht N 2010 Revista Mexicana de Física S **56** (2) 66
[7] Bereczky R, Dassanayake *B*, Das S, Ayyad A, Tőkési K and Tanis J 2012 *J. Phys.: Conf. Ser.* **388** 132024
[8] Dassanayake B, Keerthisinghe D, Wickramarachi S, Ayyad A, Stolterfoht N and Tanis J 2013 *Nucl. Instrum. Methods Phys. Res.* B **298** 1
[9] Wang W, Chen J, Yu D, Yang B, Wu Y, Zhang M, Ruan F and Cai X 2011 *Phys. Scr.* **144** 014023
[10] Vokhmyanina K, Zhukova P, Irribarra E, Kubankin A, Le T, Nazhmudinov R, Nasonov N and Pokhil G 2013 *Journal of Surface Investigation. X-ray, Synchrotron and Neutron Techniques* **7** 271

Journal of Physics: Conference Series **517** (2014) 012045 doi:10.1088/1742-6596/517/1/012045

Methods for the measurement of the refractive index of MeV photons using total internal and external reflection

M A Aginian, K A Ispirian* and M Ispiryan
Alikhanian National Scientific Laboratory (Yerevan Physics Institute)
Br. Alikhanians 2, Yerevan 0036, Armenia

E-mail: karo@mail.yerphi.am

Abstract. Recently it has been theoretically and experimentally shown that for 1–10 MeV and 1–2 MeV photons, respectively, the refractive index of Si is greater than 1. Taking into account the difficulties of the carried out experiment it is proposed to measure directly the refractive index of Si and other materials detecting the total internal and external reflections.

1. Introduction

Up to the publication of the results of the work [1] it was accepted that in the gamma ray energy region of photons with energy $\hbar\omega$ > 1 MeV the real part of the refractive index $n = 1 + \delta$ of all the materials is less than 1, $n(\omega) < 1$, and is given by the formula $n(\omega) = \sqrt{\varepsilon(\omega)} = 1 + \delta(\omega) \approx 1 - 0.5(\omega_p / \omega)^2$ where $\varepsilon(\omega)$ is the real part of the dielectric constant, ω_P is the plasma frequency, and, therefore, $\delta(\omega) < 0$. The sensational experimental results of [1] have been obtained only for photon energies 0.786, 1.165 and 1.951 MeV in a very difficult experiment measuring the nanoradian deflection of very monochromatic gamma beams with very small angular spread. The experiment has been carried out at Institut Laue-Langevin (ILL) on the facility GAMS5 [2] using a double crystal Bragg diffraction spectrometer. Experimentally it has been shown [1] that for the above energies $n(\omega) > 1$ with measured values $\delta \approx +\,(10^{-10}-10^{-9})$.

The authors of [1] carrying out complicated non-perturbative QED calculations explained their observations with the help of QED nonlinear process of Delbruck scattering which as the photon-photon scattering takes place by virtual electron-positron pair production. A comment [3] on the work [1] has been sent in September 2012 and published in March 2013 in which it has been shown that even using the Delbruck scattering cross sections given in [1] the contribution of Delbruck effect into δ is $10^{-6}-10^{-5}$ times less than it was estimated in [1]. The most amazing fact is that there is a good agreement between the theoretical and experimental results of [1].

The discovery of [1] is of fundamental importance because it opens new possibilities and directions in gamma optics, nuclear physics, etc [4–7]. Therefore, a problem in physics of the XXI century is present. It is more reasonable to solve the above described problem experimentally. Taking into account that the methods of [1] are very difficult we have carried out transparent calculations on a new phenomenon of gamma ray Cherenkov-transition radiation (GCTR) and proposed [8] a much simpler experiment. The observation and study of GCTR can confirm or decline the fundamental discovery of [1].

In this short note it is discussed and proposed a direct method for the measurement of n (ω) of various materials using total external reflection (TER) and total internal reflection (TIR). The advantages of the proposed method compared with the method of [1] are: 1) there is no need of very monochromatic MeV photons and high detection energy resolution and 2) instead of nanoradian accuracies of [1] only microradian accuracies of angles are enough. The first advantage is due to the fact that despite to the regions around K-, L- edges $n(\omega)$ in MeV region varies slowly. The second advantage is connected with the absence of Bragg diffraction spectrometers.

2. Method of Using TIR and TER for the Measurement of n(ω) in MeV Region

It is well known that if $n(\omega)>1$ (optical and softer region), and the photons intersect the boundary medium-vacuum, it can take place total internal reflection, if the incidence angle, $\theta_{inc}=\pi/2-\theta_{gl}$ (θ_{gl} is the glancing angle) is greater than a critical angle $\theta_{cr}^{TIR}=\arcsin(1/n)$. Similarly, if $n(\omega)<1$ (X-ray and harder region), and the photons intersect the boundary vacuum- medium it can take place total external reflection if the glancing angle, θ_{gl}, is less than a critical angle $\theta_{cr}^{TER}=(2\delta)^{1/2}$. In order to use TER (or TIR) for the direct measurement of $n(\omega)$, and to show: is $n(\omega)>1$ or $n(\omega)<1$ for MeV photons, it is proposed the arrangement and method shown in figure 1.

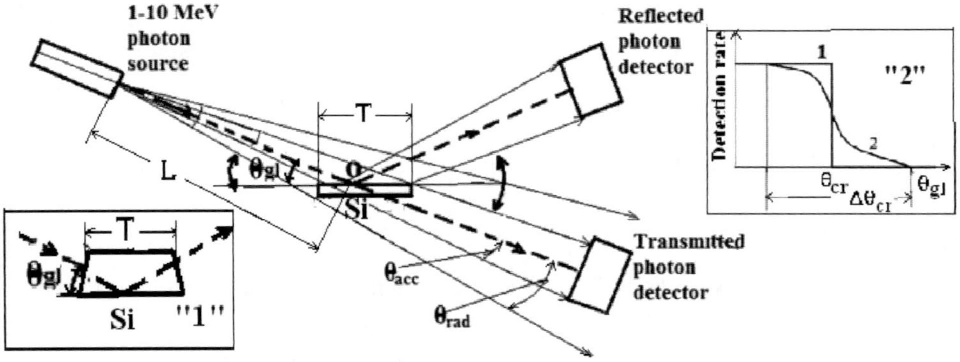

Figure 1. The proposed experimental arrangement (without collimators and shielding) and method.

As it is seen in figure 1 a photon beam with energy 1–10 MeV and with not small energy resolution (of the order of up to $\Delta\omega/\omega\sim10\%$) and angular spread, much less than the spread of the photon beam from the inverse Compton scattering source (see below), strikes a flat silicon (Si) reflector shown in figure 1. When it is expected that $n(\omega)>1$ the flat reflector of the figure 1 is used. When it is expected that $n(\omega)<1$ the table shape Si target shown in the insert "1" is used. In the case of the flat Si reflector the photons are undergone TER, if the glancing angle is less than θ_{cr}^{TER}. The photons pass through the reflector, if the glancing angle is greater than θ_{cr}^{TER}. In the case of the table shape Si reflector the photons enter freely the solid silicon, reach the bottom boundary between the silicon and vacuum, undergo TIR and come out of the reflector, if the incidence angle is large than θ_{cr}^{TIR}. The photons pass through the table shape reflector if the incidence angle is less than θ_{cr}^{TIR}. The photons passing through the reflectors are detected with the help of transmitted photon detector, while the photons undergoing TER or TIR are detected by the reflected photon detector. Both the transmitted and reflected photon detectors measure the energy of the photons with not very good (as in [1]) resolution. The angular spread of the photons from the source is $\theta_{rad}\sim1/\gamma$, where $\gamma=E/mc^2=1/[1-(V/c)^2]^{1/2}$ is the relativistic factor of the inverse Compton scattering electrons. The reflectors are placed at a distance L from the source and have longitudinal and transversal sizes T. The reflected photon and transmitted photon detectors detect the photons only in an acceptance angular interval $\theta_{acc} << \theta_{rad}$. Therefore, only a small fraction $F = (\theta_{acc} / \theta_{rad})^2 \approx (T\theta_{gl} / 2L\theta_{rad})^2$ of the primary photon beam can be detected.

Let us explain the principle of the proposed method for the case when $n(\omega) > 1$ (see insert '2'). When the angle θ_{gl} increases from 0, TER will be observed, and the reflected photon detector will continue to detect TER photons till θ_{gl} becomes greater than θ_{cr}^{TER}. Measuring at this moment θ_{cr}^{TER} and calculating δ by the above given formulae one can determine $n(\omega)$ for $n(\omega) > 1$. Similarly, one can determine $n(\omega)$ for $n(\omega) < 1$ replacing the flat reflector by the table shape one shown in the insert "1" of figure 1. The above mentioned dependence of the rate of detection of TER (or TIR) photons on the θ_{gl} have the form of the curve 1 and 2 shown in the insert "2" of the figure 1for ideal and real (with spread, etc) beams, respectively. Roughly $\Delta\theta_{cr}$ is equal to the error of the measurement of θ_{cr}^{TER} (or θ_{cr}^{TIR}). In other words, it is necessary to have photon beams with angular spread $\theta_{acc} << \theta_{cr}$, say, $\theta_{acc} \approx 0.1\,\theta_{cr}$.

In the past no such experiments at MeV energies have been carried out, though in many works (see, [9]) it has been proposed, but not realized, to use TER of MeV photons for various purposes, in particular, for monochromatization (filtration) of MeV photon beams. The reason is in the following two requirements. It is necessary to have MeV photon beams 1) with very small angular spread and 2) with sufficient intensities. However, the proposed experiments become realistic recently in connection with the advance of the method of inverse Compton scattering intense photon sources. [10, 11]. Suitable parameters for such experiments are at present available at the facility HiGS of Duke university, which provides 10^8–10^{11} photons per second in a ~10% BW, Soon it will be available at the ELI facility in Romania, which will provide ~10^{13} MeV photons per second in a ~1% BW. Of course, the intensity of the HiGS beam is much less than the primary photon beam of GAMS5 emitting into 4π solid angle (~10^{15}), but it has angular spread of $\Delta\theta_{rad}$ ~ $1/\gamma$ ~ $4 \cdot 10^{-4}$ rad. As it has been mentioned above the worse monochromaticy is not important for our application.

Let us consider the case when $n < 1$. Then assuming that the refractive index of Si is determined by the formula $n(\omega) = 1 - 0.5(\omega_P / \omega)^2$, for photons with energy $\hbar\omega$=2 MeV and Si radiator with $\hbar\omega_p \approx 30$ eV, one obtains δ (2MeV) =$-2.25\text{x}10^{-10}$, and with the help of above formula θ_{cr}^{TER}=$2.25\text{x}10^{-5}$ rad. This means that for the measurement of n(2MeV) with the help of TER it is necessary to have beam of 2 MeV photons with angular spread, say, θ_{acc}= $2.25\text{x}10^{-6}$ rad.

If one uses HiGS photon beam with angular spread Δθrad ~$1/\gamma$~$4\text{x}10^{-4}$ rad the necessary θ_{acc}=$2.25\text{x}10^{-6}$ rad is provided by the fact that only a small fraction equal to $F = (\theta_{acc} / \theta_{rad})^2$ of the beam emitted under angles smaller than $\theta_{acc} = T\theta_{gl} / 2L$ strikes the Si reflector. Taking T = 3cm, L= 10 m, one obtains θ_{acc}= $3.36\text{x}10^{-8}$ rad, i.e. less than it was required θ_{acc}= $2.25\text{x}10^{-6}$ rad. This means that the second requirement is satisfied, and one can carry out the experiment having sufficiently intense photon beam. For such parameters, $F \approx 10^{-8}$, and the number of TER photons at HiGS will be more than N_{TER}= 1 per second. With such number of photons and accuracy of the measured θ_{cr}^{TER} one can carry out the measurement in short time and determine δ at HiGS and show that δ is negative or positive. Just in the same manner following the results of [1] and assuming that δ is positive one can carry out TIR measurement of δ>0.

Above it has been shown that, in principle, the proposed experiment can be performed. In reality the HiGS beam-line has collimators after which the MeV photon beam has transverse sizes ~3x3 mm^2 in the experimental hall at a distance of about 60 m. One can collimate this beam stronger, especially, in the vertical direction, down to ~1mm. Using such collimated beam one can make the measurements, detecting with the help of reflected and transmitted photon detectors only the photons, reflected or passing the reflectors. However, some problems remain. Here the method of rotating and fixing the reflectors with microradian accuracy will not be discussed taking into account the nanoradian accuracy achieved in [1] with the help of interferometer. We will not discuss also the problem of the non-planarity and roughness of the reflectors, etc. Nevertheless, it is necessary to consider how the reflected and transmitted photon detectors work, since if $\theta_{gl} \approx 10\ \mu$rad the distance between them will be equal to ~1 cm if the detectors will have distance from the reflector equal to $L_{det} \approx 1$ km.

To overcome this difficulty and to place the reflected and transmitted detectors much closer, say, at distance $L_{det} \approx (50\text{–}100)$ m, it is proposed to use the set-up shown in figure 2 in which

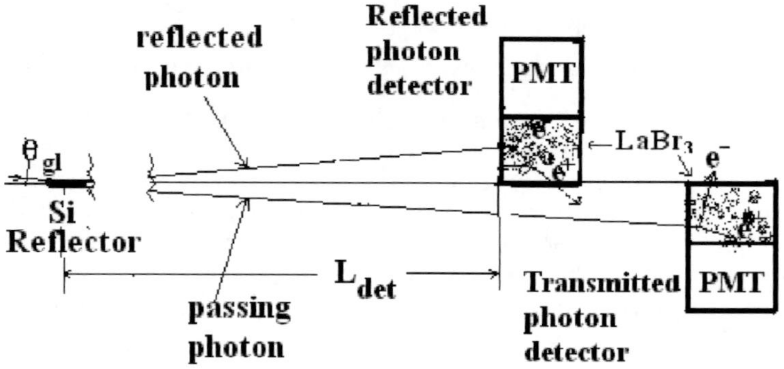

Figure 2. The proposed set-up for the reflected and transmitted photon detectors.

the recent progress in positron emission tomography, PET, technology (see [12]) is used. The reflected (transmitted) photon with known energy in the interval $\hbar\omega$=1–10 MeV produces a Compton electron or e^-e^+ pair in the top (bottom) scintillator of the reflected (transmitted) photon detector placed after a distance L_{det}, much shorter than $L_{det} \approx 1 1$ km, at a few mm from the reflection plane. Since the energy of e^- or e^+ is less than $\hbar\omega$ the range of one of them can be up to ~1 cm, and, therefore, due to the escape from the top (bottom) scintillator the energy measured by the top (bottom) photomultiplier, PMT, will be less than the energy $\hbar\omega$ of the reflected (transmitted) photon. It is necessary to use more effective scintillators, say $LaBr_3$, with better characteristics (see [12]). The scintillation light of these charged particles is detected by the top (bottom) PMTs. The transmitted photon detector is placed a little downstream and cannot detect the escaped particles. Such a set up of the detectors allows to detect separately the reflected and passing photons. Let us note that this problem can be solved also with the help of PET liquid xenon time projection chamber (LXeTPC) [13] or proposed LXeTPC on the basis of graphene [14].

3. Conclusions

It is clear that in order to confirm or decline the results of [1] which are very important for experimental and theoretical physics and various applications it is necessary to carry out new measurement better with the help of new methods. As it has been shown above the advance of the inverse Compton scattering photon sources allows at present to measure in relatively simple way $n(\omega)$ of various materials in MeV photon energy region and thus to confirm or decline the very important results of [1]. The proposed method is easier than that of [1] because it can be realized having MeV photon beam with energy spread about 10% and angular orientation microradian accuracy instead of much harder requirements of [1].

Journal of Physics: Conference Series **517** (2014) 012045 doi:10.1088/1742-6596/517/1/012045

References

[1] Habs D, Gunther M M, Jentschel M and Urban W 2012 *Phys. Rev. Lett.* **108** 184802

[2] Doll M *et al* 2000 *J. of Res. of Nat. Inst. of Stand. and Technologies* **105** 1

[3] Donohue J T 2013 *Phys. Rev. Lett.* **110** 129501

[4] Habs D, Gunther M M, Jentschel M and Thirolf P G 2012 *ArXiv*:1201.4466

[5] Jentschel M 2013*Gamma Ray Optics*, http://www.inn.jir

[6] Habs D 2013 *Delbruck scattering and non-perturbative QED and possible applications* http://www.eli-np.ro

[7] *CERN Courier* 2012, July18

[8] Aginian M A, Ispirian K A and Ispiryan M To be published in a) 2013 *European Physics Letters*; b) 2014 *Izvestya of NAS of Armenia;* c) 2014 Proc of this Conf .

[9] Kumakhov M A 1986 *Radiation of Channeled Particles in Crystal*s (Moscow Energoatomizdat, in Russian)

[10] Wu Y K 2012 *Proc. PAC*, TUXB03

[11] Raubenheimer T 2013 Lecture at SLAC Summer School on Electron and Photon Beams

[12] Lewellen T K 2008 *Phys. Med. Biol.* **53** R287

[13] Aprile E and Doke T 2010 *Rev, Mod. Phys.* **82** 2053

[14] Ispirian K A and Ispiryan R K 2011 *Nuovo Cimento* **34** 37

Setups for Tomographic Imaging with Submillimeter Spatial Resolution

S Stuchebrov, A Batranin, D Verigin, A Wagner
Tomsk Polytechnic University, Tomsk, Russia

E-mail: stuchebrov@tpu.ru

Abstract. In the work development of setups for tomographic vizualisation is presented. Detectors with resolution better than 1 mm were used. As a result of the work sinograms were obtained. These results can be used for medical purposes.

In spite of wide spreading X-ray methods for study inner structure of objects, X-ray technologies are still developing actively. It is driven by advances in computer science that allow using more sophisticated software and hardware to obtain more precise and reliable results. From the other side, researchers pay more and more attention to visualization of inner structure of objects. These factors lead to high urgency of the problem.

In this paper we present results of research on developing prototypes of X-ray setups that were used to obtain both two-dimensional shadow projection images and tomograms. A tomogram or a planar slice is a common way for studying internal structure of objects and it is a base for 3-D tomography and volume visualization. The goal of the developing prototypes is applying obtained results in medicine. In fact, X-ray tomography has made a great advance since its invention. However, dose rate decreasing in medical applications is still an urgent issue. In this paper we present results which can help to solve the dose rate decreasing problem.

In the work two X-ray prototypes were created. Pulsed X-ray tube was used as radiation source in these prototypes. Different linear detectors were chosen for different prototypes.

In the first prototype strip semiconducting GaAs detector was applied. It was chosen due to its high spatial resolution, good sensitivity and response speed.

In the second prototype multichannel linear gas-discharged detector was used. Detectors of this type have low spatial resolution compare to this of semiconducting ones. However, they are effective-cost solutions and have a large sensitive area which make them useful devices in some applications.

We used pulsed X-ray tube RAP-160-5 as radiation source. Focus spot for this tube is 1.2 mm × 1.2 mm. Power with the peak parameter is 0.6 kW. Accelerating voltage could be changing in the range from 40 to 160 kV. Values of anode current are changing from 0.4 to 5 mA. All parameters are given as integral ones due to pulse mode of X-ray tube. Frequency of pulses could be changed automatically from 60 to 700 Hz. The frequency depends on value of required current. Pulse duration is about 140 μs. The tube can be synchronized with external devices by the dedicated output [1].

Solid-state strip detector GaAs 512-0.1 was used for prototype based on semiconductor detector line. Period of line generation could be set by inner or external sync signal with duration not less than 140 μs. This is connected with procedures needed for inner processing of a cycle. The device has function of choosing time for dose acquiring. Resolution of the detector is 5 LP/mm. Width of sensitive area is 51.2 mm.

Digital linear multichannel gas-discharge detector PRIZ-1536 was used in the second prototype. Detecting plate consists of 1536 signal strips with pitch 250 μm. Width of sensitive area is 384 mm. Time of signal integrating is 2.5 msec. PRIZ-1536 can also operate with inner and external sync signal. Detecting part is arranged on a special designed mechanism, which allows to set the detecting part automatically on the translation arc with path 400 mm [2].

In the first prototype setup with GaAs line detector, both the X-ray source and the detector stay motionless on the same line while the scanned object rotates and moves to the right or to the left during the data acquisition. In the second setup, the X-ray source and the object stay on the same line while the gas-discharge detector moves to the right or to the left on its translation stage. Both setups produce continues series of line shadow projections also known as sinograms.

Synogram is a visual representation of stacked projections acquired from a layer of the object at different angles. The method used for obtaining lines of synogram image and principle of stacking lines into sinogram are shown on figure 1.

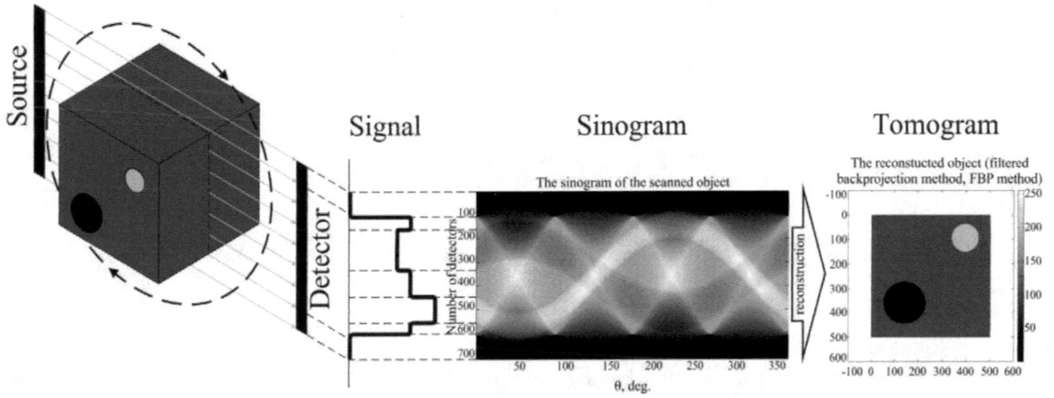

Figure 1. Schematic conception of obtaining sinogram.

The X-ray source, the scanned object and a detector are placed on the same line to obtain a single line projection. The object rotates by some definite angle step on the axis perpendicular to the "source-detector" line. Thus, shadow projections of the object on different angle positions are registered by detector. Typically, the object makes a complete revolution, 360 degrees, to produce a single sinogram. The GaAs line detector setup employs the slightly divergence beam geometry with angles not exceed 5 degrees and can be treated as pseudo-parallel, while the second one uses the fan beam geometry and should be processed properly.

Resolution of reconstructed image depends on intervals of object's rotation angle (the smaller interval the better resolution). However, there are number of disadvantages because of decreasing interval of angles. Firstly, the number of images increases with increasing number of angular projections, and consequently dose loading of the investigated object increases. Secondly, size of differential equation system, formed during reconstruction, depend on number of detectors and projections.

Computing software MATLAB was used for reconstruction of tomographic slices from sinograms in case of parallel beam of X-ray. We used function "iradon" from MATLAB's addon Image Processing Toolbox [3] for the Radon transform.

There are a number of difficulties with using divergent beam of X-ray in design of algorithm for reconstruction from sinogram was. There is function "ifanbeam" in MATLAB software for inverse Radon transform in the case of divergent beam. A number of restrictions for input parameters in the function "ifanbeam" make impossible using this function for achieving our goals [4]. We decided to use software NRecon designed by company Bruker micro-CT (former Skyscan) [5] for reconstruction of slice in case of divergent beam. There are no restrictions for geometry of a beam, intervals of an angle and total angle of scanning.

Series of projection images and sinograms of different objects were obtained using these prototypes.

The example of shadow projection image produced by the GaAs line detector setup is presented on figure 2. The object is a bar made of Pinus sylvestris 50×50×20 mm in size with a screw and a fastener in its body.

Figure 2. X-ray projection image of pine bar with screw and fastener system.

The image has a number of defects which are typical for digital detection systems. Linear artifacts caused by poor calibration are clearly seen. Different sensitivity of detectors on the intermediate intensity of signals (in range of the lowest and the highest digital accounts) prevented calibration of high quality. Moreover, there are several horizontal lines caused by bad detectors. It should be noticed that these issues are easily eliminated by software.

A sinogram of the wooden bar and its reconstructed slice (a tomogram) are presented on figure 3. The tomogram has pronounced ring artifacts due to difference in detectors sensitivity. However, the internal structure of the sample is easily revealed.

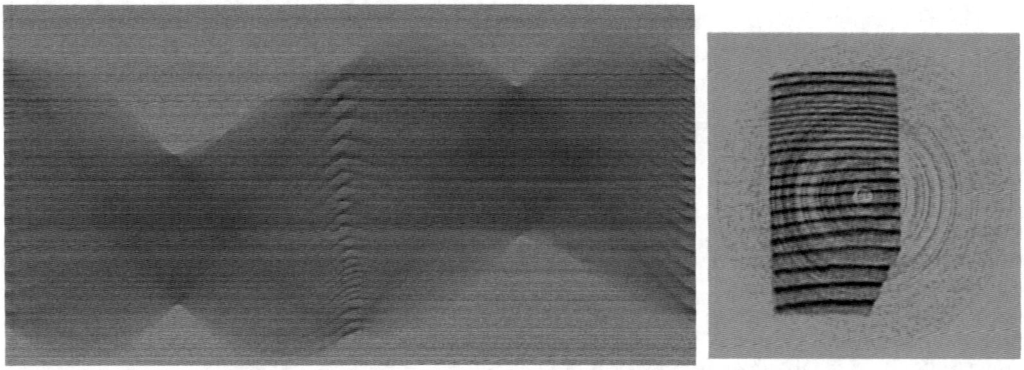

Figure 3. A sinogram of the wooden bar and its tomogram.

The defects described above define the artifacts arising during reconstruction of tomographic slice. The lines caused by a poor calibration of detectors will be transformed into ring artifacts on the sinogram. Nevertheless, the reconstructed tomographic slice allows identifying growth rings. Quality of growth rings on the tomographic slice allows defining thickness and the sizes of rings.

On figure 4 reconstructed tomographic slice of "Low Contrast insert" (the module of a tomographic phantom) is presented. The module is made from tissue-equivalent material. Diameter of the module is 203.2 mm [6].

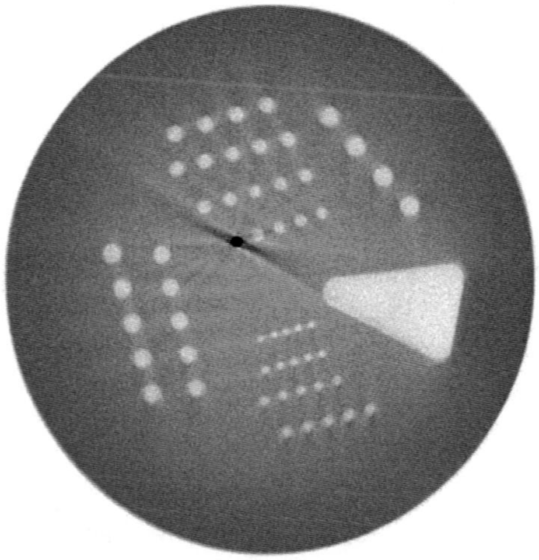

Figure 4. Reconstructed tomographic slice of "Low Contrast insert".

There are beam artifacts on figure 4 caused by opacity of a rod on which the module fastened. Fluctuations in registration of radiation by the multichannel gas-discharge detector and statistical nature of X-rays led to visible granularity. Fluctuations were caused by low level of registered radiation.

After obtaining images, the technique of a dose rate measurement during work with a pulse source was developed. Radiation monitoring with pulse sources have a number of difficulties. The RAP-160-5 tube emits only 2% during the working period. A current in a single impulse is high, approximately 50 times higher than integrated values. Besides, impusle frequency of the source is high and that requires using only high speed data-acquisition equipment.

In experiment solid-state thermoluminescent detectors were used. These dosimeters meet all requirements of a problem: working limits of dose rates and the absorbed doses correspond to measured modes [7]. Using the storage type detectors solves a problem with registration speed of a dose.

Dosimeters were irradiated in points of sample arrangement at the same modes, as at obtaining pictures. Dose rates were defined taking into account an exposition.

Using prototype with the semi-conductor detector for obtaining shadow projection images the equivalent dose was 14 μSv. Obtaining one sinogram on this installation resulted in the calculated equivalent dose 2.8 mSv.

The calculated equivalent dose was 15 μSv during obtaining shadow projection images using prototype with the gas-discharge detector. The calculated equivalent dose was 9.5 mSv for obtaining synograms.

These resulted doses are less than standard doses for medical film radiography. For example, for photofluorography the norm is considered 0.8 mSv, for radiography – 0.4 mSv, for radioscopy – 10 mSv [8].

Doses decreased mostly due to using pulse source in the described prototypes. Using synchronization between the pulse source and the detector excluded useless dose loading.

Unfortunately, application of low intensity radiation leads to decrease in quality of visualization. Contrast sensitivity decreases with using the gas-discharge detector. In addition, there are issues due to nonlinearity of detectors sensitivity in case of the GaAs line detector. However, decreasing of dose rates is crucial in medical diagnostic, therefore the further studies should be made in order to achieve higher quality of visualization.

References

[1] Information on http://www.tsnk-lab.ru/files/file42.pdf. (in Rusian)

[2] Operation manual: Mnogokanalnaja ionizatsionnaja camera dlja rentgenographicheskoi ustanovki skanirueshego tipa 2006 *IJF the Siberian Branch of the Russian Academy of Science, Novosibirsk* (in Russian)

[3] Information on http://www.mathworks.com/help/images/ref/iradon.html

[4] Information on http://www.mathworks.com/help/images/ref/ifanbeam.html

[5] Information on http://www.skyscan.be/products/downloads.htm

[6] Information on http://www.cirsinc.com/products/all/31/aapm-ct-performance-phantom/

[7] Spurny, Z. 1962 A thermoluminescent dosimeter Kernenergie (East Germany) **5**

[8] F. Mettler et al., *Effective doses in radiology and diagnostic nuclear medicine: a catalog, Radiology.* **248** (2008) 254-263

The possibility of the neutron beams formation on base of cyclotron C18

R H Avagyan, G L Bazoyan, M H Hakobyan and I A Kerobyan

Alikanian National Science Laboratory (Yerevan Physics Institute)
Br. Alikhanyan 2, Yerevan 0036, Armenia

E-mail: keropyan@mail.yerphi.am

Abstract. It is performed the investigation of the possibility of obtaining neutron beams under external proton beams of cyclotron C18. The intensity and neutron spectra of neutron beams have been investigated in the view of using them for investigation of the processes in nucleosynthesis and of the radiation resistance of the electronics.

It is suggested on the external beam of cyclotron C18 with proton energy 18 MeV and current 100 μA obtaining the monoenergetic neutron beam.

A neutron source is considered monoenergetic when the energy spectrum consists of a single line with an energy width which is much less than the energy itself. In accelerator based sources this width is mainly determined by the target thickness. In practical applications it is usually of little importance to have a narrow energy width of the line. So the "thick target yield" of a source is of interest, rather than the differential one.

Therefore the monochromaticity of a source must be viewed at with an eye on the application. Basically there are three types of applications:

a) scientific ones (physics)
b) technical ones and
c) medical ones.

In each case the demands on a monoenergetic source can differ greatly.

The yield of neutrons of the desired energy from an accelerator based monoenergetic source depends on [1]:

- the charged particle beam intensity,
- the target thickness,
- the differential cross section of the reaction
- the specific energy loss in the target material.

For instance, high-Z targets can be used to produce spallation neutrons. However, there is a recent interest in the use of a ^{7}Li target, which, when bombarded with protons, can produce a relatively high yield of quasimonoenergetic neutrons in the forward direction via the $^{7}Li(p, n)$ reaction. The quasimonoenergetic and broad neutron sources are needed for research in different areas in nuclear science and technology, e.g. Accelerator Transmutation of Waste (ATW), radiation damage studies, medical isotope production, and physics cross section experiments. In order to assess the feasibility of using lithium target to produce neutrons, accurate evaluated cross section data are needed. These data can be used to predict the neutron yield, as well as the neutron energy and angular dependencies, from both thin and thick targets. (Thin targets allow the possibility of producing quasimonoenergetic neutron sources, whereas thicker targets produce broad-spectrum neutron sources.).

Neutron production in the MeV range by the $^{7}Li(p, n)^{7}Be$ reaction is attractive for two reasons:

- emission of neutrons only into a narrow forward cone and
- the neutron production cross section is very high, especially at lower energies.

On the figure 1 is shown the cross section of the reaction $^7Li(p, n)^7Be$ in dependence of proton beam energy [2, 3]. The cross section of reaction $^7Li(p, n)^7Be$ has the peak in the region of E_p = 5-7 MeV , the value of the peak is about 250 mb. Consequently, to obtain the high yield of neutrons on the external proton beam of cyclotron C18 necessary to use degrader for decrease proton beam energy up to 5 MeV.

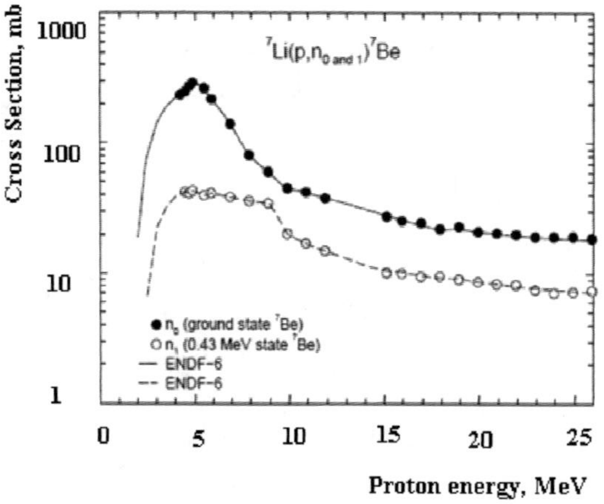

Figure 1. The cross section of reaction $^7Li(p,n)^7Be$ in dependence of incident proton energy [3].

The neutron spectra in the case of incident proton energy 5 MeV calculated by code TALYS 1.4 [4] is presented on the figure 2.

Usually only neutron emission at an angle of 0° with respect to the charged particle beam is considered.

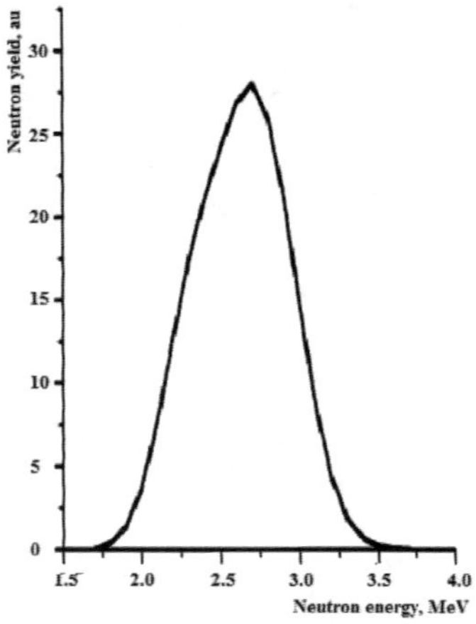

Figure 2. Neutron spectra at angle 0^0 in case of incident proton energy 5 MeV, calculated by TALYS 1.4.

The angular distribution of emitted neutrons ^{7}Li(p,n)^7 Be via neutron energy calculated by code DROSG2000 [5] is presented on the figure 3.

From the figure 3 is seen, that placing the sample at 0° has the following advantages:

- The neutron yield is usually forward peaked;
- Neutrons emitted at 0° have the highest energy.

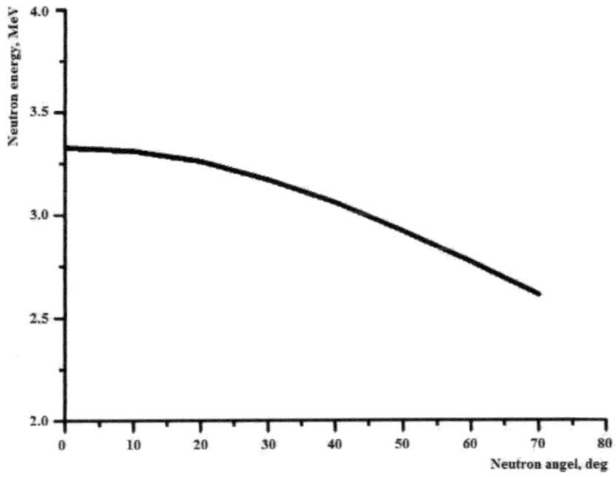

Figure 3. The dependence emitted neutrons energy via neutron angle, calculated by DROSG2000 [5].

The yield of neutrons from reaction ^{7}Li(p, n)^{7}Be under protons beam in the protons energy region 2-5.5 MeV was investigated in [6]. With the increase of the energy of the incident proton, the neutron yield increases. The neutron yield behavior in dependence of the energy of the protons is shown in figure 4.

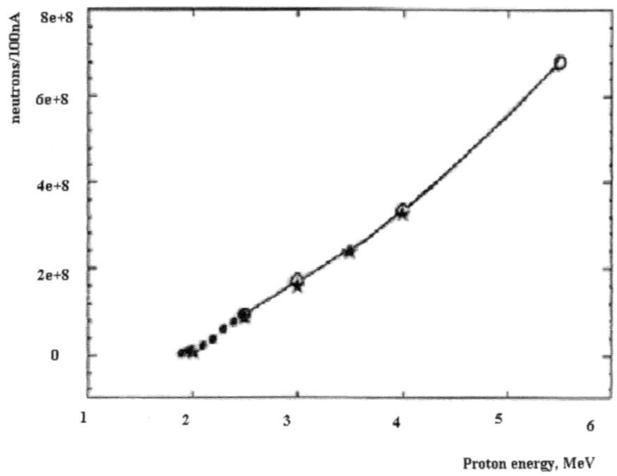

Figure 4. Neutron yield from a 5mm thick natural Lithium target for proton energies from 2.5 to 5.5 MeV. o and curve – [6] , • – [7], * – [8].

The integral yield of neutrons at incident protons energy 5.5 MeV is $6*10^8$ n/100nA, or $6*10^9$ n/μA. In the same [5] work the angular distribution of the emitted neutrons is presented also (figure 5). From this figure the neutrons yield in the angle region $0\text{-}20^0$ is estimated, $1.43*10^9$ n/μA.

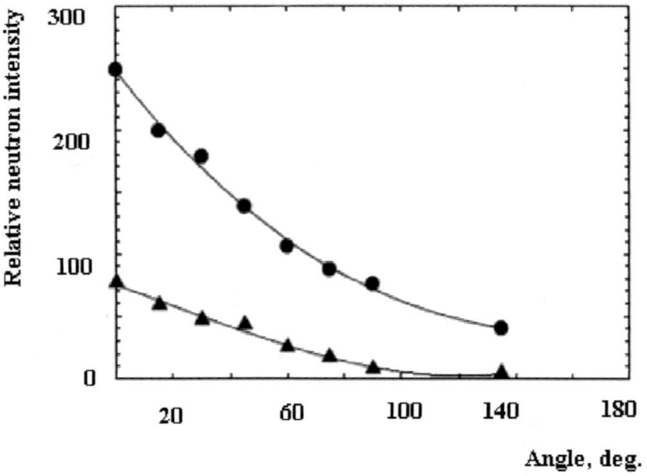

Figure 5. Angular distribution of neutron yield from thick Lithium target from 5.5 and 4.5 MeV proton energies [5]. Circles - 5.5 MeV, triangles – 4.5 MeV.

Thus, the carried out preliminary studies show that it is possible to obtain neutron beams on the extracted proton beam of the cyclotron C18 with intensity about 10^{11} n/μA which sufficient for performing research on the study of the astrophysical reactions and application problems connected with the radiation resistance of the components of spacecraft electronics.

References

[1] Drosg M, Characteristics of accelerator based neutron sources, monoenergetic neutron production by two-body reactions in the energy range from 0.0001 to 500 MeV (Overview), Institute of Experimental Physics, University of Vienna, A-1090 Wien, Austria

[2] Koning A J, Rochman D, van der Marck S, Kopecky J, Sublet J Ch, Pomp S, Sjostrand H, Forrest R, Bauge E and H Henriksson, TENDL-2012: TALYS-based evaluated nuclear data library, www.talys.eu/tendl2012.html

[3] Mashnik S G, Chadwick M B, Hughes H G, Little R C, MacFarlane R E, Waters L S, and Young P G; ^{7}Li(p,n) Nuclear data library for incident proton energies to 150 MeV, Los Alamos National Laboratory, Los Alamos, NM 87545, USA

[4] Koning A, Hilaire S, Goriely S 2011 TALYS-1.4, A nuclear reaction program

[5] Drosg M, *DROSG-2000*: Neutron Source Reactions. Institute for Experimental Physics, University of Vienna, Austria

[6] Kumawat H 2010 *Proceedings of the DAE Symp. on Nucl. Phys.* 55

[7] Lee C L and Zhou X L 1999 *Nucl. Instr. Methods Phys. Rev. B* **151** 1

[8] Randers-Pehrson G and Brener D J 1998 *Med. Phys.* **25** 894

AUTHOR INDEX

AUTHOR INDEX

Institute of Physics
Dirac House, Temple Back
Bristol BS1 6BE UK

CURRAN ASSOCIATES INC
proceedings.com